VOLUME FIVE HUNDRED AND FIFTY FIVE

METHODS IN ENZYMOLOGY

Hydrogen Sulfide in Redox Biology Part B

METHODS IN ENZYMOLOGY

VOLUME FIVE HUNDRED AND FIFTY FIVE

METHODS IN ENZYMOLOGY

Hydrogen Sulfide in Redox Biology Part B

Edited by

ENRIQUE CADENAS

Pharmacology & Pharmaceutical Sciences, School of Pharmacy, University of Southern California, USA

LESTER PACKER

Department of Molecular Pharmacology and Toxicology, School of Pharmaceutical Sciences, University of Southern California, USA

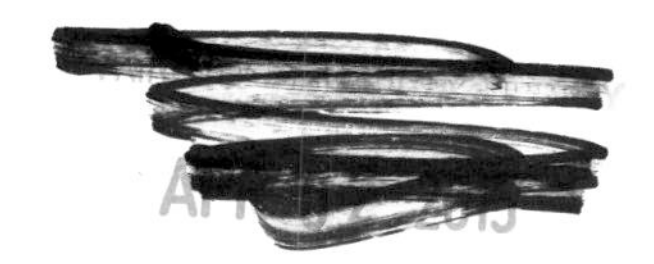

AMSTERDAM • BOSTON • HEIDELBERG • LONDON
NEW YORK • OXFORD • PARIS • SAN DIEGO
SAN FRANCISCO • SINGAPORE • SYDNEY • TOKYO
Academic Press is an imprint of Elsevier

Academic Press is an imprint of Elsevier
225 Wyman Street, Waltham, MA 02451, USA
525 B Street, Suite 1800, San Diego, CA 92101-4495, USA
125 London Wall, London, EC2Y 5AS, UK
The Boulevard, Langford Lane, Kidlington, Oxford OX5 1GB, UK

First edition 2015

Notices
Knowledge and best practice in this field are constantly changing. As new research and experience broaden our understanding, changes in research methods, professional practices, or medical treatment may become necessary.

Practitioners and researchers must always rely on their own experience and knowledge in evaluating and using any information, methods, compounds, or experiments described herein. In using such information or methods they should be mindful of their own safety and the safety of others, including parties for whom they have a professional responsibility.

To the fullest extent of the law, neither the Publisher nor the authors, contributors, or editors, assume any liability for any injury and/or damage to persons or property as a matter of products liability, negligence or otherwise, or from any use or operation of any methods, products, instructions, or ideas contained in the material herein.

ISBN: 978-0-12-801511-7
ISSN: 0076-6879

For information on all Academic Press publications
visit our website at store.elsevier.com

CONTENTS

Contributors xi
Preface xv

Section I
The Redox Biochemistry of Hydrogen Sulfide

1. Investigating the Role of H_2S in 4-HNE Scavenging **3**

Hilde Laggner and Bernhard M.K. Gmeiner

1. Introduction 4
2. Experimental Compounds and Considerations 5
3. Conclusions and Perspectives 14

Acknowledgment 16
References 16

2. Inhalation Exposure Model of Hydrogen Sulfide (H_2S)-Induced Hypometabolism in the Male Sprague-Dawley Rats **19**

Asaf Stein, David W. Kraus, Jeannette E. Doeller, and Shannon M. Bailey

1. Introduction 20
2. Exposure Protocol for H_2S-Induced Hypometabolism in Rats 21
3. Other Considerations for H_2S Exposure Studies 31
4. Summary 32

Acknowledgments 33
References 33

Section II
Mechanisms of H_2S Cell Signaling and Transcriptional Pathways

3. Use of the "Tag-Switch" Method for the Detection of Protein S-Sulfhydration **39**

Chung-Min Park, Igor Macinkovic, Milos R. Filipovic, and Ming Xian

1. Introduction 40
2. The Design of "Tag-Switch" Method 40
3. Chemistry Validation Using Small-Molecule Substrates 41
4. "Tag-Switch" Assay on Bovine Serum Albumin and GAPDH as Model Proteins 46

5. "Tag-Switch" Assay for the Detection on Intracellular Protein Persulfides 49
6. "Tag-Switch" Assay for the Detection of Intracellular S-Sulfhydration by Fluorescence Microscopy 51
7. Conclusions 54
Acknowledgments 55
References 55

4. Real-Time Assays for Monitoring the Influence of Sulfide and Sulfane Sulfur Species on Protein Thiol Redox States 57

Romy Greiner and Tobias P. Dick

1. Introduction 58
2. PTEN Activity Assay 59
3. roGFP2 Redox Assay 61
4. Application of "H_2S Donors" and Polysulfides 64
5. Quantitation of Sulfane Sulfur by Cold Cyanolysis 67
6. Elimination of Sulfane Sulfur by Cold Cyanolysis 71
Acknowledgments 76
References 76

5. Protein Sulfhydration 79

Bindu D. Paul and Solomon H. Snyder

1. Introduction 80
2. Detection of Sulfhydration Using the Modified Biotin Switch Assay 81
3. Detection of Sulfhydration Using the Maleimide Assay 86
4. Summary 88
Acknowledgments 89
References 89

Section III

H_2S in Cell Signaling in the Cardiovascular and Nervous System and Inflammatory Processes

6. Intravital Microscopic Methods to Evaluate Anti-inflammatory Effects and Signaling Mechanisms Evoked by Hydrogen Sulfide 93

Mozow Y. Zuidema and Ronald J. Korthuis

1. Introduction 94
2. Molecular Determinants of Neutrophil/Endothelial Cell Adhesive Interactions 94
3. Intravital Microscopic Approaches to Study Leukocyte/Endothelial Cell Adhesive Interactions 98

4. Assessing Leukocyte Rolling, Adhesion, and Emigration in the Intact Microcirculation 103
5. Detection of Chemokine and Adhesion Molecule Expression using Intravital Microscopy 108
6. Intravital Microscopic Methods to Assess Changes in Microvascular Permeability 109
7. Assessment of Reactive Oxygen Species Generation Using Intravital Microscopy 111
8. Fluorescence Detection of Cell Injury using Intravital Microscopy 113
9. Perfused Capillary Density Assessment with Intravital Microscopy 113
10. Acute and Preconditioning-Induced Anti-inflammatory Actions of Hydrogen Sulfide: Assessment Using Intravital Microscopy 114
11. Conclusion and Perspectives 119
Acknowledgment 121
References 121

7. Attenuation of Inflammatory Responses by Hydrogen Sulfide (H_2S) in Ischemia/Reperfusion Injury 127

Neel R. Sodha and Frank W. Sellke

1. Introduction 128
2. Ischemia–Reperfusion Injury 128
3. Central Nervous System 130
4. Respiratory System 131
5. Cardiovascular System 133
6. Gastrointestinal System 135
7. Hepatobiliary System 138
8. Renal System 139
9. Musculoskeletal 141
10. Summary 142
References 143

8. CD47-Dependent Regulation of H_2S Biosynthesis and Signaling in T Cells 145

Sukhbir Kaur, Anthony L. Schwartz, Thomas W. Miller, and David D. Roberts

1. Introduction 146
2. Regulation of H_2S Biosynthesis in T Cells 148
3. Catabolism of H_2S 149
4. Regulation of T Cell Activation by H_2S Signaling 150
5. Autocrine and Paracrine Roles of H_2S in T Cell Activation 153

6. Role of H_2S in the Cytoskeleton 155
7. T Cell Regulation by TSP1/CD47 Signaling 156
8. H_2S Regulation of Leukocyte Adhesion 157
9. Role of H_2S in Diseases Associated with Altered T Cell Immunity 158
10. Future Prospective 160
Acknowledgments 162
References 162

9. Anti-inflammatory and Cytoprotective Properties of Hydrogen Sulfide 169

Burcu Gemici and John L. Wallace

1. Introduction 170
2. Enzymatic Synthesis of H_2S 170
3. Healing and Resolution of Inflammation 172
4. Mechanisms of Anti-inflammatory Effects of H_2S 173
5. Effects of H_2S on Visceral Pain 176
6. Cytoprotective Actions of H_2S 176
7. Therapeutic Applications of H_2S-Releasing Drugs 180
Acknowledgments 188
References 188

10. H_2S and Substance P in Inflammation 195

Madhav Bhatia

1. Introduction 195
2. Disease Models Used to Study the Role of H_2S and Substance P 196
3. H_2S and Substance P—What Are They Doing Together? 201
4. Summary 202
Acknowledgments 202
References 202

11. Role of Hydrogen Sulfide in Brain Synaptic Remodeling 207

Pradip Kumar Kamat, Anuradha Kalani, and Neetu Tyagi

1. Introduction 208
2. Pharmacological and Physiological Effect of H_2S 210
3. Effect of H_2S on the CNS 212
4. Effect of H_2S on Brain Cells (Astrocyte, Microglia, and Oligodendrocyte) 214
5. Synapse 216
6. Glia and Neurons Interactions 217
7. Effect of H_2S on Neuronal Redox Stress 218

8. Effect of H_2S on Glutamate Neurotransmission 219
9. Effect of H_2S on NMDA Receptor Regulation 220
10. Effect of H_2S on GABA-Mediated Neurotransmission 221
11. Effect of H_2S on Calmodulin Kinase 222
12. Conclusion 222
Conflict of Interest 223
Acknowledgment 223
References 223

Section IV
H_2S in Plants

12. Detection of Thiol Modifications by Hydrogen Sulfide 233
E. Williams, S. Pead, M. Whiteman, M.E. Wood, I.D. Wilson, M.R. Ladomery, T. Teklic, M. Lisjak, and J.T. Hancock

1. Introduction 234
2. Hydrogen Sulfide Acts as a Signal in Cells 235
3. Modification of Thiols by Signaling Molecules 236
4. Identification of Modified Thiols by Other Methods 238
5. Experimental Protocols 239
6. *Caenorhabditis elegans* as a Model Organism 239
7. Growth of *C. elegans* 240
8. Treatment of Samples with H_2S 240
9. Estimation of Toxicity of H_2S Compounds 241
10. Treatment of Samples with Thiol Tag 242
11. Isolation and Analysis of Modified Proteins 244
12. Estimation of Protein Concentrations in Samples 245
13. Further Analysis and Identification of Modified Proteins 246
14. Concluding Remarks 247
References 248

13. Analysis of Some Enzymes Activities of Hydrogen Sulfide Metabolism in Plants 253
Zhong-Guang Li

1. Theory 254
2. Equipment 256
3. Materials 256
4. Protocol 1 260
5. Step 1: Analyze of L-/D-Cysteine Desulfhydrase Activity 261

6. Protocol 2 263
7. Step 1: Analyze of Sulfite Reductase Activity 263
8. Protocol 3 265
9. Step 1: Analyze of β-Cyano-L-Alanine Synthase Activity 266
10. Protocol 4 267
11. Step 1: Analyze of L-Cysteine Synthase Activity 268
Acknowledgment 269
References 269

14. Sulfide Detoxification in Plant Mitochondria **271**

Hannah Birke, Tatjana M. Hildebrandt, Markus Wirtz, and Rüdiger Hell

1. Introduction 272
2. Methods 275
3. Summary 283
Acknowledgments 283
References 284

Section V
Molecular Hydrogen

15. Molecular Hydrogen as a Novel Antioxidant: Overview of the Advantages of Hydrogen for Medical Applications **289**

Shigeo Ohta

1. Introduction 290
2. Comparison of H_2 with Other Medical Gasses 291
3. Oxidative Stress as Pathogenic Sources 292
4. Physiological Roles of H_2O_2 294
5. Measurement of H_2 Gas Concentration 295
6. Advantages of Hydrogen in Medical Applications 296
7. Methods of Ingesting Molecular Hydrogen 301
8. Medical Effects of H_2 304
9. Possible Molecular Mechanisms Underlying Various Effects of Molecular Hydrogen 306
10. Unresolved Questions and Closing Remarks 309
References 310

Author Index ***319***
Subject Index ***345***

CONTRIBUTORS

Shannon M. Bailey
Department of Environmental Health Sciences; Center for Free Radical Biology, and Department of Pathology, University of Alabama at Birmingham, Birmingham, Alabama, USA

Madhav Bhatia
Department of Pathology, University of Otago, Christchurch, New Zealand

Hannah Birke*
Centre for Organismal Studies Heidelberg, University of Heidelberg, Heidelberg, Germany

Tobias P. Dick
Division of Redox Regulation, German Cancer Research Center (DKFZ), DKFZ-ZMBH Alliance, Heidelberg, Germany

Jeannette E. Doeller
Department of Environmental Health Sciences, and Center for Free Radical Biology, University of Alabama at Birmingham, Birmingham, Alabama, USA

Milos R. Filipovic
Department of Chemistry and Pharmacy, University of Erlangen-Nuremberg, Erlangen, Germany

Burcu Gemici
Near East University, Nicosia, Northern Cyprus, Turkey

Bernhard M.K. Gmeiner
Department of Medical Chemistry and Pathobiochemistry, Center of Pathobiochemistry and Genetics, Medical University of Vienna, Vienna, Austria

Romy Greiner
Division of Redox Regulation, German Cancer Research Center (DKFZ), DKFZ-ZMBH Alliance, Heidelberg, Germany

J.T. Hancock
Faculty of Health and Applied Sciences, University of the West of England, Bristol, United Kingdom

Rüdiger Hell
Centre for Organismal Studies Heidelberg, University of Heidelberg, Heidelberg, Germany

Tatjana M. Hildebrandt
Institute for Plant Genetics, Leibniz University Hannover, Hannover, Germany

Anuradha Kalani
Department of Physiology and Biophysics, School of Medicine, University of Louisville, Louisville, Kentucky, USA

*Current address: CSIRO Agriculture Flagship, Black Mountain Laboratories, Canberra, Australia

Pradip Kumar Kamat
Department of Physiology and Biophysics, School of Medicine, University of Louisville, Louisville, Kentucky, USA

Sukhbir Kaur
Laboratory of Pathology, Center for Cancer Research, National Cancer Institute, National Institutes of Health, Bethesda, Maryland, USA

Ronald J. Korthuis
Department of Medical Pharmacology and Physiology, School of Medicine, One Hospital Drive, and Dalton Cardiovascular Research Center, University of Missouri, Columbia, Missouri, USA

David W. Kraus
Department of Environmental Health Sciences, and Center for Free Radical Biology, University of Alabama at Birmingham, Birmingham, Alabama, USA

M.R. Ladomery
Faculty of Health and Applied Sciences, University of the West of England, Bristol, United Kingdom

Hilde Laggner
Department of Medical Chemistry and Pathobiochemistry, Center of Pathobiochemistry and Genetics, Medical University of Vienna, Vienna, Austria

Zhong-Guang Li
School of Life Sciences, Engineering Research Center of Sustainable Development, Utilization of Biomass Energy Ministry of Education, Key Laboratory of Biomass Energy, Environmental Biotechnology Yunnan Province, Yunnan Normal University, Kunming, Yunnan, PR China

M. Lisjak
Faculty of Agriculture, University of Osijek, Osijek, Croatia

Igor Macinkovic
Department of Chemistry and Pharmacy, University of Erlangen-Nuremberg, Erlangen, Germany

Thomas W. Miller
Laboratory of Pathology, Center for Cancer Research, National Cancer Institute, National Institutes of Health, Bethesda, Maryland, USA

Shigeo Ohta
Department of Biochemistry and Cell Biology, Institute of Development and Aging Sciences, Graduate School of Medicine, Nippon Medical School, Kawasaki, Japan

Chung-Min Park
Department of Chemistry, Washington State University, Pullman, Washington, USA

Bindu D. Paul
The Solomon H. Snyder Department of Neuroscience, Johns Hopkins University School of Medicine, Baltimore, Maryland, USA

S. Pead
Faculty of Health and Applied Sciences, University of the West of England, Bristol, United Kingdom

David D. Roberts[†]
Laboratory of Pathology, Center for Cancer Research, National Cancer Institute, National Institutes of Health, Bethesda, Maryland, USA

Anthony L. Schwartz
Laboratory of Pathology, Center for Cancer Research, National Cancer Institute, National Institutes of Health, Bethesda, Maryland, USA

Frank W. Sellke
Division of Cardiothoracic Surgery, Department of Surgery, Alpert Medical School of Brown University, Providence, Rhode Island

Solomon H. Snyder
The Solomon H. Snyder Department of Neuroscience; Department of Pharmacology, and Department of Psychiatry, Johns Hopkins University School of Medicine, Baltimore, Maryland, USA

Neel R. Sodha
Division of Cardiothoracic Surgery, Department of Surgery, Alpert Medical School of Brown University, Providence, Rhode Island

Asaf Stein
Department of Environmental Health Sciences, and Center for Free Radical Biology, University of Alabama at Birmingham, Birmingham, Alabama, USA

T. Teklic
Faculty of Agriculture, University of Osijek, Osijek, Croatia

Neetu Tyagi
Department of Physiology and Biophysics, School of Medicine, University of Louisville, Louisville, Kentucky, USA

John L. Wallace
Department of Physiology & Pharmacology, University of Calgary, Calgary, Alberta, and Department of Pharmacology & Toxicology, University of Toronto, Toronto, Ontario, Canada

M. Whiteman
University of Exeter Medical School, University of Exeter, Exeter, United Kingdom

E. Williams
Faculty of Health and Applied Sciences, University of the West of England, Bristol, United Kingdom

I.D. Wilson
Faculty of Health and Applied Sciences, University of the West of England, Bristol, United Kingdom

[†]Current address: Paradigm Shift Therapeutics, Rockville, Maryland, USA

Markus Wirtz
Centre for Organismal Studies Heidelberg, University of Heidelberg, Heidelberg, Germany

M.E. Wood
Biosciences, University of Exeter, Exeter, United Kingdom

Ming Xian
Department of Chemistry, Washington State University, Pullman, Washington, USA

Mozow Y. Zuidema
Harry S. Truman Veterans Administration Hospital, Cardiology, Columbia, Missouri, USA

PREFACE

Hydrogen sulfide is viewed as the third gasotransmitter, gaseous signaling molecule, together with nitric oxide and carbon monoxide. The cellular sources of hydrogen sulfide involve enzymes of the transsulfuration pathway CBS (cystathionine β-synthase) and CSE (cystathionine γ-lyase) and 3MST (3-mercaptopyruvate sulfurtransferase). Storages of hydrogen sulfide occur in mitochondria (iron–sulfur clusters of enzymes) and cytosol (bound sulfane sulfur). Of course, the release of hydrogen sulfide from these storages is tightly regulated by several pathophysiological processes.

In addition to the myriad of effects arising from hydrogen sulfide as a signaling molecule, it also protects against oxidative stress and glutamate toxicity, inhibits the release of insulin, preserves mitochondrial function, and is a modulator of inflammatory responses. These pleiotropic effects of hydrogen sulfide have been the subject of numerous investigations in the last years and are largely accounted for by its role as a gaseous signaling molecule. Hydrogen sulfide may act alone or in conjunction with other gasotransmitters and, in doing so, it regulates a number of physiological processes and is involved in some stages of the pathogenesis of several diseases.

These volumes of *Methods in Enzymology* were designed as a compendium for hydrogen sulfide detection methods, the pharmacological activity of hydrogen sulfide donors, the redox biochemistry of hydrogen sulfide and its metabolism in mammalian tissues, the mechanisms inherent in hydrogen sulfide cell signaling and transcriptional pathways, and cell signaling in specific systems, such as cardiovascular and nervous system as well as its function in inflammatory responses. Three chapters are also devoted to hydrogen sulfide in plants and a newcomer, molecular hydrogen, its function as a novel antioxidant.

In bringing these volumes of *Methods in Enzymology* to fruition, credit must be given to the experts in various aspects of hydrogen sulfide research, whose thorough and innovative work is the basis of these *Methods in Enzymology* volumes. We hope that these volumes will be of help to both new and established investigators in the field.

Enrique Cadenas
Lester Packer

SECTION I

The Redox Biochemistry of Hydrogen Sulfide

Investigating the Role of H_2S in 4-HNE Scavenging

Hilde Laggner[1], Bernhard M.K. Gmeiner[1]
Department of Medical Chemistry and Pathobiochemistry, Center of Pathobiochemistry and Genetics, Medical University of Vienna, Vienna, Austria
[1]Corresponding authors: e-mail address: hildegard.laggner@meduniwien.ac.at; bernhard.gmeiner@meduniwien.ac.at

Contents

1. Introduction 4
2. Experimental Compounds and Considerations 5
   2.1 H_2S generation 5
   2.2 Preparation of 4-HNE solutions 6
   2.3 Preparation of 4-HNA solutions 6
   2.4 Reaction of 4-HNE with H_2S 6
   2.5 Approaches to monitor the effect of H_2S on protein modification by 4-HNE 8
   2.6 Approaches to monitor the effect of H_2S on cytotoxicity of 4-HNE 10
   2.7 Detection of 4-HNE-modified cellular proteins 13
3. Conclusions and Perspectives 14
Acknowledgment 16
References 16

Abstract

4-HNE (4-hydroxy-2-nonenal) is a highly reactive α,β-unsaturated aldehyde generated from oxidation of polyunsaturated fatty acids and has been suggested to play a role in the pathogenesis of several diseases. 4-HNE can bind to amino acids, proteins, polynucleotides, and lipids and exert cytotoxicity. 4-HNE forms adducts (Michael adducts) with cysteine, lysine, as well as histidine on proteins with the thiol function as the most reactive nucleophilic moiety. Thus, detoxification strategies by 4-HNE scavenging compounds might be of interest. Recently, hydrogen sulfide (H_2S) has been identified as an endogenous vascular gasotransmitter and neuromodulator. Assuming that the low-molecular thiol H_2S may react with 4-HNE, methods to monitor the ability of H_2S to counteract the protein-modifying and cytotoxic activity of 4-HNE are described in this chapter.

Methods in Enzymology, Volume 555
ISSN 0076-6879
http://dx.doi.org/10.1016/bs.mie.2014.11.018

1. INTRODUCTION

4-Hydroxy-2-nonenal (4-HNE) is one of the reaction products of lipid hydroperoxide break down occurring in response to oxidative stress (Esterbauer, Schaur, & Zollner, 1991; Spiteller, Kern, Reiner, & Spiteller, 2001). The chemical reactivity of this and other α,β-unsaturated aldehydes has been extensively studied in the past (Esterbauer, Ertl, & Scholz, 1976; Esterbauer et al., 1991; Schultz, Yarbrough, & Johnson, 2005; Spiteller et al., 2001).

4-HNE has been shown to be toxic to cells (Esterbauer et al., 1991). Beside its bare cytotoxic ability, 4-HNE-modified proteins may play a mechanism in the pathogenesis of human diseases and in addition, 4-HNE may act as signaling molecule (Petersen & Doorn, 2004). Most of the biochemical effects of 4-HNE may be due to its easy reaction of the C=C bond (Michael addition) with the nucleophilic thiol and amino groups of free or protein-bound amino acids (cysteine, histidine, and lysine). Lipids (phosphatidyl-ethanol amine) and nucleic acids are also targets of this highly reactive aldehyde (Schaur, 2003). The double bond can be reduced by an NAD(P)H-dependent alkenal/one oxidoreductase forming 4-HNA (4-hydroxy-nonanal), thus detoxifying 4-HNE (Dick, Kwak, Sutter, & Kensler, 2001). Epoxidation of 4-HNE can also take place in presence of hydroperoxide (Schaur, 2003).

The C=O group can undergo hemi-acetal and acetal formation with alcohols or thiols. Schiff base formation with primary amino groups (e.g., lysine) and enzymatic oxidation (aldehyde dehydrogenase/NAD) and reduction (alcohol dehydrogenase/NADH) results in 4-hydroxy-nonenoic acid and 1,4-dihydroxy-nonen formation, respectively (Schaur, 2003). Oxidation of the 4-hydroxy group results in the formation of 4-ONE (4-oxo-2-nonenal), an extremely neurotoxic derivative (Lin et al., 2005).

As 4-HNE is suggested to play a role in the pathogenesis of several diseases, molecular strategies should be developed to detoxify this highly reactive compound (Aldini, Carini, Yeum, & Vistoli, 2014; Mali & Palaniyandi, 2013). Possible approaches are (i) inhibiting 4-HNE formation, (ii) activating/upregulating detoxifying enzymes, and (iii) scavenging of 4-HNE by low-molecular-weight compounds (Aldini et al., 2014). The latter are the cysteine-mimetic, lysine-mimetic, and histidine-mimetic HNE-sequestering agents directly reacting via Michael adduct and Schiff base formation. The reaction of 4-HNE with thiol compounds has received particular attention. Glutathione (GSH) reacts readily with 4-HNE, a reaction which has been attributed to the HNE-detoxifying action of GSH (Esterbauer, Zollner, & Scholz, 1975).

Recently, H_2S (hydrogen sulfide) has been identified as the third gasotransmitter, beside NO and CO, in the vasculature (Lefer, 2007; Leffler, Parfenova, Jaggar, & Wang, 2006; Wang, 2002; Zhao, Zhang, Lu, & Wang, 2001). The enzymes cystathionine-β-synthase (CBS EC 4.2.1.22), cystathionine-γ-lyase (CSE EC 4.4.1.1), and 3-mercapto-pyruvate sulfurtransferase (3MST EC 2.8.1.2) are responsible for the endogenous production of H_2S (Kabil & Banerjee, 2014; Shibuya et al., 2009).

In the brain (human, rat, and bovine), CBS is highly expressed and the primary physiological source of H_2S (Wang, 2012). Thus, a neuromodulatory action of H_2S has been proposed (Abe & Kimura, 1996; Kimura, 2000; Moore, Bhatia, & Moochhala, 2003). Perturbed H_2S production in the brain has been linked to certain diseases. Decreased *S*-adenosylmethionine concentration has been reported for Alzheimer's disease (Morrison, Smith, & Kish, 1996) which may lead to diminished CBS activity and result in low endogenous H_2S levels. An overproduction of H_2S was found in Down syndrome patients, where the CBS gene located on chromosome 21 is overexpressed as reported by Kamoun, Belardinelli, Chabli, Lallouchi, and Chadefaux-Vekemans (2003). The synthesis of endogenous H_2S is significantly lower but 4-HNE is markedly increased in Alzheimer's disease (Butterfield et al., 2006; Liu, Raina, Smith, Sayre, & Perry, 2003; Lovell, Ehmann, Mattson, & Markesbery, 1997). We found that the low-molecular-weight thiol H_2S exerts protective activity against 4-HNE induced cytotoxicity and HNE protein-adduct formation (Schreier et al., 2010). Here we describe various approaches to monitor the interaction of 4-HNE with H_2S. We refer to scavenging reactions and inhibition of protein modifying and cytoprotective properties of H_2S using a neuroblastoma cell line (SH-SY5Y).

2. EXPERIMENTAL COMPOUNDS AND CONSIDERATIONS

2.1. H_2S generation

Sodium hydrogen sulfide (NaHS) and disodium sulfide (Na_2S) can be purchased from Aldrich. NaHS or Na_2S stock solutions (100 mmol/L) are prepared daily in distilled water and stored on ice in the dark ($\varepsilon_{230\text{nm}} = 7700\ M^{-1}\ \text{cm}^{-1}$) (Nagy et al., 2014). Stock solutions are diluted to the desired concentrations in the respective buffer.

At pH 7.4, H_2S concentration is taken as 30% of the NaHS or Na_2S concentration (Beauchamp, Bus, Popp, Boreiko, & Andjelkovich, 1984; Reiffenstein, Hulbert, & Roth, 1992). The term H_2S for the sum of the sulfur-species H_2S, HS^-, and S^{2-} is used according to Whiteman et al. (2004).

2.2. Preparation of 4-HNE solutions

4-Hydroxy-2-nonenal-dimethyl acetal (4-HNE-DMA) is supplied from Sigma-Aldrich. 4-HNE is prepared from 4-HNE-DMA by acid hydrolysis. An aliquot of hexane solution containing 4-HNE-DMA is evaporated under a gentle stream of nitrogen at room temperature. An equal volume of cold HCl (1 mmol/L) is added and the sample incubated at 4 °C for 45 min under nitrogen atmosphere and gentle agitation. At the end of incubation, the concentration of 4-HNE is determined using $\varepsilon = 13{,}600\ M^{-1}\ \mathrm{cm}^{-1}$ at 222 nm. The stock solution is further diluted into buffer solution to the desired concentration.

4-HNE is also supplied in ethanol as solvent (Cayman Chemical). In this case, an aliquot of the stock solution is evaporated under a stream of nitrogen and subsequently 4-HNE dissolved in PBS. In aqueous media, 4-HNE can be dissolved at final concentrations between 42 mmol/L (6.6 mg/mL water) (Esterbauer et al., 1991) and about 6.4 mmol/L (1 mg/mL PBS) according to the supplier. 4-HNE in buffered working solutions should be prepared daily fresh and kept at 4 °C in the dark.

2.3. Preparation of 4-HNA solutions

4-HNA can be synthesised from 4-oxohexanal (Long, Smoliakova, Honzatko, & Picklo, 2008; Picklo, Amarnath, McIntyre, Graham, & Montine, 1999). 4-HNA was a generous gift of Dr. Matthew J. Picklo, Dept of Pharmacology, Physiology and Therapeutics, University of North Dakota School of Medicine, Grand Forks, USA. Present address: USDA-ARS Grand Forks Human Nutrition Research Center, Grand Forks, North Dakota, USA.

2.4. Reaction of 4-HNE with H_2S

All reactions are carried out in 50 mmol/L phosphate buffer pH 7.4 at 37 °C. The reaction is monitored as the decrease of 4-HNE (absorbance at 222 nm) according to Esterbauer et al. (1975). As NaHS in solution absorbs at 230 nm (Nagy et al., 2014), which is to close to the A_{max} of 4-HNE, HCl is added to the incubations (final concentration 50 mmol/L, pH <1) destroying 94–99% of NaHS, and the absorbance of the samples is recorded after 5 min.

Figure 1 shows the time- (A) and concentration- (B) dependent reaction of H_2S with 4-HNE (50 μmol/L) at 37 °C and pH 7.4. At a molar ratio of 1:20 (4-HNE:NaHS), 50% of the aldehyde is scavenged after about 40 min (Fig. 1A). As seen in Fig. 1B at a molar ratio of 1:60 and after 90 min of incubation, the carbonyl is completely scavenged by the thiol.

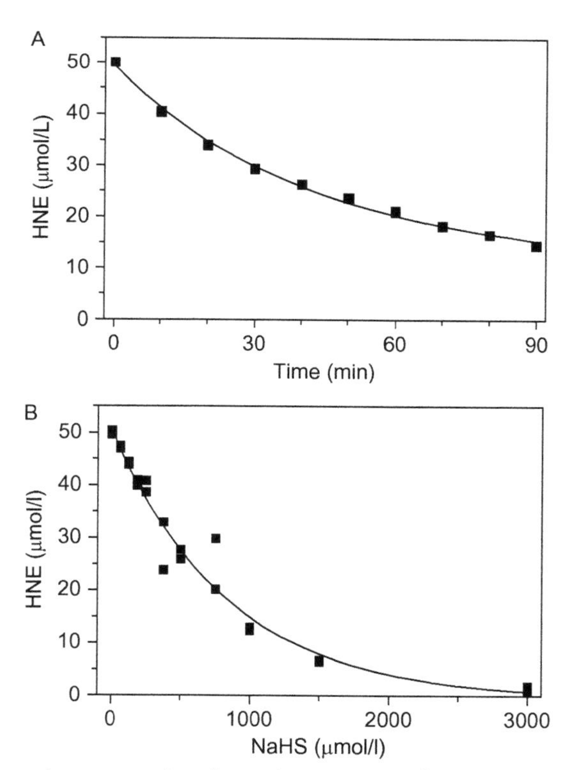

Figure 1 Time- and concentration-dependent reaction of 4-HNE with H_2S. (A) Reaction of 4-HNE (50 μmol/L) with NaHS (1 mmol/L) was followed for up to 90 min and the decrease in absorbance of the unsaturated carbonyl was monitored at 222 nm. (B) Reaction of 4-HNE (50 μmol/L) with increasing concentrations of NaHS was measured after an incubation time of 90 min. Individual points of two estimations are depicted (Schreier et al., 2010).

2.5. Approaches to monitor the effect of H_2S on protein modification by 4-HNE

2.5.1 Electrophoretic mobility shift

Protein modification by 4-HNE can be studied using human serum albumin (HSA) as a model protein. HSA can be supplied by Sigma-Aldrich. HSA (1 mg/mL) is incubated at 37 °C for 24 h with 2 mmol/L 4-HNE and the sulfur-containing compounds (NaHS, cysteine, methionine, *N*-acetylcysteine, GSH) ranging from 1.25 to 20 mmol/L in 0.1 mol/L phosphate buffer pH 7.0. After incubation, samples are subjected to native polyacrylamide (8% w/v) gel electrophoresis. 3 μg of protein sample is loaded per lane and separated at 200 V for 60 min. The gel is stained with Coomassie blue and protein modification is detected by electrophoretic mobility shift.

As seen in Fig. 2, 4-HNE causes a greater cathodic mobility indicating aldehyde modification of positively charged amino acid residues. H_2S (NaHS between 2.5 and 20 mmol/L) clearly inhibits the mobility shift. Very similar results are obtained with Na_2S, which is sometimes used to generate H_2S. Cysteine, GSH, and NAC run as a positive control inhibit HSA modification in contrast to methionine, which has no free SH-group.

2.5.2 Immunoblotting

HSA (1 mg/mL) is incubated at 37 °C for 24 h with 2 mmol/L 4-HNE and the sulfur-containing compounds (NaHS, cysteine, methionine,

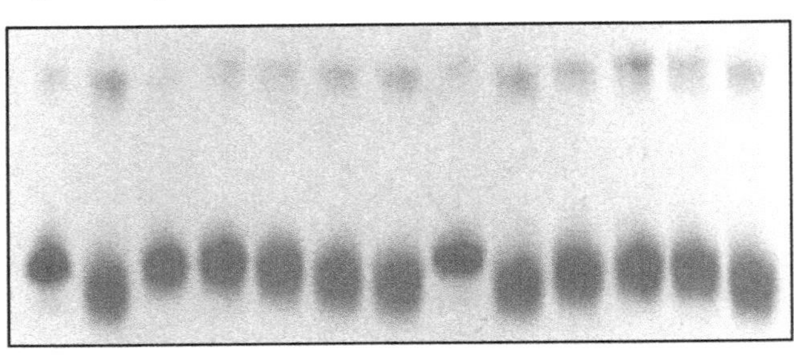

4-HNE (2 mmol/L)	–	+	+	+	+	+	+	–	+	+	+	+	+
NaHS (mmol/L)	–	–	20	10	5	2.5	1.25	–	–	–	–	–	–
Cysteine (5 mmol/L)										+			
GSH (5 mmol/L)											+		
NAC (5 mmol/L)												+	
Methionine (5 mmol/L)													+

Figure 2 Electrophoretic mobility shift of 4-HNE-modified HSA. HSA in PBS was incubated for 24 h in the absence or presence of the respective compound and analyzed on nondenaturing PAGE. Proteins were stained with Coomassie blue (Schreier et al., 2010).

N-acetylcysteine, GSH) ranging from 1.25 to 20 mmol/L in 0.1 mol/L phosphate buffer pH 7.0. After incubation, samples are subjected to native polyacrylamide (8% w/v) gel electrophoresis. 3 μg of protein sample is loaded per lane and separated at 200 V for 60 min. The gel is transferred onto nitrocellulose membrane (0.45 μm; Bio-Rad, Vienna, Austria). Membranes are blocked with 5% (w/v) nonfat milk in TBST solution [20 mmol/L Tris pH 7.4, 145 mmol/L NaCl, 0.1% (v/v) Tween-20] for 1 h at room temperature. The resulting blots are incubated with rabbit polyclonal antibody against 4-HNE (1:2500 dilution; Abcam) overnight at 4 °C. After three washes with TBST, membranes are incubated with goat anti-rabbit IgG-horseradish peroxidase-linked secondary antibody (1:7500 dilution; Sigma-Aldrich) for 1 h at room temperature, again followed by three washes with TBST for 1 h. The immunoreactive bands are visualized with Super Signal Pico (Pierce) in a chemiluminescence imager system (ChemiImager 4400 with AlphaEaseFC Software, InnoTech Corporation, Biozym).

The inhibition of 4-HNE protein modification is depicted in Fig. 3. At higher H_2S/4-HNE molar ratios, no or less immunostaining is observed, as is also seen in presence of the thio-amino acids with the exception of methionine.

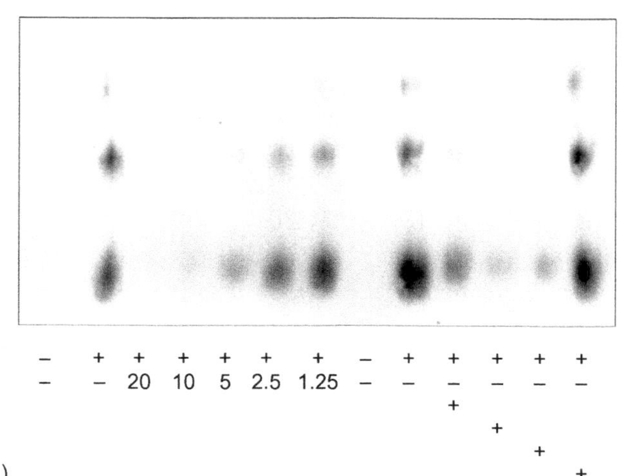

Figure 3 Western blot of 4-HNE-modified HSA. HSA in PBS was incubated for 24 h in the absence or presence of the respective compound and analyzed on nondenaturing PAGE. Separated proteins were transferred to nitrocellulose and detected by a 4-HNE-specific antibody (Schreier et al., 2010).

2.6. Approaches to monitor the effect of H_2S on cytotoxicity of 4-HNE

It is of note that when measuring cytotoxicity of various compounds in the presence of H_2S, the 3-(4,5-dimethylthiazol-2-yl)-2,5-diphenyltetrazolium bromide (MTT) test should not be used as H_2S interferes with the assay. Therefore, cytotoxicity is preferentially evaluated by neutral red retention or release of lactate dehydrogenase (LDH). Human neuroblastoma SH-SY5Y cells (ATCC CRL-2266) can be used as a cell model for neuro-degenerative diseases, e.g., Alzheimer's disease, and SH-SY5Y cells have been shown to be sensitive to 4-HNE (Lin et al., 2005).

SH-SY5Y cells are grown in RPMI 1640 medium containing 10% FCS (Gibco), 2 mmol/L glutamine, 50 U/mL streptomycin and 0.1 mg/mL penicillin (PAA Laboratories), and tissue culture flat tubes with 10 cm^2 growth areas (TPP). After washing with serum-free RPMI 1640 medium, cells are incubated with different concentrations of 4-HNE or 4-HNA in the presence or absence of thiol compound in serum-free RPMI 1640 medium for defined times in culture tubes with caps screwed to avoid loss of H_2S.

After incubation cells are washed, 2 mL of serum-free RPMI 1640 medium with 0.005% (w/v) neutral red (Sigma-Aldrich) is added and cells are incubated for 3 h at 37 °C. Then, cells are washed twice with PBS, lysed in 50% (v/v) ethanol/1% (v/v) acetic acid for 30 min, and neutral red retention is measured photometrically at 540 nm.

LDH assay can be performed in 96-well plates containing 50 μL of cell supernatant, 1 mmol/L sodium pyruvate, and 0.2 mmol/L NADH in Tris–HCl buffer (40 mmol/L, pH 7.4). Enzyme activity is recorded by measuring the rate of decrease of NADH (Δ 340 nm) for 5 min at 25 °C in a kinetic mode in a plate reader (Wallac, Victor^2 1420 multilabel counter) and expressed as U/L.

Incubating neuronal cells for 4 h with 4-HNE (5–50 μmol/L) results in a dose-dependent increase in cytotoxicity (reduction of neutral red retention; Fig. 4A). NaHS at 100 μmol/L completely inhibits the 4-HNE-induced (5–10 μmol/L) cytotoxicity. At 25 μmol/L 4-HNE, still a significant H_2S effect can be observed ($p < 0.01$), whereas at the highest dose (50 μmol/L) of 4-HNE, H_2S does not counteract the cytotoxic effect (Fig. 4A). The saturated hydroxy aldehyde 4-HNA, which is far less cytotoxic potential compared to 4-HNE (Picklo et al., 1999), shows no cytotoxic effect on SH-SY5Y cells under the conditions employed (Fig. 4B).

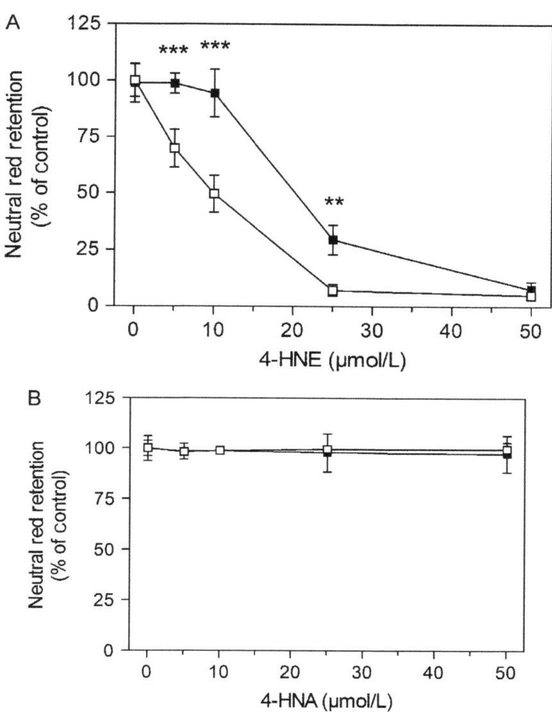

Figure 4 Effect of 4-HNE and 4-HNA on cell integrity. SH-SY5Y neuroblastoma cells were incubated in the absence (open symbols) or presence of 100 μmol/L NaHS (closed symbols) and with up to 50 μmol/L 4-HNE (A) or 4-HNA (B) for 4 h at 37 °C, and neutral red retention was measured. Data are expressed as mean ± SD ($n = 3$), ** $p < 0.01$, *** $p < 0.001$ by Newman–Keuls' test comparing NaHS-treated cells to corresponding cells without NaHS (Schreier et al., 2010).

Figure 5A shows the dose-dependent influence of NaHS (6.25–200 μmol/L) on the cytotoxic effect of 4-HNE (10 μmol/L). NaHS up to 200 μmol/L tested shows no influence on cell viability which is comparable with the findings of Hu, Lu, Wu, Wong, and Bian (2009) and Whiteman et al. (2005), who reported no effect on cell viability of up to 300 μmol/L NaHS for 24 h in SH-SY5Y cells. The cytotoxic effect of 4-HNE (55.2 ± 7.8% of control) is reversed in the presence of NaHS at a concentration as low as 12.5 μmol/L to 77 ± 3.2% of controls ($p < 0.05$ compared to 4-HNE alone). 100 μmol/L NaHS almost completely protects the cells from the 4-HNE action (89.7 ± 8.3% of control).

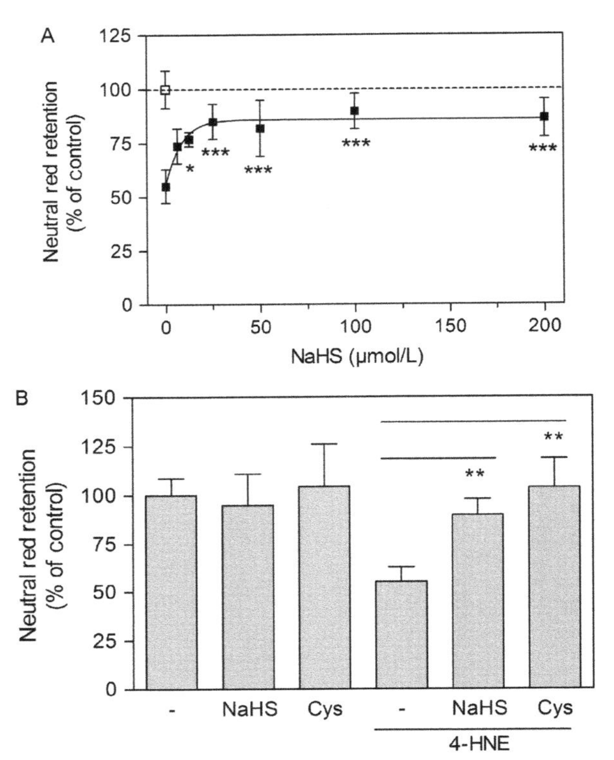

Figure 5 Effect of NaHS on 4-HNE-mediated cytotoxicity. (A) SH-SY5Y neuroblastoma cells were incubated with up to 200 µmol/L NaHS without (□) or with 10 µmol/L 4-HNE (■) for 4 h at 37 °C and neutral red retention was measured. Data are expressed as mean ± SD ($n=3$), * $p<0.05$, *** $p<0.001$ by Dunnett's test compared to cells treated with 4-HNE alone. (B) SH-SY5Y neuroblastoma cells were incubated with NaHS (100 µmol/L) or cysteine (150 µmol/L) in the absence or presence of 4-HNE (10 µmol/L) for 4 h at 37 °C and neutral red retention was measured. Data are expressed as mean ± SD ($n=3$), ** $p<0.01$ compared by Newman–Keuls' test to cells treated with 4-HNE alone (Schreier et al., 2010).

Our data of the 4-HNE/H_2S reaction suggest a biological significance at physiologically relevant concentrations of 1–10 µmol/L HNE (Subramaniam et al., 1997) and H_2S.

Cysteine (150 µmol/L) known to react readily with 4-HNE can be run as a positive control in the culture system and shows very similar cytoprotective action (see Fig. 5B) against 4-HNE (10 µmol/L) like NaHS (100 µmol/L). It is interesting to note that mature primary cortical neurons seem to be more sensitive to NaHS than immature neurons (Cheung, Peng, Chen, Moore, & Whiteman, 2007).

2.7. Detection of 4-HNE-modified cellular proteins

Beside model proteins like HSA, the inhibiting effect of H_2S on the protein-modifying activity of 4-HNE can also be tested using intact neuronal cells. SH-SY5Y cells are incubated with 4-HNE and NaHS for 4 h, washed twice with ice-cold PBS, and homogenized in lysis buffer [50 mmol/L HEPES pH 7.5, 150 mmol/L NaCl, 1.5 mmol/L $MgCl_2$, 1 mmol/L EGTA, 100 μmol/L NaF, 10 μmol/L Na_3VO_4, 10% (v/v) glycerol, 1% (v/v) NP40, and 1% (v/v) protease inhibitor cocktail from Sigma-Aldrich], and after 60 min on ice, homogenates are centrifuged at 20,800 × *g* for 10 min at 4 °C and protein concentrations within the supernatants are determined with the Bradford protein assay (Bio-Rad). Equal amounts of protein (30 μg) are separated by 8% SDS–polyacrylamide gel under reducing conditions and blotted on nitrocellulose membrane. Membranes are blocked with 5% (w/v) nonfat milk in TBST solution [20 mmol/L Tris pH 7.4, 145 mmol/L NaCl, 0.1% (v/v) Tween-20] for 1 h at room temperature. The resulting blots are incubated with rabbit polyclonal antibody against 4-HNE (1:2500 dilution; Abcam) overnight at 4 °C. After three washes with TBST, membranes are incubated with goat anti-rabbit IgG-horseradish peroxidase-linked secondary antibody (1:7500 dilution; Sigma-Aldrich) for 1 h at room temperature, again followed by three washes with TBST for 1 h. The immunoreactive bands are visualized with Super Signal Pico (Pierce) in a chemiluminescence imager system (ChemiImager 4400 with AlphaEaseFC Software, InnoTech Corporation, Biozym).

When cells were incubated with 4-HNE (10 μmol/L) and cell extracts were analyzed for 4-HNE-modified proteins, the results reveal that a variety of proteins is modified by the aldehyde under the conditions employed (see Fig. 6, lane 3). However, in the presence of NaHS (100 μmol/L), an almost complete inhibition of the 4-HNE protein-adduct formation can be observed (see Fig. 6, lane 4).

H_2S may counteract the protein-modifying and cytotoxic activity of 4-HNE considering the relatively high concentrations of H_2S found in the brain (Abe & Kimura, 1996; Kimura, 2000). However, recent work has suggested that levels of free H_2S *in vivo* are much lower than previously thought (Furne, Saeed, & Levitt, 2008; Kimura, Shibuya, & Kimura, 2012). On the other hand, as H_2S can be stored as bound sulfur (sulfane sulfur) and released (Ishigami et al., 2009; Kabil & Banerjee, 2014), one may speculate that the overall amount could be adequate to modulate pathophysiological reactions.

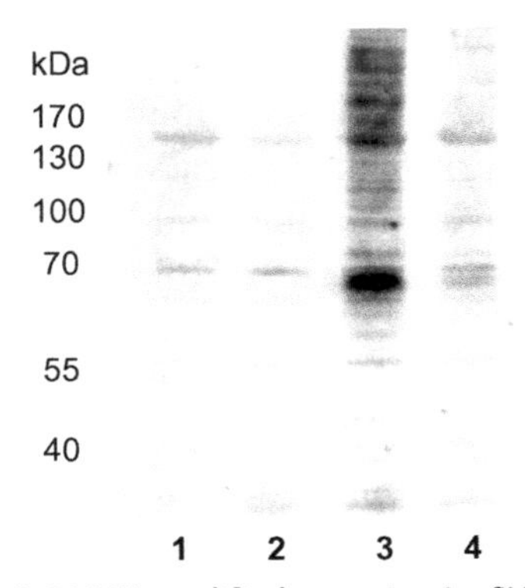

Figure 6 Western blot of 4-HNE-modified proteins in SH-SY5Y neuroblastoma cells. Cells were incubated for 4 h at 37 °C, and subsequently cells were lysed and 4-HNE-modified proteins detected by a 4-HNE-specific antibody. Control (1), 100 μmol/L NaHS (2), 10 μmol/L 4-HNE (3), and 10 μmol/L 4-HNE together with 100 μmol/L NaHS (4) (Schreier et al., 2010).

3. CONCLUSIONS AND PERSPECTIVES

The advanced lipid peroxidation end product 4-HNE has been shown to be toxic to cells (Esterbauer et al., 1991). Beside its bare cytotoxic ability, 4-HNE-modified proteins may play a mechanism in the pathogenesis of human diseases.

GSH is known to react with conjugated carbonyls (Esterbauer et al., 1975; Schultz et al., 2005), and this reaction may detoxify 4-HNE. The reaction of GSH or cysteine with 4-HNE in aqueous solution is straightforward resulting in the formation of a saturated aldehyde with the thio-compound linked to carbon 3 (see scheme A in Fig. 7), which can further react to cyclic products (Esterbauer et al., 1991). In analogy to this reaction and the reaction of 2-butenal with hydrogen sulfide (Badings et al., 1976; Kleipool, Tas, Maarse, Neeter, & Badings, 1976), one may hypothesize that 3-mercapto-4-hydroxy-nonanal could be an initial product of the reaction of H_2S with 4-HNE (see proposed scheme B in Fig. 7).

Thus H_2S can be counted to the so-called HNE-sequestering agents, i.e., direct scavenging low-molecular-weight compounds like the cysteine mimetics (Aldini et al., 2014). In this respect, compounds with defined H_2S-releasing property would be needed to use as therapeutic agents (Guo, Cheng, & Zhu, 2013; Szabo, 2007). The currently available H_2S

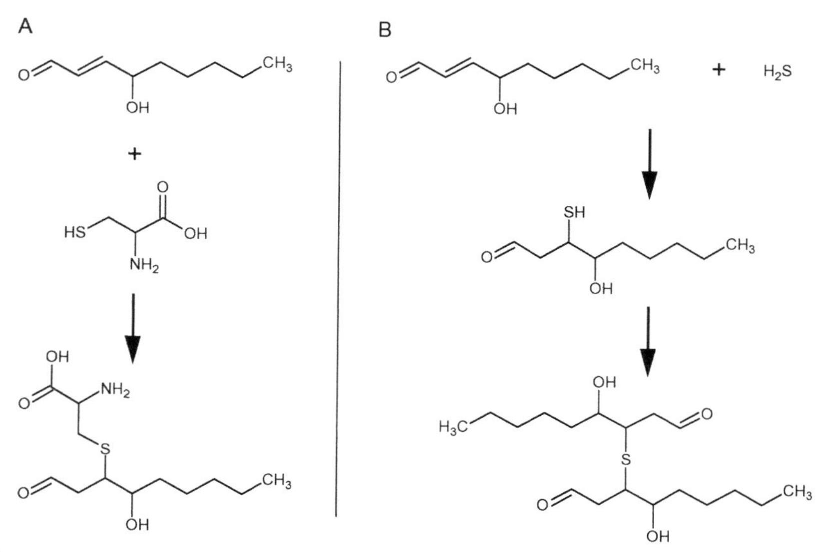

Figure 7 Proposed scheme for the reaction of 4-HNE with cysteine (A) or H_2S (B) (Schreier et al., 2010).

donors can be divided in inorganic and organic compounds and fast- and slow-releasing molecules (Papapetropoulos, Whiteman, & Cirino, 2014). The mostly used compound for in *in vitro* and *in vivo* experiments is the synthetic slow rate releasing H_2S donor GYY4137 (*p*-methoxyphenyl morpholino-phosphinodithioic acid), which is commercially available in contrast to other H_2S-releasing compounds reported in the literature (Papapetropoulos et al., 2014).

The present results add to the chemical foundations of hydrogen sulfide biology (Li & Lancaster, 2013), suggesting that H_2S may be an additional important protective factor against carbonyl stress by directly inactivating/modulating the action of highly reactive α,β-unsaturated aldehydes like 4-HNE (Schreier et al., 2010). H_2S is a lipophilic molecule (Li & Moore, 2008) and may therefore be an excellent scavenger of 4-HNE, which preferentially distributes in biomembranes rather than in the aqueous phase (Schaur, 2003). However, one must keep in mind that this action and possible therapeutic potential of H_2S in respect to degenerative diseases (e.g., AD) still needs more demonstration. The approaches described here provide an experimental basis for further studies on the reactivity of the gasotransmitter H_2S with unsaturated aldehydes.

ACKNOWLEDGMENT

The authors gratefully thank F. H. Slama for her help in preparing the manuscript.

REFERENCES

Abe, K., & Kimura, H. (1996). The possible role of hydrogen sulfide as an endogenous neuromodulator. *Journal of Neuroscience, 16*(3), 1066–1071.

Aldini, G., Carini, M., Yeum, K.-J., & Vistoli, G. (2014). Novel molecular approaches for improving enzymatic and nonenzymatic detoxification of 4-hydroxynonenal: Toward the discovery of a novel class of bioactive compounds. *Free Radical Biology and Medicine, 69*, 145–156.

Badings, H. T., Maarse, H., Kleipool, R. J., Tas, A. C., Neeter, R., & Noever de Brauw, M. C. (1976). Formation of odorous compounds from hydrogen sulphide and 2-butenal. *Zeitschrift für Lebensmittel-Untersuchung und -Forschung, 161*(1), 53–59.

Beauchamp, R. J., Bus, J., Popp, J., Boreiko, C., & Andjelkovich, D. A. (1984). A critical review of the literature on hydrogen sulfide toxicity. *Critical Reviews in Toxicology, 13*, 25–97.

Butterfield, D. A., Reed, T., Perluigi, M., De Marco, C., Coccia, R., Cini, C., et al. (2006). Elevated protein-bound levels of the lipid peroxidation product, 4-hydroxy-2-nonenal, in brain from persons with mild cognitive impairment. *Neuroscience Letters, 397*(3), 170–173.

Cheung, N. S., Peng, Z. F., Chen, M. J., Moore, P. K., & Whiteman, M. (2007). Hydrogen sulfide induced neuronal death occurs via glutamate receptor and is associated with calpain activation and lysosomal rupture in mouse primary cortical neurons. *Neuropharmacology, 53*(4), 505–514.

Dick, R. A., Kwak, M.-K., Sutter, T. R., & Kensler, T. W. (2001). Antioxidative function and substrate specificity of NAD(P)H-dependent alkenal/one oxidoreductase: A new role for leukotriene B412-hydroxydehydrogenase/15-oxoprostaglandin 13-reductase. *Journal of Biological Chemistry, 276*(44), 40803–40810.

Esterbauer, H., Ertl, A., & Scholz, N. (1976). The reaction of cysteine with [alpha], [beta]-unsaturated aldehydes. *Tetrahedron, 32*(2), 285–289.

Esterbauer, H., Schaur, R., & Zollner, H. (1991). Chemistry and biochemistry of 4-hydroxynonenal, malonaldehyde and related aldehydes. *Free Radical Biology and Medicine, 11*(1), 81–128.

Esterbauer, H., Zollner, H., & Scholz, N. (1975). Reaction of glutathione with conjugated carbonyls. *Zeitschrift für Naturforschung. Section C, 30*(4), 466–473.

Furne, J., Saeed, A., & Levitt, M. D. (2008). Whole tissue hydrogen sulfide concentrations are orders of magnitude lower than presently accepted values. *American Journal of Physiology. Regulatory, Integrative and Comparative Physiology, 295*(5), R1479–R1485.

Guo, W., Cheng, Z. Y., & Zhu, Y. Z. (2013). Hydrogen sulfide and translational medicine. *Acta Pharmacologica Sinica, 34*(10), 1284–1291.

Hu, L.-F., Lu, M., Wu, Z.-Y., Wong, P. T.-H., & Bian, J.-S. (2009). Hydrogen sulfide inhibits rotenone-induced apoptosis via preservation of mitochondrial function. *Molecular Pharmacology, 75*(1), 27–34.

Ishigami, M., Hiraki, K., Umemura, K., Ogasawara, Y., Ishii, K., & Kimura, H. (2009). A source of hydrogen sulfide and a mechanism of its release in the brain. *Antioxidants & Redox Signaling, 11*(2), 205–214.

Kabil, O., & Banerjee, R. (2014). Enzymology of H2S biogenesis, decay and signaling. *Antioxidants & Redox Signaling, 20*(5), 770–782.

Kamoun, P., Belardinelli, M.-C., Chabli, A., Lallouchi, K., & Chadefaux-Vekemans, B. (2003). Endogenous hydrogen sulfide overproduction in down syndrome. *American Journal of Medical Genetics Part A, 116A*(3), 310–311.

Kimura, H. (2000). Hydrogen sulfide induces cyclic AMP and modulates the NMDA receptor. *Biochemical and Biophysical Research Communications, 267*(1), 129–132.

Kimura, H., Shibuya, N., & Kimura, Y. (2012). Hydrogen sulfide is a signaling molecule and a cytoprotectant. *Antioxidants & Redox Signaling, 17*(1), 45–57.

Kleipool, R. J., Tas, A. C., Maarse, H., Neeter, R., & Badings, H. T. (1976). Reaction of hydrogen sulphide with 2-alkenals. *Zeitschrift für Lebensmittel-Untersuchung und -Forschung, 161*(3), 231–238.

Lefer, D. J. (2007). A new gaseous signaling molecule emerges: Cardioprotective role of hydrogen sulfide. *Proceedings of the National Academy of Sciences of the United States of America, 104*(46), 17907–17908.

Leffler, C. W., Parfenova, H., Jaggar, J. H., & Wang, R. (2006). Carbon monoxide and hydrogen sulfide: Gaseous messengers in cerebrovascular circulation. *Journal of Applied Physiology, 100*(3), 1065–1076.

Li, Q., & Lancaster, J. R., Jr. (2013). Chemical foundations of hydrogen sulfide biology. *Nitric Oxide, 35*, 21–34.

Li, L., & Moore, P. K. (2008). Putative biological roles of hydrogen sulfide in health and disease: A breath of not so fresh air? *Trends in Pharmacological Sciences, 29*(2), 84–90.

Lin, D., Lee, H.-G., Liu, Q., Perry, G., Smith, M. A., & Sayre, L. M. (2005). 4-Oxo-2-nonenal is both more neurotoxic and more protein reactive than 4-hydroxy-2-nonenal. *Chemical Research in Toxicology, 18*(8), 1219–1231.

Liu, Q., Raina, A. K., Smith, M. A., Sayre, L. M., & Perry, G. (2003). Hydroxynonenal, toxic carbonyls, and Alzheimer disease. *Molecular Aspects of Medicine, 24*(4–5), 305–313.

Long, E. K., Smoliakova, I., Honzatko, A., & Picklo, M. J. (2008). Structural characterization of α, β-unsaturated aldehydes by GC/MS is dependent upon ionization method. *Lipids, 43*(8), 765–774.

Lovell, M. A., Ehmann, W. D., Mattson, M. P., & Markesbery, W. R. (1997). Elevated 4-hydroxynonenal in ventricular fluid in Alzheimer's disease. *Neurobiology of Aging, 18*(5), 457–461.

Mali, V. R., & Palaniyandi, S. S. (2013). Regulation and therapeutic strategies of 4-hydroxy-2-nonenal metabolism in heart disease. *Free Radical Research, 48*(3), 251–263.

Moore, P. K., Bhatia, M., & Moochhala, S. (2003). Hydrogen sulfide: From the smell of the past to the mediator of the future? *Trends in Pharmacological Sciences, 24*(12), 609–611.

Morrison, L. D., Smith, D. D., & Kish, S. J. (1996). Brain S-adenosylmethionine levels are severely decreased in Alzheimer's disease. *Journal of Neurochemistry, 67*(3), 1328–1331.

Nagy, P., Pálinkás, Z., Nagy, A., Budai, B., Tóth, I., & Vasas, A. (2014). Chemical aspects of hydrogen sulfide measurements in physiological samples. *Biochimica et Biophysica Acta, 1840*(2), 876–891.

Papapetropoulos, A., Whiteman, M., & Cirino, G. (2014). Pharmacological tools for hydrogen sulfide research: A brief, introductory guide for beginners. *British Journal of Pharmacology*. http://dx.doi.org/10.1111/bph.12806.

Petersen, D. R., & Doorn, J. A. (2004). Reactions of 4-hydroxynonenal with proteins and cellular targets. *Free Radical Biology & Medicine, 37*(7), 937–945.

Picklo, M. J., Amarnath, V., McIntyre, J. O., Graham, D. G., & Montine, T. J. (1999). 4-Hydroxy-2(E)-nonenal inhibits CNS mitochondrial respiration at multiple sites. *Journal of Neurochemistry, 72*(4), 1617–1624.

Reiffenstein, R. J., Hulbert, W. C., & Roth, S. H. (1992). Toxicology of hydrogen sulfide. *Annual Review of Pharmacology and Toxicology, 32*, 109–134.

Schaur, R. J. (2003). Basic aspects of the biochemical reactivity of 4-hydroxynonenal. *Molecular Aspects of Medicine, 24*(4–5), 149–159.

Schreier, S. M., Muellner, M. K., Steinkellner, H., Hermann, M., Esterbauer, H., Exner, M., et al. (2010). Hydrogen sulfide scavenges the cytotoxic lipid oxidation product 4-HNE. *Neurotoxicity Research, 17*(3), 249–256.

Schultz, T. W., Yarbrough, J. W., & Johnson, E. L. (2005). Structure-activity relationships for reactivity of carbonyl-containing compounds with glutathione. *SAR and QSAR in Environmental Research, 16*(4), 313–322.

Shibuya, N., Tanaka, M., Yoshida, M., Ogasawara, Y., Togawa, T., Ishii, K., et al. (2009). 3-Mercaptopyruvate sulfurtransferase produces hydrogen sulfide and bound sulfane sulfur in the brain. *Antioxidants & Redox Signaling, 11*(4), 703–714.

Spiteller, P., Kern, W., Reiner, J., & Spiteller, G. (2001). Aldehydic lipid peroxidation products derived from linoleic acid. *Biochimica et Biophysica Acta, 1531*(3), 188–208.

Subramaniam, R., Roediger, F., Jordan, B., Mattson, M. P., Keller, J. N., Waeg, G., et al. (1997). The lipid peroxidation product, 4-hydroxy-2-*trans*-nonenal, alters the conformation of cortical synaptosomal membrane proteins. *Journal of Neurochemistry, 69*(3), 1161–1169.

Szabo, C. (2007). Hydrogen sulphide and its therapeutic potential. *Nature Reviews. Drug Discovery, 6*(11), 917–935.

Wang, R. U. I. (2002). Two's company, three's a crowd: Can H2S be the third endogenous gaseous transmitter? *FASEB Journal, 16*(13), 1792–1798.

Wang, R. (2012). Physiological implications of hydrogen sulfide: A whiff exploration that blossomed. *Physiological Reviews, 92*(2), 791–896.

Whiteman, M., Armstrong, J. S., Chu, S. H., Jia-Ling, S., Wong, B.-S., Cheung, N. S., et al. (2004). The novel neuromodulator hydrogen sulfide: An endogenous peroxynitrite 'scavenger'? *Journal of Neurochemistry, 90*(3), 765–768.

Whiteman, M., Cheung, N. S., Zhu, Y.-Z., Chu, S. H., Siau, J. L., Wong, B. S., et al. (2005). Hydrogen sulphide: A novel inhibitor of hypochlorous acid-mediated oxidative damage in the brain? *Biochemical and Biophysical Research Communications, 326*(4), 794–798.

Zhao, W., Zhang, J., Lu, Y., & Wang, R. (2001). The vasorelaxant effect of H_2S as a novel endogenous gaseous K(ATP) channel opener. *EMBO Journal, 20*(21), 6008–6016.

CHAPTER TWO

Inhalation Exposure Model of Hydrogen Sulfide (H_2S)-Induced Hypometabolism in the Male Sprague-Dawley Rat

Asaf Stein*,†, David W. Kraus*,†, Jeannette E. Doeller*,†, Shannon M. Bailey*,†,‡,1

*Department of Environmental Health Sciences, University of Alabama at Birmingham, Birmingham, Alabama, USA
†Center for Free Radical Biology, University of Alabama at Birmingham, Birmingham, Alabama, USA
‡Department of Pathology, University of Alabama at Birmingham, Birmingham, Alabama, USA
[1]Corresponding author: e-mail address: sbailey@uab.edu

Contents

1. Introduction 20
2. Exposure Protocol for H_2S-Induced Hypometabolism in Rats 21
   2.1 Special gases and equipment 21
   2.2 Animals 22
   2.3 H_2S exposure protocol 22
   2.4 Physiological measurements 25
   2.5 Lung histopathology assessment 27
   2.6 Tissue and blood collection after H_2S exposure for experimental measurements 28
   2.7 Comments on measuring H_2S and H_2S-derived metabolites in blood and tissues from H_2S-exposed rats 29
3. Other Considerations for H_2S Exposure Studies 31
4. Summary 32
Acknowledgments 33
References 33

Abstract

Hydrogen sulfide (H_2S) has been accepted as a physiologically relevant cell-signaling molecule with both toxic and beneficial effects depending on its concentration in mammalian tissues. Notably, exposure to H_2S in breathable air has been shown to decrease aerobic metabolism and induce a reversible hypometabolic-like state in laboratory rodent models. Herein, we describe an experimental exposure setup that can be used to define the reversible cardiovascular and metabolic physiology of rodents (rats) during H_2S-induced hypometabolism and following recovery.

Methods in Enzymology, Volume 555
ISSN 0076-6879
http://dx.doi.org/10.1016/bs.mie.2014.11.019

1. INTRODUCTION

Hydrogen sulfide (H_2S), an environmental toxicant, has gained significant attention in the biomedical research community in recent years following its discovery by Abe and Kimura (1996) as an endogenously produced cell-signaling molecule in mammalian tissues (see reviews by Kabil, Motl, & Banerjee, 2014; Kimura, Shibuya, & Kimura, 2012; Wang, 2012). Interest in H_2S for therapeutic uses is also quite high as pharmacological application of H_2S and other H_2S-donating drugs have shown promise in preventing and/or reducing injury and disease in experimental animal model systems (Elrod et al., 2007; King et al., 2014; Li, Salto-Tellez, Tan, Whiteman, & Moore, 2009; Polhemus et al., 2013). Importantly, the study of H_2S in mammalian physiology intensified following publication of the landmark paper by Roth and coworkers who reported that H_2S induced a "suspended animation"-like or hypometabolic state in normal healthy laboratory mice (Blackstone, Morrison, & Roth, 2005). When mice were exposed to 80 ppm H_2S in their breathable air for 6 h, heart rate, CO_2 production, and O_2 consumption were significantly depressed, and body temperature dropped to just above ambient room temperature (Blackstone et al., 2005). It should be mentioned that ambient temperature in these studies was 13 °C. This low temperature may have exaggerated decreases in these metabolic parameters. Importantly, when H_2S was removed from the air, metabolism returned to pre-exposure baseline levels in mice, suggesting reversibility of H_2S-mediated hypometabolism. Following this study, Volpato et al. (2008) confirmed the hypometabolic effects of H_2S in mice that were maintained at higher ambient temperatures; e.g., 35 °C. While the mechanism underlying H_2S-induced hypometabolism remains unknown, it is proposed that hypometabolism may be attributed to an inhibitory effect on mitochondrial respiration through the binding of H_2S (or SH^-) to cytochrome *c* oxidase (Nicholls, Marshall, Cooper, & Wilson, 2013), inducing a reversible "pause" in mitochondrial function at low concentrations of H_2S (Elrod et al., 2007; Szabo et al., 2014). In light of this, reversible metabolic inhibition by H_2S might be highly useful in trauma treatment and disease prevention as depression of mitochondrial respiration would lower reactive oxygen species production and oxidative tissue injury.

Although the phenomenon of H_2S-induced hypometabolism is largely accepted in the field, the ability of H_2S to function as a ubiquitous trigger of metabolic suppression has not been consistently found in all mammalian species tested. Notably, H_2S-induced hypometabolism has been shown to be less pronounced or even absent in larger mammalian species; e.g., sheep and pigs (Drabek et al., 2011; Haouzi et al., 2008; Li, Zhang, Cai, & Redington, 2008; Wagner et al., 2011). Our work highlighted in this chapter shows that rats are also less sensitive to H_2S-induced hypometabolism (Stein et al., 2012) as compared with mice (Blackstone et al., 2005). There are many possible reasons to explain differential susceptibilities to H_2S-induced hypometabolism, including differences in experimental conditions (H_2S concentration, duration of exposure, use of awake vs. sedated animals), the form of H_2S used (H_2S gas vs. NaHS or Na_2S), route of administration (inhaled vs. injected), a dilution effect linked to body mass (higher ventilation rate to body mass ratio in mice vs. larger mammals), and species differences in H_2S detoxifying systems. Notably, some biological effects of H_2S in *in vitro* systems have been shown to be enhanced by lowering the O_2 tension (Koenitzer et al., 2007). As rates of mitochondrial oxygen consumption have been shown to be altered by changes in O_2 concentration (Gnaiger, Steinlechner-Maran, Mendez, Eberl, & Margreiter, 1995; Wilson, Owen, & Erecinska, 1979), we speculated that O_2 tension may also be a critical factor determining the extent of H_2S-induced hypometabolism in mammals. This is supported by studies showing that mitochondria incubated under low O_2 are more sensitive to H_2S-mediated inhibition of respiration (Gnaiger & Kuznetsov, 2002; Gnaiger et al., 1995; Wilson et al., 1979). Therefore, we developed an inhalation exposure system to examine the interaction of H_2S and O_2 on metabolism *in vivo* using a rat model. Specifically, this system was used to test the novel hypothesis that reducing the O_2 tension in breathable air would enhance H_2S-induced hypometabolism in rats (Stein et al., 2012). In this chapter, we describe the experimental setup used for performing H_2S exposures, provide an example of a typical experimental protocol, and show the effects of inhaled H_2S on some physiological parameters. We also include helpful suggestions on how to prepare tissues and blood for various experimental endpoint measurements, and include a brief discussion regarding measurement of H_2S in biological samples using an H_2S-specific polarographic sensor developed by Drs. D.W. Kraus and J.E. Doeller (Doeller et al., 2005).

2. EXPOSURE PROTOCOL FOR H_2S-INDUCED HYPOMETABOLISM IN RATS

2.1. Special gases and equipment

A gas mixture of 5% H_2S:95% N_2 was used to generate the desired H_2S concentration for inhalation exposure studies. This special gas mixture was obtained from Scott Specialty Gases (Plumsteadville, PA), which is now Air Liquide America Specialty Gases LLC (Houston, TX). The N_2 and air tanks used for experiments were obtained from a local gas supply company that provides standard gas mixtures for research laboratories. The digital mass flow controllers used to blend gases were obtained from Sierra Instruments (Monterey, CA). The H_2S gas analyzer used to confirm the H_2S concentration in the exposure chamber was a Rack-Mount configured gas analyzer from Interscan Corporation (Simi Valley, CA). A 50 ppm H_2S gas standard (Scott Specialty Gases) was used to calibrate the H_2S analyzer. An O_2 analyzer (S-3A AEI Technologies Inc., Pittsburgh, PA) was also used to monitor O_2 concentration.

2.2. Animals

The experimental animal model used in this protocol was the male Sprague-Dawley rat (Harlan Laboratories, Indianapolis, IN). Male rats (300–350 g) were allowed to acclimate for at least 1 week before use in exposure studies. The acclimation period is necessary to allow rats to return back to metabolic homeostasis before their use in experimental studies. Rats used in this protocol were housed under a standard 12 h light:12 h dark cycle with normal laboratory rat chow and water provided *ad libitum*. Further, the rats used in these studies were viral antigen-free (VAF) rats. VAF rats are raised and maintained in tightly controlled isolators to prevent exposure to infectious agents, including bacteria, viruses, and parasites known to produce respiratory disorders, which could confound experimental endpoints of interest; e.g., various respiratory and cardiovascular physiological parameters, lung histopathology, cell-signaling responses, and inflammatory markers.

2.3. H_2S exposure protocol

For exposure experiments, one male VAF rat was placed in a custom-built 6.0 L glass exposure chamber similar to the setup described in Ballinger et al. (2005) and acclimated for at least 30 min with 100% air flowing through the chamber at a rate of 3.0 L/min. A diagram illustrating the basic experimental setup system for these studies is provided in Fig. 1. To achieve the desired O_2 and H_2S concentrations for these experiments, digital mass flow controllers were used to mix the gas from three separate gas tanks containing (1) air alone; (2) N_2 gas alone; or, (3) 5% H_2S:95% N_2 gas mixture. Importantly, the inflow rate of H_2S into the system did not affect the gas phase

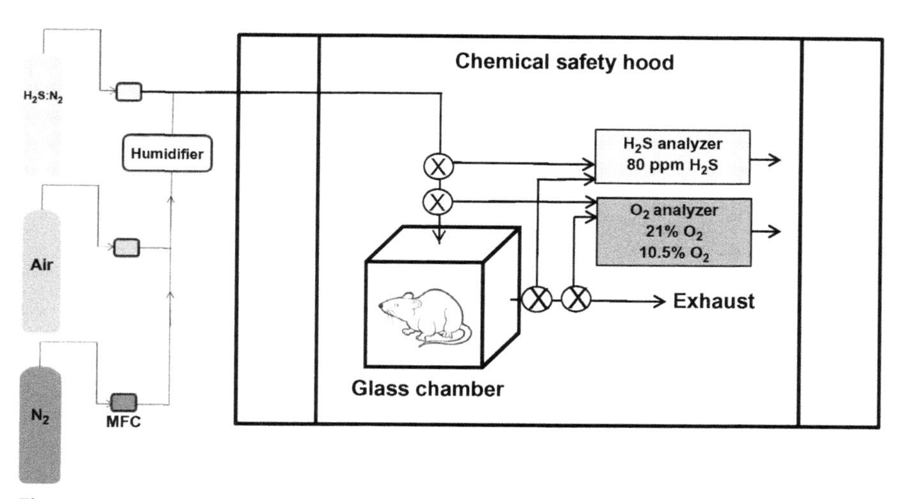

Figure 1 Diagram of the H_2S exposure setup. Rats are placed in the glass exposure chamber fitted with input and output ports. The system is housed in a ventilated chemical safety hood. Digital mass flow controllers (MFC) mix input gases from three tanks containing 5% H_2S:95% N_2, air, and N_2 to create a 3.0 L/min stream of 80 ppm H_2S in 21% or 10.5% O_2 balance N_2 that is humidified before reaching the glass chamber. An O_2 analyzer and an H_2S analyzer (calibrated with a precision H_2S standard) monitor O_2 and H_2S concentrations of both input and output chamber gas, respectively.

concentration of O_2. The exposure system was also designed so that air and N_2 gases were humidified prior to injection of H_2S into the flow system. The input and output chamber gas concentrations of O_2 and H_2S were continuously monitored during experiments using the gas analyzers described in Section 2.1. With this type of setup, the O_2 concentration in the exposure chamber inflow and outflow air stream can also be used to determine the animal O_2 consumption rate before, during, and after gas exposures. Moreover, the concentration of H_2S in the exposure chamber was also monitored during H_2S exposures using a H_2S gas analyzer. The experimental exposures were conducted at the same time each day from 9 AM to 3 PM, which corresponds to the less active (sleep) period of the day for rats, and rat chow was not present during the exposure period. The entire exposure setup (Fig. 1) was contained within a chemical safety hood with the ambient temperature maintained at 22–23 °C.

Exposures were started at time zero by adjusting the mass flow controllers to deliver a specific $H_2S:O_2$ gas mixture using the same flow rate. The equilibration time for the exposure chamber volume used in these studies

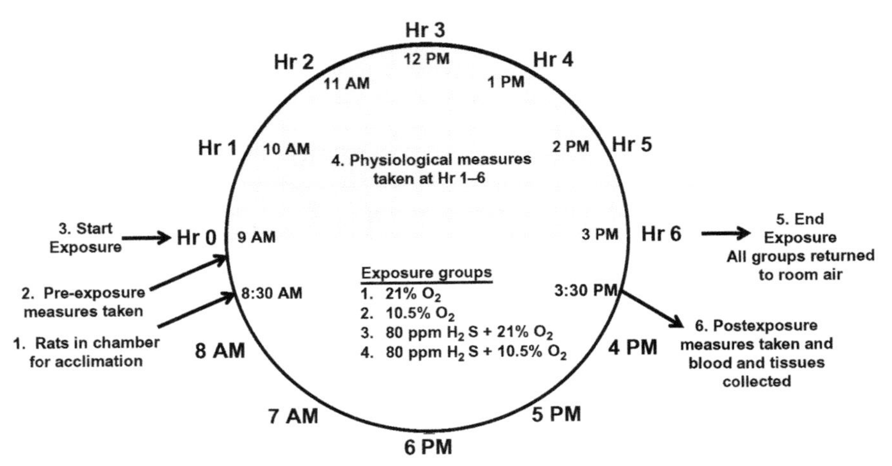

Figure 2 A typical H_2S exposure experimental protocol. Rats are placed into the glass exposure chamber and allowed to acclimate (1) before baseline physiological parameters are measured (2). At 9 AM, the exposure is started (3) with the rat exposed to one of the following exposure conditions for 6 h: (1) 21% O_2; (2) 10.5% O_2, (3) 80 ppm H_2S +21% O_2, or (4) 80 ppm H_2S+10.5% O_2. Physiological parameters (heart rate, breathing rate, body temperature, and hemoglobin saturation) are taken every hour (on the hour) until 3 PM (4). The rat continues to breathe the appropriate gas mixture through a nose cone when physiological measures are being taken. At 3 PM (after 6 h of exposure), the protocol is stopped and the rat is allowed to recover and breathe air (5). After the desired recovery period is completed, a final measure is taken for all physiological parameters before the rat is anesthetized for collection of blood and tissues for various biochemical and molecular measures.

was approximately 5 min. The typical protocol used to study H_2S-mediated hypometabolism in rats involved the following treatment groups: (1) 21% O_2 (i.e., air control group); (2) 10.5% O_2 (i.e., the low O_2 control group); (3) 80 ppm H_2S+21% O_2; and (4) 80 ppm H_2S+10.5% O_2. Rats were maintained under these exposure conditions for 6 h to investigate the potential impact inhaled H_2S had on various physiological parameters and biochemical endpoints. Using this system, the H_2S concentration can be adjusted; however, higher concentrations of H_2S (e.g., up to 160 ppm) did not induce a greater hypometabolic effect in rats (unpublished data). After the exposure protocol was completed, rats were allowed to recover in room air before tissue and blood were collected for experimental measures. A schematic of the exposure protocol for these studies is provided in Fig. 2. This exposure protocol was approved by the

IACUC and studies were conducted in accordance with the National Research Council *Guide for the Care and Use of Laboratory Animals* (8th Edition, 2011).

2.4. Physiological measurements

A MouseOx® Pulse Oximeter (STARR Life Sciences, Oakmont, PA) was used to noninvasively monitor and record heart rate, ventilation rate, and arterial oxygen saturation (hemoglobin saturation) before, during, and after exposures. These parameters serve two purposes: (1) as *indicators* of the physiological state of exposed animals for adjustment if needed to achieve a desired degree of hypometabolism and (2) as *reporters* of the metabolic status of animals during the exposure and recovery periods. For experiments, these physiological parameters were measured before the exposure (pre-exposure baseline assessment), once per hour during the exposure period (Hr 1–6), and at 30 min intervals after the exposure (Fig. 2). To measure these physiological parameters during the exposure period, the rat was gently removed from the exposure chamber and a custom-made face cone was placed over the rat's nose to maintain continuous exposure to the gas. The rat foot sensor was placed on the right hind foot for 5 min. This provided sufficient time for physiological readings to stabilize before data were recorded. During this time, the inner ear temperature was also measured with an infrared thermometer as a reliable indicator of core body temperature. After these physiological responses were recorded, the animal was carefully placed back into the exposure chamber until the next measurement time point, at which time the same procedure was performed. At the end of the exposure protocol, the animal was allowed to recover under room air conditions for a designated period of time (typically 30 min or 1 h) before the animal was anesthetized for blood and tissue collection.

An example of the type of physiological data that can be collected using this experimental system is provided in Fig. 3. These results show the effects that H_2S (80 ppm), low O_2 (10.5% O_2), or a combination exposure have on heart rate (Fig. 3A), ventilation rate (Fig. 3B), body temperature (Fig. 3C), and hemoglobin saturation (Fig. 3D) during a 6 h exposure period followed by a 30 min recovery period. These results were collected with a cohort of male VAF Sprague-Dawley rats ($n = 5$–12 animals/group) and illustrate the ability of H_2S to induce a hypometabolic state. Notably, the overall hypometabolic effect of 80 ppm H_2S was larger when H_2S

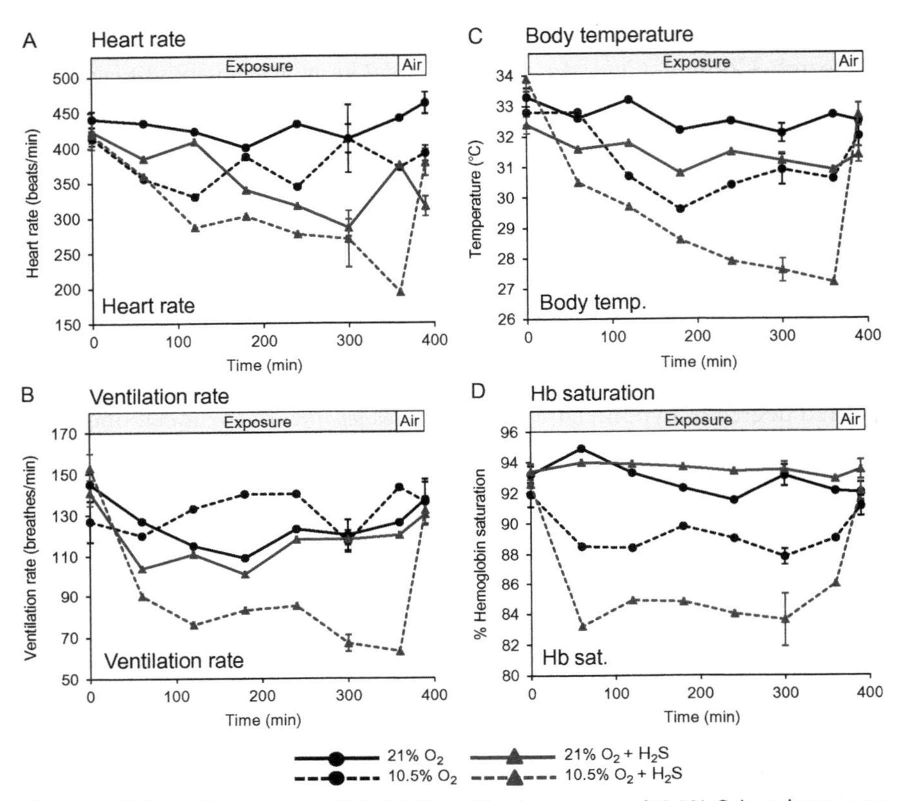

Figure 3 Effect of hydrogen sulfide (H_2S) and/or low oxygen (10.5% O_2) on heart rate (A), ventilation rate (B), body temperature (C), and % hemoglobin saturation (D) in rats subjected to a 6 h exposure protocol. The results shown in these panels illustrate the effect 80 ppm H_2S in combination with either 21% O_2 (red solid line (dark gray in the print version), triangles) or 10.5% O_2 (red dashed line (light gray in the print version), triangles) has on key physiological responses in male Sprague-Dawley rats. The control groups for these exposures are 21% O_2 alone (black solid line, circles) and 10.5% O_2 alone (black dashed line, circles). Measures were taken at baseline (pre-exposure, time zero) and every hour during the 6 h exposure. At the end of the exposure, animals were maintained under room air conditions for 30 min (recovery period) before the last measure was taken. Results represent the average responses from 5 to 12 rats/exposure group with SEM provided at select time points (time zero, 300 and 400 min).

was administered (inhaled) in combination with low O_2 concentration (i.e., 10.5% O_2) as compared to results obtained with 80 ppm H_2S + 21% O_2. This was particularly the case with regard to ventilation rate, body temperature, and hemoglobin saturation, but not with heart rate. Similar impacts on heart rate were observed in both H_2S exposure groups (21% O_2 and

10.5% O_2), the only exception was at the 6 h time point where there was a large disparity in heart rate between the two groups; the reason for this is not known. Interestingly, the impact of H_2S + low O_2 was more immediate for ventilation rate and hemoglobin saturation as significant depression of these parameters was observed at the first (Hr 1) exposure time point. In contrast, heart rate and body temperature gradually decreased over the entire 6 h exposure. Once exposures to H_2S and/or low O_2 were ended, these physiological parameters largely returned to baseline (i.e., pre-exposure) levels with one exception: the heart rate. The failure of heart rate to return to pre-exposure levels within the first 30 min of recovery suggests a potential persistent hypometabolic or toxic effect of H_2S on the heart. Additional experimental results (e.g., cardiac signaling events) collected from rats subjected to these exposure protocols are included in Stein et al. (2012).

2.5. Lung histopathology assessment

Because H_2S is a toxic gas (at high concentrations; e.g., >300 ppm), it is important to assess the impact H_2S has on lung histopathology even when used at the lower nonlethal concentrations suggested in this protocol or in other experimental protocols designed to induce a hypometabolic state. Airway injury can be assessed by several well-established histological methods, with one provided here for use in studies. Following administration of anesthesia, the trachea is cannulated and a mid-line thoracotomy is performed to expose the lungs. Saline is perfused to clear the lungs of blood after which the lungs are resected from the thoracic cavity and formalin-fixed under 25 cm H_2O hydrostatic pressure for 1 h (Hayatdavoudi, Crapo, Miller, & O'Neil, 1980). After fixation, a cross-section of the lung tissue (e.g., a cut that is perpendicular to the main airways) is removed and embedded in paraffin for the preparation of tissue sections that can be stained with hemotoxylin and eosin (H&E) for histopathology scoring. The H&E-stained tissue sections should be labeled with a numeric code so that they can be evaluated by a pathologist blinded to the experimental design and status of each individual animal used in the study. A standard histopathological scoring system is provided here for assessing airway damage to the lungs and is based in methods used in Fanucchi et al. (2012). The H&E-stained lung sections can also be evaluated for the presence (increased numbers) of cell nuclei in the alveolar regions and for the presence of extravascular red blood cells, an additional indicator of epithelial damage.

Lung histopathological scoring system

Score	Histopathological description
0	Normal columnar airway epithelium with a majority of ciliated cells and a smaller number of nonciliated cells
1	Up to 25% of the airway epithelium with some evidence of epithelial cell injury, defined as cell swelling, vacuolization, and/or presence of a very thin epithelium
2	25–50% of the airway epithelium with cell injury markers
3	50–75% of the airway epithelium with cell injury markers
4	>75% of the airway epithelium with cell injury markers

Using these methods and scoring system, we previously reported that exposure to 80 ppm H_2S under both normoxic (21% O_2) and low O_2 (10.5%) conditions damaged the lungs as epithelial cell exfoliation and swelling were observed in both the large and small airways of the lung (Stein et al., 2012). Thus, the potential for tissue damage should not be ignored in models of low exposure (inhaled) H_2S.

2.6. Tissue and blood collection after H_2S exposure for experimental measurements

Following exposures and recovery, blood can be collected by standard laboratory protocols to generate serum or plasma for a wide variety of biochemical measurements. Blood can be collected from the heart via a cardiac puncture procedure. Following full anesthesia (i.e., no response following a toe-pinch), a mid-line thoracotomy is performed to open the chest cavity. Using a 3–10 mL syringe with a 20 gauge needle attached, blood is withdrawn from the right ventricle by applying a small amount of negative pressure. It is important to note that some experimental measures may be altered by several commonly used anesthesia agents. For example, barbiturates (sodium pentobarbital) and inhalation anesthetics (isoflurane) inhibit mitochondrial respiratory chain enzymes (Brody & Bain, 1954; Hanley, Ray, Brandt, & Daut, 2002); thus, these drugs should not be used if functional mitochondrial assays (e.g., respiration, mitochondrial permeability transition, and calcium uptake) are planned. In this case, decapitation is a preferred method of euthanasia. If decapitation is performed, trunk blood can easily be collected for assays. Following collection, blood is then transferred to centrifuge tubes containing an anticoagulant for plasma preparation, or untreated tubes for serum preparation. Blood is typically allowed to set on ice for 30 min (but no more than 1.5 h) and then subjected to low-speed centrifugation (3000 rpm) to prepare plasma or serum. It is important to remove the needle from the syringe after collecting blood by cardiac puncture. Expulsion of blood through the needle into the centrifuge tubes will result in significant hemolysis. Whole blood can also be maintained for assays and/or preparation of red blood cells, if desired for H_2S studies. For our studies, we have used serum or plasma samples to determine the activities of creatine-kinase-MB (CK-MB) and alanine aminotransferase as indicators of possible H_2S-mediated toxicity to heart and liver, respectively. Commercial spectrophotometric kits for both these markers enzymes are available from Pointe Scientific (Canton, MI). Studies from our lab show no effect of 6 h H_2S exposure on liver enzyme release; however, rats exposed to 80 ppm H_2S under low O_2 conditions (10.5% O_2) had elevated plasma levels of CK-MB indicating a mild cardiotoxic effect (Stein et al., 2012).

Organs of interest (heart, liver, kidney, brain, adipose, etc.) should be collected quickly under aseptic conditions for molecular and biochemical assays after H_2S exposures. Tissues can be placed in 10% buffered formalin solution for most standard histopathological assessments, e.g., H&E staining. For most biochemical assays and measurements, tissues can be excised, flash-frozen in liquid N_2, and then stored at −80 °C until used. Tissues stored under these conditions are suitable for protein gels/Western blots, gene expression, and various stable metabolites, e.g., glycogen and lipid species. Tissues can also be immediately placed in RNAlater® solution to stabilize RNA for a wide variety of gene-based measures (RT-PCR, microarrays). On the other hand, measurement of highly labile metabolites including adenine and pyridine nucleotides requires that tissues be rapidly freeze-clamped, stored in liquid N_2, and processed immediately by assay-specific conditions, e.g., acid extraction under liquid N_2 for adenine nucleotide measurements. If mitochondrial preparations are needed for studies following exposures, fresh tissues should be immediately placed in ice-cold mitochondrial isolation buffer (i.e., sucrose-based buffers), tissue homogenates prepared, and differential centrifugation used to isolate the mitochondrial fraction. For a detailed protocol on isolation of mitochondria from liver, please see our previous publication (King & Bailey, 2010).

2.7. Comments on measuring H_2S and H_2S-derived metabolites in blood and tissues from H_2S-exposed rats

Determining the circulating blood and tissue concentrations of H_2S that may contribute to H_2S-induced hypometabolism (i.e., the model described herein) and in other physiological and pathological conditions has been challenging due to the complexity of H_2S chemistry and high reactivity of H_2S in biological systems. Moreover, most of the standard analytical methods used to measure H_2S in biological samples (blood and/or tissue) employ chemical methods that have a high potential to generate spurious results (e.g., H_2S levels > 100 μM) that have largely been deemed non-physiological. Indeed, there has been much written on these subjects in recent years and as such these issues will not be discussed here. Readers are encouraged to seek out the many excellent reviews on these topics with a few included here (Olson, 2011, 2012; Wang, 2012). However, with the advent of several new technologies (amperometric sensors, fluorescent probes, monobromobimane-HPLC), accurate measurement of physiological levels of H_2S and H_2S-derived metabolites is now becoming feasible. To address problems associated with H_2S stability and kinetics, and to avoid potential artifacts from chemical detection methods, former members of our redox biology research group at UAB, Dr. David W. Kraus and Dr. Jeannette E. Doeller, developed a polarographic hydrogen sulfide sensor (PHSS) that can be used in combination with high resolution multisensor respirometry to measure physiological levels of H_2S in blood and tissues (Doeller et al., 2005). Modifications to the original PHSS were more recently described by Olson and colleagues (Whitfield, Kreimier, Verdial, Skovgaard, & Olson, 2008). The PHSS has the advantage of measuring free or dissolved H_2S in biological samples (e.g., blood), is sufficiently sensitive to detect H_2S at low physiological levels, and requires little to no sample manipulation. Moreover, the PHSS operates over a broad concentration range with a response time that is ideal for kinetic studies of H_2S metabolism in intact or disrupted cell suspensions, isolated organelle (mitochondria) systems, and tissue slices and/or homogenates. Detailed descriptions of, and experimental results generated with, the PHSS are provided in Benavides et al. (2007), Doeller et al. (2005), Elrod et al. (2007), Koenitzer et al. (2007), Kraus & Doeller (2004), and Teng et al. (2008) and therefore are not discussed herein. Current suppliers of H_2S sensors suitable for biological samples include World Precision Instruments, Sarasota, FL (model: ISO-H2S-2) and Innovative Instruments, Inc., Tampa, FL (H_2S electrochemical solid-state sensor). Using a novel flow through PHSS system, preliminary measurement of H_2S concentration in rat blood (30 μL) immediately following a 6 h exposure to 80 ppm H_2S (in 15% O_2) showed a slight increase from 0.5 to 0.8 μM (unpublished results). Other initial studies show that dissolved H_2S in the blood remains in the 0.5–2.0 μM range throughout the H_2S exposure (unpublished results). Use of these novel PHSS devices will allow investigators to detect small changes in dissolved H_2S in blood before, during, and after the induction of the hypometabolic state.

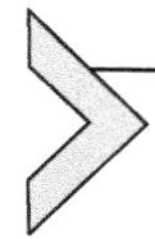

3. OTHER CONSIDERATIONS FOR H_2S EXPOSURE STUDIES

Results presented in this chapter (Fig. 3) and in Stein et al. (2012) illustrate that the biological actions of H_2S in an *in vivo* rat model are differentially altered by O_2 tension. The ability of H_2S to decrease various metabolic parameters in the male Sprague-Dawley rat is much greater when 80 ppm H_2S is administered in combination with low O_2 (10.5% O_2) as compared to room air (21% O_2) exposure (Fig. 3). Little effect of H_2S is observed when 80 ppm H_2S is inhaled under room air (21% O_2) concentration. This contrasts with results observed in mice that seem to be more sensitive to H_2S and enter a deeper state of hypometabolism in terms of key physiological readouts at 80 ppm H_2S and room air (Blackstone et al., 2005). This supports the notion that metabolic responses to H_2S may be species-dependent. Reasons for these differences were discussed in Stein et al. (2012) and briefly highlighted above; thus, they will not be restated here, except with regard to possible species differences in H_2S detoxification enzymes. Experiments show that H_2S can be oxidized to thiosulfate in the mitochondrion via an enzyme complex composed of three units, including a sulfide-quinone oxidoreductase (SQR), a sulfur dioxygenase (SDO), and a sulfur transferase, with rhodanese suggested as the putative sulfur transferase in this H_2S removal pathway (reviewed in Bouillaud & Blachier, 2011). Rhodanese is an abundant protein in most mammalian mitochondria with diverse functions including cyanide detoxification and transfer of sulfane sulfur (Cipollone, Ascenzi, & Visca, 2007). Studies show that rhodanese activity (defined as cyanide detoxification) is decreased under lower O_2 tensions (Isom & Way, 1984). Further, rhodanese activity in rats is quite high being comparable to activities measured in lug worms that live in sulfide-rich mud environments (Hildebrandt & Grieshaber, 2008). Thus, rats may be less sensitive to the hypometabolic effects of H_2S than mice and other mammals due to higher detoxification capacity of H_2S and therefore require the combination of low O_2 tension to induce hypometabolism. Differences in tissue expression of SQR, SDO, and/or rhodanese may also explain differences in biological responses to H_2S in different organ systems. For example, lower levels of rhodanese in heart versus liver mitochondria may underpin the ability of inhaled H_2S to activate Akt signaling in heart and not liver following

H_2S exposure (Stein et al., 2012). Future studies directed at understanding differences in H_2S metabolism among different species and/or tissues are needed to determine sensitivity to the metabolic effects of H_2S under both normal health and disease conditions.

Finally, one additional factor to consider with regard to H_2S-induced hypometabolism and other biological responses to H_2S is the time-of-day. Typically, experiments with laboratory animals are performed during the day (normal work day, 8 AM–5 PM), which corresponds to the sleep (less active) period for the most commonly used laboratory rodent species (rats and mice) as they are nocturnal animals. It is widely recognized that up to 10% of the transcriptome of most tissues varies over the course of the 24 h day (Miller et al., 2007; Panda et al., 2002), and this variation is regulated by an intrinsic molecular circadian clock mechanism (see review Bailey, Udoh, & Young, 2014). Notably, gene expression for rhodanese (*Tst*) and sulfur dioxygenase (*Ethe1*) shows daily rhythms in activity in liver (Pizarro, Hayer, Lahens, & Hogenesch, 2013), suggesting that H_2S disposal may vary during the day, which is predicted to affect H_2S-mediated hypometabolism in a time-of-day dependent manner. Further, initial studies from our laboratories also show that H_2S exposure *in vivo* significantly increases cardiac expression of *Period1*, a core component of the circadian clock mechanism (unpublished results). Taken together, these data suggest that the metabolic actions of H_2S may be markedly different at different times of the day and/or may also alter circadian regulation of metabolism. We speculate that the hypometabolic action of H_2S may be more pronounced if exposures are conducted in the more active (night) period in nocturnal laboratory rats and mice. While conducting exposures in the dark would be more difficult, use of telemetry to record physiological parameters would facilitate these studies (Volpato et al., 2008), along with use of infrared night vision goggles that are typically used in circadian-focused investigations.

4. SUMMARY

Herein, we have described an experimental exposure system that can be used to increase understanding of H_2S-induced hypometabolism and to more fully develop various laboratory rodent models to study the physiological, biochemical, and molecular effects of inhaled H_2S. The ability to titrate the H_2S and O_2 concentrations during exposure is also important, as this will allow investigators to define the depth and the kinetics of entering and exiting a state of H_2S-induced hypometabolism in rodents as a function of the $H_2S:O_2$ ratio. This basic experimental setup can also be adapted to investigations of other physiologically relevant gaseous signaling molecules (e.g., carbon monoxide) that have been shown to have beneficial biological actions. Studies can also be expanded to determine disease-mediated differences in conditions required for, and physiological responses to, H_2S-induced hypometabolism in acute and chronic disease models.

ACKNOWLEDGMENTS

The work described in this chapter was supported by National Heart, Lung, and Blood Institute grants T32 HL007918 (A. Stein) and R01 HL092857 (J. E. D. and S. M. B.) and served as the basis of Dr. A. Stein's doctoral dissertation project. The authors would like to thank Dr. Edward M. Postlethwait, University of Alabama at Birmingham, for assistance in developing the experimental exposure setup utilized in the studies described herein and Dr. Michelle V. Fanucchi, University of Alabama at Birmingham, for expertise in developing lung histopathology protocols and for measurements. We also thank Ms. Telisha Millender-Swain and Dr. Zhengkuan Mao, University of Alabama at Birmingham, and Dr. Joanna P. Morrison, Longwood University, for technical assistance in the experimental studies described in this chapter and in Stein et al. (2012).

REFERENCES

Abe, K., & Kimura, H. (1996). The possible role of hydrogen sulfide as an endogenous neuromodulator. *The Journal of Neuroscience, 16*(3), 1066–1071.

Bailey, S. M., Udoh, U. S., & Young, M. (2014). Circadian regulation of metabolism. *The Journal of Endocrinology, 222*, R75–R96.

Ballinger, C. A., Cueto, R., Squadrito, G., Coffin, J. F., Velsor, L. W., Pryor, W. A., et al. (2005). Antioxidant-mediated augmentation of ozone-induced membrane oxidation. *Free Radical Biology & Medicine, 38*(4), 515–526.

Benavides, G. A., Squadrito, G. L., Mills, R. W., Patel, H. D., Isbell, T. S., Patel, R. P., et al. (2007). Hydrogen sulfide mediates the vasoactivity of garlic. *Proceedings of the National Academy of Sciences of the United States of America, 104*(46), 17977–17982.

Blackstone, E., Morrison, M., & Roth, M. B. (2005). H_2S induces a suspended animation-like state in mice. *Science, 308*(5721), 518.

Bouillaud, F., & Blachier, F. (2011). Mitochondria and sulfide: A very old story of poisoning, feeding, and signaling? *Antioxidants & Redox Signaling, 15*(2), 379–391.

Brody, T. M., & Bain, J. A. (1954). Barbiturates and oxidative-phosphorylation. *The Journal of Pharmacology and Experimental Therapeutics, 110*(2), 148–156.

Cipollone, R., Ascenzi, P., & Visca, P. (2007). Common themes and variations in the rhodanese superfamily. *IUBMB Life, 59*(2), 51–59.

Doeller, J. E., Isbell, T. S., Benavides, G., Koenitzer, J., Patel, H., Patel, R. P., et al. (2005). Polarographic measurement of hydrogen sulfide production and consumption by mammalian tissues. *Analytical Biochemistry, 341*(1), 40–51.

Drabek, T., Kochanek, P. M., Stezoski, J., Wu, X., Bayir, H., Morhard, R. C., et al. (2011). Intravenous hydrogen sulfide does not induce hypothermia or improve survival from hemorrhagic shock in pigs. *Shock, 35*(1), 67–73.

Elrod, J. W., Calvert, J. W., Morrison, J., Doeller, J. E., Kraus, D. W., Tao, L., et al. (2007). Hydrogen sulfide attenuates myocardial ischemia-reperfusion injury by preservation of

mitochondrial function. *Proceedings of the National Academy of Sciences of the United States of America*, *104*(39), 15560–15565.
Fanucchi, M. V., Bracher, A., Doran, S. F., Squadrito, G. L., Fernandez, S., Postlethwait, E. M., et al. (2012). Post-exposure antioxidant treatment in rats decreases airway hyperplasia and hyperreactivity due to chlorine inhalation. *American Journal of Respiratory Cell and Molecular Biology*, *46*(5), 599–606.
Gnaiger, E., & Kuznetsov, A. V. (2002). Mitochondrial respiration at low levels of oxygen and cytochrome c. *Biochemical Society Transactions*, *30*(2), 252–258.
Gnaiger, E., Steinlechner-Maran, R., Mendez, G., Eberl, T., & Margreiter, R. (1995). Control of mitochondrial and cellular respiration by oxygen. *Journal of Bioenergetics and Biomembranes*, *27*(6), 583–596.
Hanley, P. J., Ray, J., Brandt, U., & Daut, J. (2002). Halothane, isoflurane and sevoflurane inhibit NADH:ubiquinone oxidoreductase (complex I) of cardiac mitochondria. *The Journal of Physiology*, *544*(Pt. 3), 687–693.
Haouzi, P., Notet, V., Chenuel, B., Chalon, B., Sponne, I., Ogier, V., et al. (2008). H_2S induced hypometabolism in mice is missing in sedated sheep. *Respiratory Physiology & Neurobiology*, *160*(1), 109–115.
Hayatdavoudi, G., Crapo, J. D., Miller, F. J., & O'Neil, J. J. (1980). Factors determining degree of inflation in intratracheally fixed rat lungs. *Journal of Applied Physiology: Respiratory, Environmental and Exercise Physiology*, *48*(2), 389–393.
Hildebrandt, T. M., & Grieshaber, M. K. (2008). Three enzymatic activities catalyze the oxidation of sulfide to thiosulfate in mammalian and invertebrate mitochondria. *The FEBS Journal*, *275*(13), 3352–3361.
Isom, G. E., & Way, J. L. (1984). Effects of oxygen on the antagonism of cyanide intoxication: Cytochrome oxidase, in vitro. *Toxicology and Applied Pharmacology*, *74*(1), 57–62.
Kabil, O., Motl, N., & Banerjee, R. (2014). H2S and its role in redox signaling. *Biochimica et Biophysica Acta*, *1844*(8), 1355–1366.
Kimura, H., Shibuya, N., & Kimura, Y. (2012). Hydrogen sulfide is a signaling molecule and a cytoprotectant. *Antioxidants & Redox Signaling*, *17*(1), 45–57.
King, A. L., & Bailey, S. M. (2010). Assessment of mitochondrial dysfunction arising from treatment with hepatotoxicants. *Current Protocols in Toxicology*, *44*, 14.18.11–14.18.29.
King, A. L., Polhemus, D. J., Bhushan, S., Otsuka, H., Kondo, K., Nicholson, C. K., et al. (2014). Hydrogen sulfide cytoprotective signaling is endothelial nitric oxide synthase-nitric oxide dependent. *Proceedings of the National Academy of Sciences of the United States of America*, *111*(8), 3182–3187.
Koenitzer, J. R., Isbell, T. S., Patel, H. D., Benavides, G. A., Dickinson, D. A., Patel, R. P., et al. (2007). Hydrogen sulfide mediates vasoactivity in an O_2-dependent manner. *American Journal of Physiology. Heart and Circulatory Physiology*, *292*(4), H1953–H1960.
Kraus, D. W., & Doeller, J. E. (2004). Sulfide consumption by mussel gill mitochondria is not strictly tied to oxygen reduction: Measurements using a novel polarographic sulfide sensor. *The Journal of Experimental Biology*, *207*(Pt. 21), 3667–3679.
Li, L., Salto-Tellez, M., Tan, C. H., Whiteman, M., & Moore, P. K. (2009). GYY4137, a novel hydrogen sulfide-releasing molecule, protects against endotoxic shock in the rat. *Free Radical Biology & Medicine*, *47*(1), 103–113.
Li, J., Zhang, G., Cai, S., & Redington, A. N. (2008). Effect of inhaled hydrogen sulfide on metabolic responses in anesthetized, paralyzed, and mechanically ventilated piglets. *Pediatric Critical Care Medicine*, *9*(1), 110–112.
Miller, B. H., McDearmon, E. L., Panda, S., Hayes, K. R., Zhang, J., Andrews, J. L., et al. (2007). Circadian and CLOCK-controlled regulation of the mouse transcriptome and cell proliferation. *Proceedings of the National Academy of Sciences of the United States of America*, *104*(9), 3342–3347.

Nicholls, P., Marshall, D. C., Cooper, C. E., & Wilson, M. T. (2013). Sulfide inhibition of and metabolism by cytochrome c oxidase. *Biochemical Society Transactions*, *41*(5), 1312–1316.

Olson, K. R. (2011). The therapeutic potential of hydrogen sulfide: Separating hype from hope. *American Journal of Physiology. Regulatory, Integrative and Comparative Physiology*, *301*(2), R297–R312.

Olson, K. R. (2012). A practical look at the chemistry and biology of hydrogen sulfide. *Antioxidants & Redox Signaling*, *17*(1), 32–44.

Panda, S., Antoch, M. P., Miller, B. H., Su, A. I., Schook, A. B., Straume, M., et al. (2002). Coordinated transcription of key pathways in the mouse by the circadian clock. *Cell*, *109*(3), 307–320.

Pizarro, A., Hayer, K., Lahens, N. F., & Hogenesch, J. B. (2013). CircaDB: A database of mammalian circadian gene expression profiles. *Nucleic Acids Research*, *41*(Database issue), D1009–D1013.

Polhemus, D. J., Kondo, K., Bhushan, S., Bir, S. C., Kevil, C. G., Murohara, T., et al. (2013). Hydrogen sulfide attenuates cardiac dysfunction after heart failure via induction of angiogenesis. *Circulation. Heart Failure*, *6*(5), 1077–1086.

Stein, A., Mao, Z., Morrison, J. P., Fanucchi, M. V., Postlethwait, E. M., Patel, R. P., et al. (2012). Metabolic and cardiac signaling effects of inhaled hydrogen sulfide and low oxygen in male rats. *Journal of Applied Physiology*, *112*(10), 1659–1669.

Szabo, C., Ransy, C., Modis, K., Andriamihaja, M., Murghes, B., Coletta, C., et al. (2014). Regulation of mitochondrial bioenergetic function by hydrogen sulfide. Part I. Biochemical and physiological mechanisms. *British Journal of Pharmacology*, *171*(8), 2099–2122.

Teng, X., Scott Isbell, T., Crawford, J. H., Bosworth, C. A., Giles, G. I., Koenitzer, J. R., et al. (2008). Novel method for measuring S-nitrosothiols using hydrogen sulfide. *Methods in Enzymology*, *441*, 161–172.

Volpato, G. P., Searles, R., Yu, B., Scherrer-Crosbie, M., Bloch, K. D., Ichinose, F., et al. (2008). Inhaled hydrogen sulfide: A rapidly reversible inhibitor of cardiac and metabolic function in the mouse. *Anesthesiology*, *108*(4), 659–668.

Wagner, K., Georgieff, M., Asfar, P., Calzia, E., Knoferl, M. W., & Radermacher, P. (2011). Of mice and men (and sheep, swine etc.): The intriguing hemodynamic and metabolic effects of hydrogen sulfide (H_2S). *Critical Care*, *15*(2), 146.

Wang, R. (2012). Physiological implications of hydrogen sulfide: A whiff exploration that blossomed. *Physiological Reviews*, *92*(2), 791–896.

Whitfield, N. L., Kreimier, E. L., Verdial, F. C., Skovgaard, N., & Olson, K. R. (2008). Reappraisal of H_2S/sulfide concentration in vertebrate blood and its potential significance in ischemic preconditioning and vascular signaling. *American Journal of Physiology. Regulatory, Integrative and Comparative Physiology*, *294*(6), R1930–R1937.

Wilson, D. F., Owen, C. S., & Erecinska, M. (1979). Quantitative dependence of mitochondrial oxidative phosphorylation on oxygen concentration: A mathematical model. *Archives of Biochemistry and Biophysics*, *195*(2), 494–504.

Mechanisms of H_2S Cell Signaling and Transcriptional Pathways

Use of the "Tag-Switch" Method for the Detection of Protein S-Sulfhydration

Chung-Min Park*[,1], Igor Macinkovic†[,1], Milos R. Filipovic†[,2], Ming Xian*[,2]

*Department of Chemistry, Washington State University, Pullman, Washington, USA
†Department of Chemistry and Pharmacy, University of Erlangen-Nuremberg, Erlangen, Germany
[1]These authors contributed equally.
[2]Corresponding authors: e-mail address: milos.filipovic@fau.de; mxian@wsu.edu

Contents

1. Introduction 40
2. The Design of "Tag-Switch" Method 40
3. Chemistry Validation Using Small-Molecule Substrates 41
   3.1 Materials 41
   3.2 Model reactions of MSBT with thiol and persulfide substrates 44
   3.3 Model reaction of cyanoacetate with R-S-S-BT 45
4. "Tag-Switch" Assay on Bovine Serum Albumin and GAPDH as Model Proteins 46
   4.1 Materials 46
   4.2 Preparation of BSA–SOH, BSA–SSG, and BSA–SSH 46
   4.3 "Tag-switch" labeling of model proteins for MS and dot blot 47
   4.4 Studying the mechanism of protein S-sulfhydration by "tag-switch" assay 49
5. "Tag-Switch" Assay for the Detection on Intracellular Protein Persulfides 49
   5.1 Materials 49
   5.2 Tag-switch labeling of Jurkat cell extracts 50
   5.3 Tag-switch labeling during immunoprecipitation 51
6. "Tag-Switch" Assay for the Detection of Intracellular S-Sulfhydration by Fluorescence Microscopy 51
   6.1 Materials 51
   6.2 Protocol 53
   6.3 Colocalization of intracellular S-sulfhydration with different organelles 54
7. Conclusions 54
Acknowledgments 55
References 55

Abstract

Protein S-sulfhydration (i.e., converting protein cysteines –SH to persulfides –SSH) is a redox-based posttranslational modification. This reaction plays an important role in

Methods in Enzymology, Volume 555
ISSN 0076-6879
http://dx.doi.org/10.1016/bs.mie.2014.11.033

signaling pathways mediated by hydrogen sulfide or other reactive sulfane sulfur species. Recently, our laboratories developed a "tag-switch" method which can be used to selectively label and detect protein S-sulfhydrated residues. In this chapter, we provide a comprehensive summary of this method, including the design of the method, preparation of the reagents, validation on small-molecule substrates, as well as applications in protein labeling. Experimental protocols for the use of the method are described in details.

1. INTRODUCTION

Protein S-sulfhydration (forming –SSH adducts from –SH residues) is a newly defined oxidative posttranslational modification and is receiving fast-growing attention in the context of hydrogen sulfide (H_2S)-mediated signaling pathways involved in many physiological and pathophysiological processes (Abe & Kimura, 1996; Eberhardt et al., 2014; Elrod et al., 2007; Greiner et al., 2013; Kabil & Banerjee, 2010; Krishnan, Fu, Pappin, & Tonks, 2011; Li, Rose, & Moore, 2011; Mustafa et al., 2009, 2011; Paul & Snyder, 2012; Sen et al., 2012; Szabo, 2007; Vandiver et al., 2013; Yang et al., 2008, 2013; Zhao, Zhang, Lu, & Wang, 2001). To date, the underlying mechanisms of S-sulfhydration are still a subject of active investigation and several possibilities have been suggested. H_2S may react with cysteine-modified proteins such as –S–S–, –S–OH, and –S–NO to form S-sulfhydrated products. Protein cysteines may also react with reactive sulfane sulfurs such as hydrogen polysulfides to form the same adducts (Greiner et al., 2013; Kabil, Motl, & Banerjee, 2014; Kimura, 2014; Kimura et al., 2013; Koike, Ogasawara, Shibuya, Kimura, & Ishii, 2013; Nagy et al., 2014). Nevertheless, the detection of protein S-sulfhydration plays an important role in understanding the biological functions of sulfhydration. The major challenge associated with sulfhydration detection is that cysteine persulfides (–SSH) have very similar reactivity as thiols (–SH), especially as good nucleophiles to react with –SH-blocking reagents (Pan & Carroll, 2013; Paulsen & Carroll, 2013). Thus to accurately detect sulfhydration, the key is to differentiate persulfides in the presence of thiols. Recently, we have developed a "tag-switch" method for selective detection of sulfhydration (Zhang et al., 2014). In this chapter, we report the detailed protocols of this method.

2. THE DESIGN OF "TAG-SWITCH" METHOD

As illustrated in Scheme 1, this method employs two reagents to label protein persulfides in a stepwise process. In the first step, a –SH-blocking

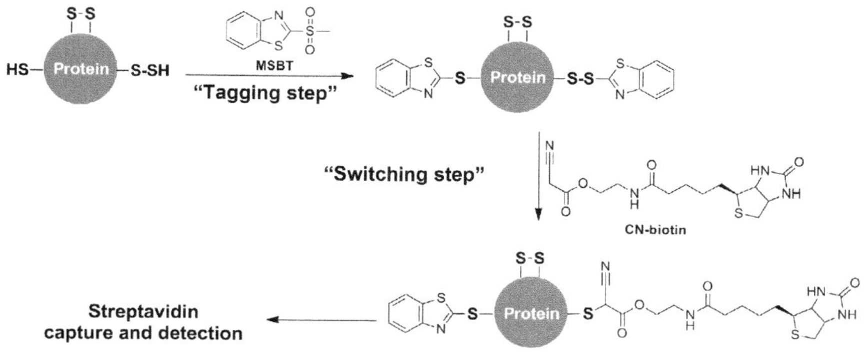

Scheme 1 The design of "tag-switch" method.

reagent—methylsulfonyl benzothiazole (MSBT)—is introduced (Zhang, Devarie-Baez, Li, Lancaster, & Xian, 2012). Given the very similar nucleophilicity of persulfides (–SSH) and thiols (–SH), MSBT should react with both –SH and –SSH to form corresponding blocked adducts. However, these two adducts should have very different reactivity toward nucleophiles. The adducts resulted from thiols are thiol ethers (–S–BT), which should be unreactive to nucleophiles. In contrast, the adducts resulted from persulfides are disulfides (–S–S–BT). It is known that heterocycle-conjugated disulfides like R–S–S–BT are very reactive (Pan & Xian, 2011). They could react with certain carbon-based nucleophiles to form stable thioether linkages. As such, a cyanoacetate-based reagent CN-biotin was identified and used successfully to "switch tag." Using this method, the persulfide residues can be converted to biotin-labeled adducts. It should be noted that regular disulfides (R–S–S–R) in proteins are stable to cyanoacetates. Therefore, this method should only label persulfides.

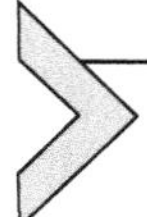

3. CHEMISTRY VALIDATION USING SMALL-MOLECULE SUBSTRATES

3.1. Materials

Methyl cyanoacetate (99%), purchased from Aldrich.

Ethyl cyanoacetate (98%), purchased from Aldrich.

MSBT (this reagent was synthesized using the following procedure): Step 1: A solution of 2-mercaptobenzothiazole (10.0 g, 59.8 mmol) in dry THF (240 mL) was cooled to 0 °C and NaH (60% in min. oil) (2.63 g, 65.8 mmol) was slowly added within 10 min. The resulting solution was stirred at 0 °C for 30 min followed by the addition of methyl iodide

(4.50 mL, 71.8 mmol) slowly. The mixture was warmed to room temperature and stirred for additional 10 h. The reaction was quenched by adding saturated NH_4Cl (100 mL). The crude product was extracted with ethyl acetate (EtOAc) (3 × 250 mL). The combined organic layers were washed with water, brine, dried over $MgSO_4$, filtered, and concentrated. The yellowish solid was used without further purification for the next step. Step 2: A solution of the product from Step 1 (10.0 g, 55.2 mmol) in EtOH (276 mL, 0.20 *M*) was cooled to 0 °C and 10 mol% of $Na_2WO_4 \cdot 2H_2O$ (1.82 g, 5.51 mmol) was added. After 5 min at 0 °C, 40 mL of H_2O_2 (30% aqueous solution) was slowly added and stirring continued at room temperature. After 10 h additional 25 mL of H_2O_2 was added into the mixture and the reaction was heated to 80 °C for 1 h to give a single product. Upon completion, 100 mL of $Na_2S_2O_3$ solution was added and the mixture was extracted with EtOAc (3 × 250 mL). The combined organic layers were washed with brine, dried over $MgSO_4$, filtered, and concentrated. The crude product was recrystallized by ether/hexane to yield the desired MSBT as a white solid (9.88 g, 84% yield).

MSBT-A: This reagent is a water-soluble version of MSBT. It has the same reactivity as MSBT (Zhang et al., 2012). MSBT-A was synthesized using the following procedure: Step 1: To a solution of 2-mercaptobenzothiazole (2.00 g, 12.0 mmol) in THF (70 mL) was added K_2CO_3 (3.30 g, 24.0 mmol) at room temperature. To this mixture was added bromoacetic acid (1.67 g, 12.0 mmol) in one portion. The mixture was stirred for 20 h and then acidified with 1 N HCl to pH 2. The precipitates formed were filtered and washed with water and hexanes to give the product as a pale yellow solid (85%). Step 2: This product (1.50 g, 6.66 mmol) was dissolved in EtOH (60 mL) and cooled to 0 °C. To this solution was added $(NH_4)_6Mo_7O_{24} \cdot 4H_2O$ (0.823 g, 0.666 mmol) and the mixture was stirred for 10 min. Then H_2O_2 (46 mL, 30%) was added slowly and the mixture was stirred at room temperature. After 20 h, additional H_2O_2 (9 mL) was added and the reaction was further stirred for 10 h at room temperature. After that, H_2O (80 mL) was added to the mixture and the aqueous layer was extracted with CH_2Cl_2 (5 × 40 mL). The organic layers were combined and washed with water (5 × 30 mL) and brine, dried over with $MgSO_4$, and concentrated. The solid residue was dissolved in CH_2Cl_2 (10 mL) and crystallized by adding hexanes (90 mL). The desired product MSBT-A was obtained by filtration (1.42 g, 82%).

CN-biotin: This reagent was synthesized from commercially available biotin (**1**) in three steps (Scheme 2).

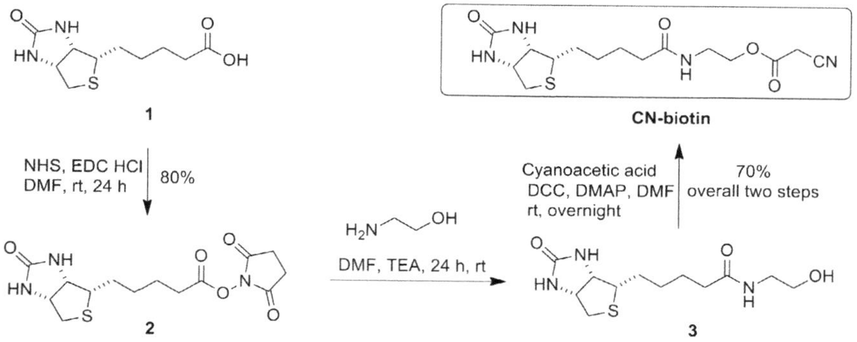

Scheme 2 Preparation of CN-biotin.

Compound **2**. To a suspension of biotin **1** (990 mg, 4.06 mmol) in anhydrous DMF (60 mL) was added *N*-hydroxysuccinimide (514 mg, 4.46 mmol) at room temperature under argon atmosphere. 1-Ethyl-3-(3-dimethylaminopropyl)carbodiimide hydrogen chloride EDC (933 mg, 4.87 mmol) was then added and the reaction was stirred for 24 h. The resulting mixture turned clear. The solvent was removed under reduced pressure to provide a white solid, which was washed thoroughly with anhydrous methanol, filtered, and dried to give compound **2** (80%). This product was used directly in the next step without any further purification.

Compound **3**. To a solution of compound **2** (1.1 g, 3.24 mmol) in anhydrous DMF (66 mL) was added 2-aminoethanol (0.29 mL, 4.86 mmol) dropwise at room temperature followed by the addition of triethylamine (0.9 mL, 6.48 mmol). The resulting mixture was stirred for 24 h. Solvent was then removed under reduced pressure and the resultant crude product was purified by flash column chromatography (gradient elution, 50:1 DCM/MeOH to 7:1 DCM/MeOH) to afford product **3** which was used for the next step.

CN-biotin. To a solution of compound **3** (3.24 mmol) in dry DMF (50 mL) was added cyanoacetic acid (331 mg, 3.89 mmol). Dicyclohexyl carbodiimide (867 mg, 4.2 mmol) was then added followed by 4-dimethylaminopyridine (40 mg, 0.32 mmol). The reaction mixture was stirred 24 h and solvent was removed under reduced pressure to give the crude product. Purification by flash column chromatography (gradient elution, 50:1 DCM/MeOH then 7:1 DCM/MeOH) afforded the final product as a white solid (804 mg, 70% over two steps). mp 132–134 °C; ^{1}H NMR (300 MHz, DMSO-d_6) δ 7.96 (t, J=5.3 Hz, 1H), 6.43 (s, 1H), 6.37 (s, 1H), 4.30 (dd, J=7.7, 4.8 Hz, 1H), 4.11 (m, 3H), 3.97 (s, 2H), 3.36–3.23

(m, 2H), 3.16–3.03 (m, 1H), 2.82 (dd, $J = 12.4$, 5.0 Hz, 1H), 2.57 (m, 1H), 2.06 (t, $J = 7.3$ Hz, 2H), 1.68–1.37 (m, 4H), 1.29 (m, 2H); ^{13}C NMR (75 MHz, DMSO-d$_6$) δ 172.3, 164.3, 162.6, 114.9, 64.4, 60.9, 59.1, 55.3, 39.7, 37.1, 34.9, 28.1, 27.9, 25.0, 24.5. FT-IR (thin film, cm^{-1}) 3270.6, 3079.7, 2931.3, 2264.3, 2200.7, 1742.4, 1696.7, 1549.2, 1264.7, 1032.1, 726.8. MS (ESI) *m/z*, calcd for $C_{15}H_{22}N_4NaO_4S$ $[M+Na]^+$ 377.1, found 377.1.

3.2. Model reactions of MSBT with thiol and persulfide substrates

To prove MSBT can effectively block thiols and persulfides, two substrates (**4** and **5**) were used to test the reactions. As small molecules of persulfides are known to be very unstable, compound **5** was used to *in situ* generate a persulfide model and further react with MSBT to form the blocked product **6** (Scheme 3).

Reaction of MSBT with thiol **4**: to a stirring solution of **4** (50.6 mg, 0.2 mmol) in THF (8 mL) and phosphate buffer (200 m*M*, pH 7.4, 16 mL) was added MSBT (86 mg, 0.4 mmol). The reaction was stirred at room temperature for 20 min and then quenched with saturated NaCl solution. The mixture was extracted with EtOAc two times. The combined organic layers were dried over anhydrous $MgSO_4$ and concentrated. The resulting residue was subjected to flash chromatography (EtOAc: hexane = 1:2 to 2:1) to isolate the desired product **6** in 95% yield (71 mg, 0.190 mmol). m.p. 192–195 °C. ^{1}H NMR (300 MHz, $CDCl_3$) δ 8.10 (d, $J = 1.2$ Hz, 1H), 7.63–7.71 (m, 2H), 7.18–7.35 (m, 7H); 4.76–4.77 (br, 1H); 4.38–4.41 (m, 2H); 3.59–3.79 (m, 2H); 1.93 (s, 3H); ^{13}C NMR (75 MHz, $CDCl_3$) δ 171.7, 170.1, 168.3, 152.5, 138.0, 135.6, 128.9, 127.9, 127.7, 126.7, 125.1, 121.5, 121.2, 55.0, 43.8, 35.4, 23.4;

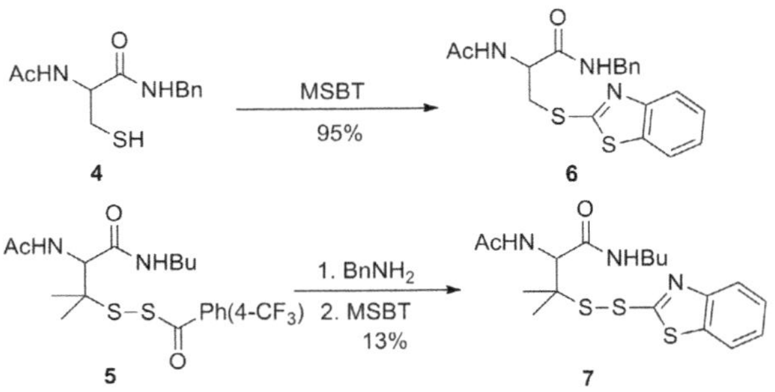

Scheme 3 Model reactions of MSBT.

IR (cm^{-1}) 3277.6, 3065.4, 2926.8, 1632.0, 1538.5, 1462.3, 1427.8, 1368.9, 990.7; mass spectrum (ESI/MS) m/z 408.1 $[M+Na]^+$.

Reaction of MSBT with persulfide model **5**: to a solution of *S*-(4-trifluoromethylbenzothioacidic)-*N*-Ac-penicillamine-NHBu **5** (73.0 mg, 0.163 mmol) in dry THF (12 mL) was added MSBT (139 mg, 0.652 mmol) under argon atmosphere. Then a solution of benzylamine (52.4 mg, 0.489 mmol) was added dropwise into the mixture at room temperature. The resulting mixture was stirred overnight and the solvent was removed under vacuum. The crude material was subjected to flash column chromatography (2% MeOH in DCM) to afford the desired product **7** as a brownish solid in 13% yield (9 mg). mp 167–168 °C; ^1H NMR (300 MHz, $CDCl_3$) δ 8.04 (s, 1H), 7.78 (t, J=8.9 Hz, 2H), 7.42 (t, J=7.7 Hz, 1H), 7.32 (t, J=7.5 Hz, 1H), 6.90 (d, J=8.9 Hz, 1H), 4.87 (d, J=9.1 Hz, 1H), 3.43 (m, 1H), 3.27–3.09 (m, 1H), 2.02 (s, 3H), 1.52 (m, 5H), 1.39 (m, 5H), 0.89 (t, J=7.2 Hz, 3H); ^{13}C NMR (75 MHz, $CDCl_3$) δ 170.4, 168.9, 153.9, 136.1, 126.6, 125.1, 121.9, 121.5, 58.7, 55.5, 39.7, 31.6, 31.2, 25.2, 23.6, 20.5, 14.0; IR (thin film, cm^{-1}) 3273, 3080, 2951, 2864, 1683, 1634, 1564, 1455, 1424, 1380, 1363, 1002, 749, 721; MS (ESI) 434.1 $[M+Na]^+$.

3.3. Model reaction of cyanoacetate with R-S-S-BT

To prove MSBT-blocked persulfide products (R-S-S-BT) can effectively undergo "tag-switch" with cyanoacetate, a model substrate **8** was used to react with methyl cyanoacetate (Scheme 4). The desired product **9** was obtained in high yield.

Protocol: To a stirring solution of **8** (0.2 mmol) in THF (8 mL) and NaPi (20 m*M*, 8 mL) was added methyl cyanoacetate (0.4 mmol). The reaction mixture was then stirred for 20 min at room temperature and quenched with 1 N HCl (16 mL). The solution was extracted with EtOAc (10 mL ×3). The combined organic layers were washed with brine, dried over with anhydrous $MgSO_4$, and concentrated. The resulting residue was subjected to flash column chromatography to provide product **9** in 93% yield as 1:1 diastereomers. ^1H NMR (300 MHz, $CDCl_3$) δ 6.38 (s, 1H), 4.90

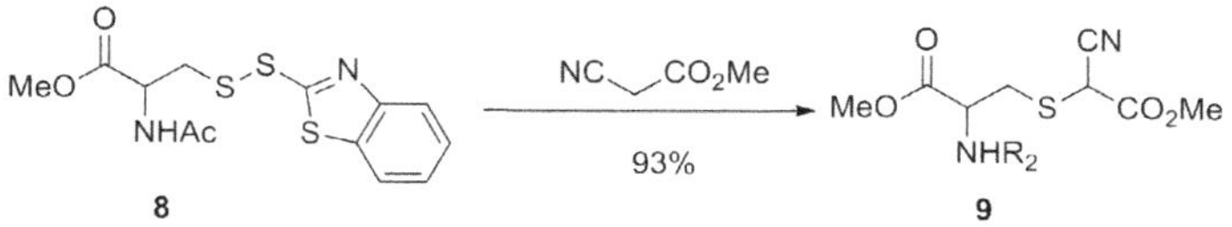

Scheme 4 Model reaction of the "tag-switch" step.

(q, J=5.9 Hz, 1H), 4.43 (d, J=27.1 Hz, 1H), 3.87 (s, 3H), 3.81 (s, 3H), 3.53–3.31 (m, 1H), 3.32–3.15 (m, 1H), 2.07 (s, 3H); ^{13}C NMR (75 MHz, $CDCl_3$) δ 170.7, 170.5, 164.6, 164.3, 114.1, 113.9, 54.6, 53.3, 51.8, 51.7, 36.1, 35.9, 34.7, 23.3; FT-IR (thin film, cm^{-1}) 3374.8, 3280.8, 2963.9, 2917.1, 2843.6, 2247.0, 1740.5, 1654.5, 1531.9, 1429.7, 1368.4, 1213.2, 1008.9; MS (ESI) m/z calcd for $C_{10}H_{14}N_2NaO_5S$ $[M+Na]^+$ 297.1, found 297.1.

4. "TAG-SWITCH" ASSAY ON BOVINE SERUM ALBUMIN AND GAPDH AS MODEL PROTEINS

Next the "tag-switch" method was validated by S-sulfhydrated proteins. Bovine serum albumin (BSA) and glyceraldehyde 3-phosphate dehydrogenase (GAPDH) were used as the models. In these experiments, S-sulfhydrated proteins were first generated and then treated with "tag-switch" assay.

4.1. Materials

BSA (≥98% (agarose gel electrophoresis), lyophilized powder, essentially fatty acid free, essentially globulin free, Sigma Aldrich)
2-Mercaptoethanol (≥99.0%, Sigma Aldrich)
5,5′-Dithiobis(2-nitrobenzoic acid) (≥98%, Sigma Aldrich)
H_2O_2 (30 wt.% in H_2O, ACS reagent, Sigma Aldrich)
Na_2S anhydrous (Sigma Aldrich)
Na_2S stock solutions (3 mL, 100 mM) in nano-pure water treated with Chelex-100 resins were prepared as recently reported (Wedmann et al., 2014).
Glutathione (≥98.0%, Sigma Aldrich)
Catalase (from bovine liver, aqueous suspension, 10,000–40,000 units/mg protein, Sigma Aldrich)
Micro Bio-Spin™ P-6 Gel Columns, Tris Buffer (BioRad, USA)

4.2. Preparation of BSA-SOH, BSA-SSG, and BSA-SSH

Preparation of BSA sulfenic acid (BSA–SOH) was done by modifying the published protocol (Carballal et al., 2003).

1. 0.65 g of BSA was dissolved in 10 mL nano-pure water (to give 0.98 mM BSA solution) and 4.3 μL of 2-mercaptoethanol was added to reduce the oxidized cysteines.

2. After 45 min, the solution was placed in the dialysis bag and dialyzed under argon in 50 m*M* phosphate buffer pH 7.4, overnight. The efficiency of thiol reduction was determined by Ellman's reaction. Only the samples containing 1:1 free thiol to protein concentration ratio were used.
3. The solution was transferred into glass vials containing PTFE septa and kept under argon for further use. To prepare BSA–SOH, BSA–SH solution (0.6 m*M*) was incubated with 10 m*M* H_2O_2 for 15 min at 37 °C and the reaction was stopped by the addition of catalase (1221 U, Sigma Aldrich, USA) for 10 min at 37 °C.

   Note: Formation of BSA sulfenic acid was confirmed by treating the freshly prepared BSA–SOH with 50 m*M* dimedone (final concentration) for 30 min at 37 °C. Dimedone-labeled protein was digested with proteomic-grade trypsin (trypsin:protein, 1:100, mg/mg) for 18 h at 37 °C. The tryptic digest was then injected into ESI-TOF-MS (maXis, Bruker Daltonics) and analyzed for the presence of dimedone labeled peptide GLVLIAFSQYLQQ^{34}CPFDEHVK.

4a. For the preparation of BSA–SSG, 10 min after the addition of catalase into BSA–SOH sample, 2 m*M* glutathione (final concentration) was added and the reaction mixture incubated for 30 min at room temperature.

5a. The samples were then desalted two times by passing through the Micro Bio-Spin™ P-6 Gel Columns (BioRad).

4b. For the preparation of BSA–SSH, 10 min after the addition of catalase into BSA–SOH sample (see above), 2 m*M* Na_2S solution (final concentration) was added and the reaction mixture incubated for 15 min at room temperature.

5b. The samples were then desalted two times by passing through the Micro Bio-Spin™ P-6 Gel Columns (BioRad).

4.3. "Tag-switch" labeling of model proteins for MS and dot blot

1. BSA–SH, BSA–SOH, BSA–SSG, and BSA–SSH (50 μL each sample) were mixed with 5.5 μL of 25% sodium dodecyl sulfate (SDS) (final concentration 2.5%), vortexed, and the solution then treated with 2.9 μL of 200 m*M* MSBT-A (dissolved in DMSO) for 15 min at room temperature.

2. The samples were desalted by passing through the Micro Bio-Spin™ P-6 Gel Columns (BioRad) and then treated with 2 m*M* CN-biotin for 1 h at room temperature.
3. To remove the excess of CN-biotin, Micro Bio-Spin™ P-6 Gel Columns (BioRad) were used.
4. For the ESI-TOF-MS analysis (maXis, Bruker Daltonics) (Fig. 1A), the samples were dialyzed against the nano-pure water overnight and then mixed with acetonitrile (1:1, v/v) containing 0.1% formic acid.
5. For the immuno-dot blotting, 10 μL of each sample was spotted on the nitrocellulose membrane. After drying, the membrane was blocked with 5% nonfat milk for 1 h at room temperature and incubated for 2 h with

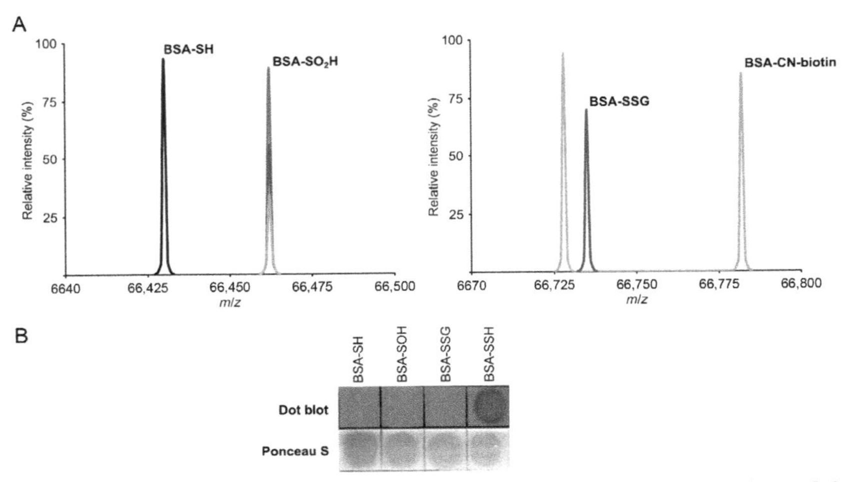

Figure 1 Tag-switch assay on BSA as a model protein. (A) ESI-TOF MS analysis of the BSA–SOH, BSA–SSG, and BSA–SSH treated with tag-switch assay. Figure shows redrown deconvoluted MS spectra obtained using Data Analysis software (Bruker Daltonics). Untreated BSA served as a control, giving the peak at *m/z* 66,430 ± 2 Da. BSA–SOH sample gave only the peak at *m/z* 66,462 ± 2 Da which we assigned to BSA–SO_2H (mass change is 32 when compared to the control spectrum). BSA–SO_2H was inevitable end product after treating the samples of BSA–SOH few hours with all the reagents of tag-switch assay. The same peak could be seen in BSA–SSG and BSA–SSH samples. However, BSA–SSG sample clearly shows the peak at *m/z* 66,735 ± 2 Da (addition of a glutathione moiety), while the BSA–SSH sample has a peak at *m/z* 66,782 ± 2 Da, which is assigned to the BSA labeled with CN-biotin. The peak at *m/z* 66,728 could be the decomposition product of BSA–CN-biotin derivative, suggesting the loss of NC(O)N moiety of biotin structure. (B) Immuno-dot blot detection of successful biotinylation (upper lane) and Ponceau S staining for the protein load (lower lane). *This figure is reprinted (with modification) from Zhang et al. (2014). Copyright 2014* Angewandte Chemie International Edition.

HRP-labeled mouse monoclonal anti-biotin antibody (Sigma Aldrich, St. Louis, MO) according to the manufacture recommendation. Signal was visualized using Pierce ECL reagent (Thermo Scientific, IL, USA) and exposing membrane to X-ray films (Fig. 1B).

4.4. Studying the mechanism of protein S-sulfhydration by "tag-switch" assay

In addition to BSA, we have successfully applied "tag-switch" assay on GAPDH as the model system, in order to study the mechanism of protein S-sulfhydration. To do so, 2 mg/mL of GAPDH (from rabbit muscle, lyophilized powder, ≥75 units/mg protein, Sigma Aldrich) was incubated with 100 μ*M* H_2S, 100 μ*M* H_2S with rigorous vortexing, 100 μ*M* Angeli's salt, and 100 μ*M* H_2O_2 for 30 min at room temperature. GAPDH was also treated with 100 μ*M* Angeli's salt and 100 μ*M* H_2O_2 for 15 min followed by additional 15 min with 100 μ*M* H_2S. Additionally, 2 mg/mL GAPDH was mixed with 100 μ*M* H_2S and 20 μ*M* water-soluble iron porphyrin (Miljkovic, Kenkel, Ivanovic-Burmazovic, & Filipovic, 2013) for 30 min. Samples were then desalted on Micro Bio-Spin™ P-6 Gel Columns (BioRad) and further labeled by tag-switch assay as described above.

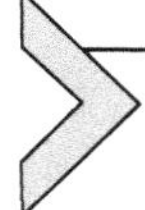

5. "TAG-SWITCH" ASSAY FOR THE DETECTION ON INTRACELLULAR PROTEIN PERSULFIDES

5.1. Materials

Jurkat E6.1 Cell Line human (ECACC)
HEPES (≥99.5%, Sigma Aldrich)
Neocuproine hydrate (99%, Sigma Aldrich)
EDTA (>99.0%, Sigma Aldrich)
NP-40 (Tergitol® solution, 70% in H_2O, Sigma Aldrich)
HEN buffer (250 m*M* HEPES, 50 m*M* NaCl, 1 m*M* EDTA, 0.1 m*M* neocuproine, 1% NP-40)
Protease inhibitor cocktail (DMSO solution, Sigma Aldrich)
SDS (≥98.5%, Sigma Aldrich)
HRP-labeled mouse monoclonal anti-biotin antibody (Sigma Aldrich, St. Louis, MO)
Nitrocellulose membrane (Carl Roth GmbH, Germany)
Nonfat-dried milk bovine (Sigma Aldrich)
Pierce ECL reagent (Thermo Scientific, IL, USA)

MSBT-A stock solution (300 m*M*, 7.7 mg in 80 μL of DMSO with addition of 20 μL of 25% SDS solution)
CN-biotin (200 m*M*, 7.1 mg in 100 μL 25% SDS solution)

5.2. Tag-switch labeling of Jurkat cell extracts

1. Jurkat cells were cultured in RPMI supplemented with 10% FCS, 1% nonessential amino acid mix and 1% Streptomycin–Penicillin at 37 °C, and 5% CO_2.
2. Cells were centrifuged at $1000 \times g$ for 5 min and resuspended in the fresh medium to get 5×10^6 cells.
3. Jurkat cells were then treated with 200 μ*M* Na_2S (30 min, 37 °C, and 5% CO_2), centrifuged at $800 \times g$ for 5 min, washed with PBS, and after another round of centrifugation pellet was resuspended in 3 volumes of HEN buffer supplemented with 1% protease inhibitors.
4. Cells were kept on ice and lysed by 5×20 s cycles of ultrasonication.
5. 50 μL of the cell lysate was mixed with 10 μL of 300 m*M* MSBT-A and incubated for 1 h at 37 °C.
6. Samples (60 μL) were desalted on Micro Bio-Spin™ P-6 Gel Columns (BioRad) and 6.6 μL of 200 m*M* CN-biotin (in 25% SDS) was added.
7. Cell extracts were incubated for 1 h at 37 °C.
8. The excess of CN-biotin was removed by passing the samples through Micro Bio-Spin™ P-6 gel columns.
9. Prior to electrophoresis, 15 μL samples were mixed with 5 μL 4 × non-reducing Laemmli buffer (BioRad, USA) and boiled at 95 °C for 1 min.
10. Proteins were resolved on either 8% or 10% SDS polyacrylamide gels.
11. After the electrophoresis, proteins were transferred on 0.2 μm nitrocellulose membrane using Trans-Blot® SD semi-dry blotter (Bio Rad, USA, 1.5 h, 17 V). Membrane was blocked with 5% nonfat milk for 1 h at room temperature, and incubated overnight with HRP-labeled mouse monoclonal anti-biotin antibody (1:10,000, Sigma Aldrich, St. Louis, MO). Signal was visualized using Pierce ECL reagent (Thermo Scientific, IL, USA) and exposing membrane to X-ray films (Fig. 2A).

Note: We observed that the treatment of the cell lysates with H_2S gives much higher yields of protein S-sulfhydration (Fig. 2B) than when the intact cells were treated with the same amount of H_2S. This suggests that some artificial protein modification occurs and thus the treatment of cell lysates should be avoided when studying the intracellular levels of S-sulfhydration.

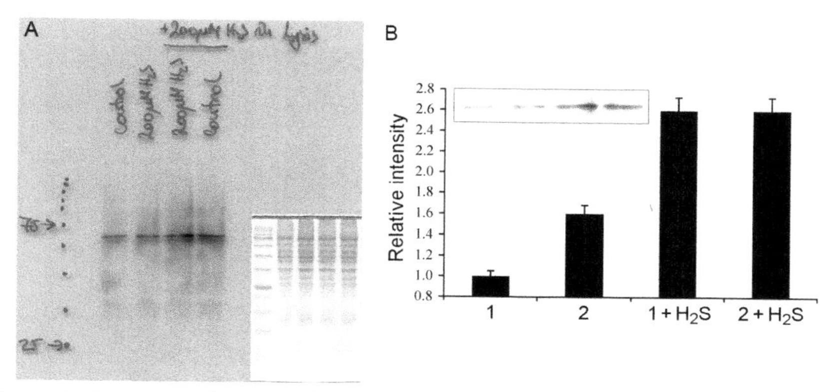

Figure 2 Detection of intracellular S-sulfhydration in Jurkat cell lysates. (A) Original, unprocessed scan of the X-ray film. Inset: Coomassie blue staining for the protein load. (B) Quantification of the S-sulfhydration based on the intensity of the 70 kDa band. 1: control, 2: H_2S-treated cells, 1 + H_2S: control cell lysates treated with H_2S, 2 + H_2S: cell lysates from H_2S-treated cells additionally treated with 200 μ*M* H_2S. *Reprinted from Zhang et al. (2014). Copyright 2014* Angewandte Chemie International Edition.

5.3. Tag-switch labeling during immunoprecipitation

In order to study the S-sulfhydration of particular protein of interest, tag-switch assay in combination with immunoprecipitation could be used. In that case, it is essential to block free thiols and persulfides with MSBT-A during the cell lysis as described in the previous protocol. The protein of interest can be then pulled down by immunoprecipitation. After the pull down, the immunoprecipitate is washed and then treated with 2 m*M* CN-biotin in PBS containing 2% SDS for 1 h at room temperature and the sample is ready for the electrophoresis.

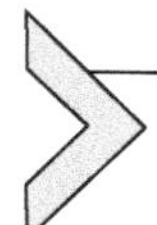

6. "TAG-SWITCH" ASSAY FOR THE DETECTION OF INTRACELLULAR S-SULFHYDRATION BY FLUORESCENCE MICROSCOPY

To be able to detect intracellular S-sulfhydration and visualize it by the means of fluorescence microscopy, we adapted the "tag-switch" assay as shown in Fig. 3. Essentially, additional step was introduced and biotin labeling visualized by fluorescently labeled streptavidin (Scheme 5).

6.1. Materials

Human umbilical vein endothelial cells (HUVECs) (Promo Cell, Heidelberg, Germany)

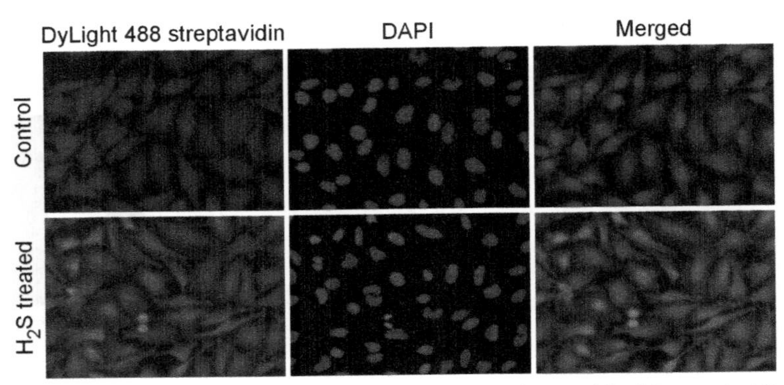

Figure 3 *In situ* labeling of intracellular S-sulfhydration by modified "tag-switch" assay. Cells (HUVEC) were either treated with 100 μ*M* Na_2S or not and then prepared for the microscopy, following the described protocol. Nuclei were stained with DAPI. Fluorescence microscopy was carried out using Carl Zeiss Axiovert 40 CLF inverted microscope, equipped with monochromatic RGB CoolLed light source (Andover, UK) and monochromatic AxioCam Icm1 camera. All experiments were performed at least in triplicate. Images were postprocessed in ImageJ software (NIH, USA). (See the color plate.)

Scheme 5 "Tag-switch" assay for *in situ* detection of intracellular S-sulfhydration.

DL-Propargylglycine (Sigma Aldrich)
2-Ketobutyric acid (97% Sigma Aldrich)
Na_2S stock solution (100 m*M*) prepared as previously described
Methanol (HPLC grade, Sigma Aldrich)
Acetone (HPLC grade, Sigma Aldrich)
4% Paraformaldehyde
Triton X-100 (for molecular biology, Sigma Aldrich)
50 m*M* solution of MSBT-A (300 m*M* stock solution, prepared as described above, was diluted 6 × with PBS)
20 m*M* solution of CN-biotin (200 m*M* stock solution, prepared as described above, was diluted 10 × with PBS)
Dimedone (95%, Sigma Aldrich)
DyLight 488 Conjugated streptavidin (Thermo Scientific Pierce; 1 mL; 1 mg/mL protein in buffered saline solution containing BSA)
DyLight 405 Conjugated streptavidin (Thermo Scientific Pierce; 1 mL; 1 mg/mL protein in buffered saline solution containing BSA)

MitoTracker® Red CMXRos (Molecular Probes, Invitrogen, Germany)
CellLight™ ER-GFP (Molecular Probes, Invitrogen, Germany)

6.2. Protocol

1. HUVECs (passage 2–3) were cultured in 35 mm μ-dishes (ibidi, Martinsried, Germany) in endothelial cell growth medium 2 (PromoCell GmbH) at 37 °C and 5% CO_2.
2. To study the effect of endogenous H_2S production, cells were incubated with 2 m*M* propargylglycine to inhibit cystathionine gamma lyase, or with 2 m*M* 2-ketobutyric acid to inhibit mercaptopyruvate sulfur transferase, for 2 h.
3. To study the effect of exogenous H_2S, cells were exposed to 100 μ*M* Na_2S for 30 min.
4. After the corresponding treatments, the cells were washed three times with PBS.
5. Cell fixation was done in two ways, with methanol and with paraformaldehyde. The quality of the detected signal differed, with the less diffused signal observed in the samples fixed with methanol.
   - **5a.** For the methanol fixation, ice-cold methanol (−40 °C) was added into the cells and they were kept at −20 °C for the following 15 min. Methanol was removed and ice-cold acetone (−40 °C) was added to the cells for 5 min and then removed and washed three times with ice-cold PBS.
   - **5b.** For the paraformaldehyde fixation, 1 mL of 4% paraformaldehyde was added into the dishes containing 1 mL of PBS and cells incubated for 10 min at room temperature. The solution was then removed and replaced with 4% paraformaldehyde and incubated additional 10 min at room temperature. After the fixation samples were washed three times with PBS, to remove excess paraformaldehyde and then permeabilized with 1% Triton X-100 (v/w) for 2 h at room temperature. The samples were then washed three times with PBS.
6. PBS was removed from the dishes and the cells were overlaid with 300 μL of 50 m*M* MSBT-A (dissolved in PBS). Incubation period could span from 1 h to overnight incubation. We observed that 1 h incubation at 37 °C is sufficient to efficiently block all the free thiols and persulfides. Samples were then thoroughly washed (5 × 10 min) with PBS.

7. PBS was removed from the dishes and the cells where covered with 300 μL of 2 m*M* CN-biotin (in PBS) and incubated for 2 h at room temperature. Samples were then thoroughly washed (5 × 10 min) with PBS.
8. Finally 1 μL of DyLight 488 conjugated streptavidin (Thermo Fischer Scientific, IL, USA) was added into each dish containing 1 mL of PBS. Cells were incubated for 1 h at room temperature and then washed three times with PBS prior to the visualization by fluorescence microscopy.

   Note: Due to the slight possibility that some of the intracellular sulfenic acids could directly react with CN-biotin, it is advisable to introduce additional step in sample preparation. After the fixation (Step 5), the cells could be treated with 50 m*M* dimedone for 1 h, thoroughly washed with PBS and then exposed to MSBT-A (as described in Step 6). It is worth mentioning that in our hands dimedone-pretreatment had no effect on overall fluorescence detected with "tag-switch" assay.

6.3. Colocalization of intracellular S-sulfhydration with different organelles

The modified "tag-switch" assay allows for the colocalization studies in combination with the fluorescent probes for different organelles. For example, for the visualization of endoplasmic reticulum, HUVECs were transfected with CellLight™ ER-GFP following the instruction of the manufacturer (Molecular Probes, Invitrogen, Germany). Mitochondria were visualized by treating the cell with 500 n*M* MitoTracker® Red CMXRos (Molecular Probes, Invitrogen) for 45 min prior the fixation. Cells were treated with 100 μ*M* Na_2S (30 min), washed, and fixed with ice-cold methanol as described above with the only difference that in the last step, DyLightTM405 conjugated streptavidin (Thermo Fischer Scientific, IL, USA) was used.

7. CONCLUSIONS

In this chapter, we provide detailed experimental protocols for the use of "tag-switch" method for the detection of S-sulfhydrated proteins. The method was proved to be selective for S-sulfhydration on protein samples. It can be used to identify and isolate S-sulfhydrated proteins in cell lysates.

ACKNOWLEDGMENTS

M. X. thanks NIH (R01HL116571) and the American Chemical Society (Teva USA Scholar Award). M. R. F. is grateful to Emerging Field Initiative intramural grant (MRIC) from FAU Erlangen-Nuremberg.

REFERENCES

Abe, K., & Kimura, H. (1996). The possible role of hydrogen sulfide as an endogenous neuromodulator. *Journal of Neuroscience*, *16*(3), 1066–1071.

Carballal, S., Radi, R., Kirk, M. C., Barnes, S., Freeman, B. A., & Alvarez, B. (2003). Sulfenic acid formation in human serum albumin by hydrogen peroxide and peroxynitrite. *Biochemistry*, *42*, 9906–9914.

Eberhardt, M., Dux, M., Namer, B., Miljkovic, J., Cordasic, N., Will, C., et al. (2014). H_2S and NO cooperatively regulate vascular tone by activating a neuroendocrine HNO-TRPA1-CGRP signalling pathway. *Nature Communications*, *5*, 4381.

Elrod, J. W., Calvert, J. W., Morrison, J., Doeller, J. E., Kraus, D. W., Tao, L., et al. (2007). Hydrogen sulfide attenuates myocardial ischemia-reperfusion injury by preservation of mitochondrial function. *Proceedings of the National Academy of Sciences of the United States of America*, *104*(39), 15560–15565.

Greiner, R., Palinkas, Z., Basell, K., Becher, D., Antelmann, H., Nagy, P., et al. (2013). Polysulfides link H_2S to protein thiol oxidation. *Antioxidants & Redox Signaling*, *19*(15), 1749–1765.

Kabil, O., & Banerjee, R. (2010). Redox biochemistry of hydrogen sulfide. *Journal of Biological Chemistry*, *285*(29), 21903–21907.

Kabil, O., Motl, N., & Banerjee, R. (2014). H_2S and its role in redox signaling. *Biochimica et Biophysica Acta*, *1844*(8), 1355–1366.

Kimura, H. (2014). The physiological role of hydrogen sulfide and beyond. *Nitric Oxide: Biology and Chemistry*, *41*, 4–10.

Kimura, Y., Mikami, Y., Osumi, K., Tsugane, M., Oka, J., & Kimura, H. (2013). Polysulfides are possible H_2S-derived signaling molecules in rat brain. *FASEB Journal*, *27*(6), 2451–2457.

Koike, S., Ogasawara, Y., Shibuya, N., Kimura, H., & Ishii, K. (2013). Polysulfide exerts a protective effect against cytotoxicity caused by t-buthylhydroperoxide through Nrf2 signaling in neuroblastoma cells. *FEBS Letters*, *587*(21), 3548–3555.

Krishnan, N., Fu, C., Pappin, D. J., & Tonks, N. K. (2011). H_2S-induced sulfhydration of the phosphatase PTP1B and its role in the endoplasmic reticulum stress response. *Science Signaling*, *4*(203), ra86.

Li, L., Rose, P., & Moore, P. K. (2011). Hydrogen sulfide and cell signaling. *Annual Review of Pharmacology and Toxicology*, *51*, 169–187.

Miljkovic, J. Lj., Kenkel, I., Ivanovic-Burmazovic, I., & Filipovic, M. R. (2013). Generation of HNO and HSNO from nitrite by heme-iron-catalyzed metabolism with H_2S. *Angewandte Chemie International Edition*, *52*, 12061–12064.

Mustafa, A. K., Gadalla, M. M., Sen, N., Kim, S., Mu, W., Gazi, S. K., et al. (2009). H_2S signals through protein S-sulfhydration. *Science Signaling*, *2*(96), ra72.

Mustafa, A. K., Sikka, G., Gazi, S. K., Steppan, J., Jung, S. M., Bhunia, A. K., et al. (2011). Hydrogen sulfide as endothelium-derived hyperpolarizing factor sulfhydrates potassium channels. *Circulation Research*, *109*(11), 1259–1268.

Nagy, P., Palinkas, Z., Nagy, A., Budai, B., Toth, I., & Vasas, A. (2014). Chemical aspects of hydrogen sulfide measurements in physiological samples. *Biochimica et Biophysica Acta*, *1840*(2), 876–891.

Pan, J., & Carroll, K. S. (2013). Persulfide reactivity in the detection of protein S-sulfhydration. *ACS Chemical Biology*, *8*(6), 1110–1116.

Pan, J., & Xian, M. (2011). Disulfide formation via sulfenamides. *Chemical Communications*, *47*(1), 352–354.

Paul, B. D., & Snyder, S. H. (2012). H_2S signalling through protein sulfhydration and beyond. *Nature Reviews. Molecular Cell Biology*, *13*(8), 499–507.

Paulsen, C. E., & Carroll, K. S. (2013). Cysteine-mediated redox signaling: Chemistry, biology, and tools for discovery. *Chemical Reviews*, *113*(7), 4633–4679.

Sen, N., Paul, B. D., Gadalla, M. M., Mustafa, A. K., Sen, T., Xu, R., et al. (2012). Hydrogen sulfide-linked sulfhydration of NF-kappaB mediates its antiapoptotic actions. *Molecular Cell*, *45*(1), 13–24.

Szabo, C. (2007). Hydrogen sulphide and its therapeutic potential. *Nature Reviews. Drug Discovery*, *6*(11), 917–935.

Vandiver, M. S., Paul, B. D., Xu, R., Karuppagounder, S., Rao, F., Snowman, A. M., et al. (2013). Sulfhydration mediates neuroprotective actions of parkin. *Nature Communications*, *4*, 1626.

Wedmann, R., Bertlein, S., Macinkovic, I., Böltz, S., Miljkovic, J. Lj., Muñoz, L. E., et al. (2014). Working with "H_2S": Facts and apparent artifacts. *Nitric Oxide: Biology and Chemistry*, *41*, 85–96.

Yang, G., Wu, L., Jiang, B., Yang, W., Qi, J., Cao, K., et al. (2008). H_2S as a physiologic vasorelaxant: Hypertension in mice with deletion of cystathionine γ-lyase. *Science*, *322*(5901), 587–590.

Yang, G., Zhao, K., Ju, Y., Mani, S., Cao, Q., Puukila, S., et al. (2013). Hydrogen sulfide protects against cellular senescence via S-sulfhydration of Keap1 and activation of Nrf2. *Antioxidants & Redox Signaling*, *18*(15), 1906–1919.

Zhang, D., Devarie-Baez, N. O., Li, Q., Lancaster, J. R., Jr., & Xian, M. (2012). Methylsulfonyl benzothiazole (MSBT): A selective protein thiol blocking reagent. *Organic Letters*, *14*(13), 3396–3399.

Zhang, D., Macinkovic, I., Devarie-Baez, N. O., Pan, J., Park, C.-M., Carroll, K. S., et al. (2014). Detection of protein S-sulfhydration by a tag-switch technique. *Angewandte Chemie International Edition*, *53*, 575–581.

Zhao, W., Zhang, J., Lu, Y., & Wang, R. (2001). The vasorelaxant effect of H_2S as a novel endogenous gaseous K(ATP) channel opener. *EMBO Journal*, *20*(21), 6008–6016.

CHAPTER FOUR

Real-Time Assays for Monitoring the Influence of Sulfide and Sulfane Sulfur Species on Protein Thiol Redox States

Romy Greiner, Tobias P. Dick[1]
Division of Redox Regulation, German Cancer Research Center (DKFZ), DKFZ-ZMBH Alliance, Heidelberg, Germany
[1]Corresponding author: e-mail address: t.dick@dkfz.de

Contents

1. Introduction 58
2. PTEN Activity Assay 59
   2.1 Principle 59
   2.2 Reagents and equipment 60
   2.3 Procedure 60
   2.4 Comments 61
   2.5 Expected results 61
3. roGFP2 Redox Assay 61
   3.1 Principle 61
   3.2 Reagents and equipment 62
   3.3 Procedure 63
   3.4 Comments 63
   3.5 Expected results 64
4. Application of "H_2S Donors" and Polysulfides 64
   4.1 General considerations 64
   4.2 Expected results 67
5. Quantitation of Sulfane Sulfur by Cold Cyanolysis 67
   5.1 Principle 67
   5.2 Reagents and equipment 69
   5.3 Procedure 70
   5.4 Comments 70
   5.5 Expected results 71
6. Elimination of Sulfane Sulfur by Cold Cyanolysis 71
   6.1 Principle 71
   6.2 Elimination of pre-existing sulfane sulfur 72
   6.3 Prevention of the *de novo* formation of sulfane sulfur 73
Acknowledgments 76
References 76

Methods in Enzymology, Volume 555
ISSN 0076-6879
http://dx.doi.org/10.1016/bs.mie.2014.11.020

Abstract

Hydrogen sulfide (H_2S) is known to induce persulfidation of protein thiols. However, the process of H_2S-induced persulfidation is not fully understood as it requires an additional oxidant. There are several mechanistic possibilities and it is of interest to determine which pathway is kinetically most relevant. Here, we detail *in vitro* assays for the real-time monitoring of thiol redox states in two model proteins with oxidizable cysteines, PTEN, and roGFP2. These allow kinetic measurements of the response of defined protein thiols (or disulfides) to sulfide and sulfane sulfur species. The combination of these assays with cold cyanolysis reveals the role of intermediary sulfane sulfur species in H_2S-induced protein thiol oxidation.

1. INTRODUCTION

H_2S is considered a physiologically relevant signaling molecule (Kabil, Motl, & Banerjee, 2014). It is believed to lead to the posttranslational modification of protein thiols (R–S–H), in particular to persulfidation (R–S–S–H). However, it is mechanistically unclear how H_2S leads to persulfidation. The isolated reaction of H_2S with a thiol or thiolate is chemically impossible, as two electrons are left unaccounted for:

$$H_2S^{-II} + R - S^{-II} - H \rightarrow R - S^{-I} - S^{-I} - H + 2e^- + 2H^+.$$

Thus, to explain persulfidation by H_2S, we are left with three main mechanistic possibilities: (1) H_2S^{-II} is oxidized first, to either H_2S_2 ($H–S^{-I}–S^{-I}–H$), a polysulfide ($–S–S^0–S–$), or other sulfane sulfur species, which then reacts with the thiol to yield the persulfide; (2) the thiol is first oxidized to a sulfenic acid ($R–S^0–O–H$), which then condenses with H_2S to yield the persulfide; and (3) the thiol is oxidized first to a disulfide ($R–S^{-I}–S^{-I}–R$), which then reacts with H_2S to yield a persulfide and a thiol.

One strategy to identify kinetically relevant mechanisms of persulfidation is the use of recombinant model proteins harboring cysteines for which the redox state (reduced vs. oxidized) can be monitored in real time. Here, we present protocols for two such assays based on different model proteins. First, PTEN (phosphatase and tensin homolog) is a redox-regulated lipid phosphatase with a low pK_a catalytic cysteine (Denu & Dixon, 1998). The thiolate is the nucleophile in the dephosphorylation reaction. Thus, a small-molecule substrate that becomes fluorescent upon dephosphorylation can be used to monitor the availability of the thiolate, reflecting the kinetics of oxidation and reduction at the catalytic cysteine. Second, roGFP2 is a modified version of enhanced green fluorescent protein which has been engineered to carry two adjacent cysteines on its surface (Hanson et al., 2004). Formation of a disulfide bond changes roGFP2 fluorescence in a ratiometric manner and allows exact quantitation of the degree of oxidation in the roGFP2 population. Both assays, PTEN activity and roGFP2 redox state, allow the quantitation of the influence of sulfur-containing species on specific protein thiols and disulfide bonds. Importantly, to dissect the reaction mechanism, these assays can be combined with additional chemical or enzymatic procedures. As an example, we show how the cyanolysis reaction can be used to identify the involvement of intermediary sulfane sulfur species in protein thiol oxidation by H_2S. Furthermore, the described assays can be combined with various additional analytic tools to further characterize the resulting thiol modifications, including mass spectrometry, gel shift analysis and the detection of released protein-bound sulfane sulfur.

The assays presented here, in combination with additional experiments, have been used recently to demonstrate that H_2S-induced oxidation of PTEN and roGFP2 is mediated by sulfane sulfur species which unavoidably form in H_2S solutions. In addition, these assays have confirmed that the reduction of disulfide bonds by hydrogen sulfide is a kinetically insignificant process (Greiner et al., 2013).

In the following protocol, for simplicity, the term H_2S will be used to refer to the total sulfide pool, i.e., H_2S, HS^-, and S^{2-}. Moreover, the term "H_2S donors" will be used to indicate all sources of H_2S, including true H_2S donors such as GYY4137, but also sulfide salts (NaHS, Na_2S) and H_2S gas.

2. PTEN ACTIVITY ASSAY

2.1. Principle

In this continuous *in vitro* assay, the artificial substrate 6,8-difluoro-4-methylumbelliferyl phosphate (DiFMUP) becomes fluorescent upon its dephosphorylation by recombinant PTEN; hence, the rate of fluorescence increase, as measured using a fluorescence microplate reader, directly reflects PTEN activity. As PTEN activity strictly depends on the free thiolate of the low pK_a active site cysteine (Cys-124), this assay enables specific real-time monitoring of the redox state of Cys-124. It is thus ideally suited to assess rapid reduction or oxidation processes brought about by small molecules, including sulfide and sulfane sulfur species. Note that the catalytic Cys-124 can form a disulfide bond with a nearby second cysteine (Cys-71), which typically happens upon reaction with hydrogen peroxide (H_2O_2)

(Lee et al., 2002). Also persulfidation of Cys-124 appears to be followed by formation of the Cys-124/Cys-71 disulfide (Greiner et al., 2013). Importantly, while this assay allows the sensitive monitoring of the Cys-124 redox state, it is not able to discriminate between different oxidative modifications of that cysteine. To identify the specific nature of the oxidative modification, other analytical tools need to be applied such as gel shift analysis, mass spectrometry, or the detection of H_2S released from protein-bound sulfane sulfur.

In the protocol detailed below, reaction conditions are 380 n*M* (2 μg) recombinant PTEN and 200 μ*M* DiFMUP in a total volume of 100 μl.

2.2. Reagents and equipment

- Recombinant PTEN-SBP-His

  We use a PTEN expression construct encoding two C-terminal tags, a streptavidin binding peptide (SBP) tag followed by a hexahistidine (His_6) tag. Recombinant PTEN-SBP-His is purified from *Escherichia coli* strain M15 (Qiagen) via consecutive metal chelate and streptavidin affinity chromatography, dialyzed against 50 m*M* Tris–HCl, 150 m*M* NaCl, 2 m*M* EDTA, 2 m*M* DTT, pH 7.4, and supplemented with 10% glycerol for long-term storage at −80 °C.
- PTEN assay buffer (100 m*M* Tris–HCl, 2 m*M* EDTA, 0.05% Nonidet P40, pH 8.0)
- DiFMUP as fluorogenic substrate (10 m*M* stock in dimethylformamide; store at −20 °C)
- 0.5 ml Zeba Desalt Spin Columns, 40 K MWCO (Thermo Scientific)
- Black flat-bottomed 96-well plate (e.g., BD Falcon #353219)
- Fluorescence microplate reader (such as FLUOstar Omega, BMG Labtech)

2.3. Procedure

1. Degas the PTEN assay buffer for a minimum of 2 h before use by applying vacuum and stirring vigorously at 37 °C.
2. Reduce freshly thawed recombinant PTEN-SBP-His with 20 m*M* DTT for 10 min at room temperature.
3. Prepare a 400 μ*M* DiFMUP working solution by diluting the 10 m*M* DiFMUP stock solution into degassed PTEN assay buffer.
4. Pre-equilibrate a Zeba gel filtration column 3× with degassed PTEN assay buffer (as detailed by manufacturer)

5. Pass prereduced PTEN-SBP-His through the pre-equilibrated Zeba gel filtration column to remove DTT and adjust the concentration of the dialyzed PTEN-SBP-His to 40 ng/μl with degassed PTEN assay buffer.
6. Dispense 50 μl of the diluted PTEN-SBP-His solution per well into a black flat-bottomed 96-well plate (for 2 μg protein/well); use 50 μl degassed PTEN assay buffer as blank.
7. Add 50 μl of 400 μ*M* DiFMUP to sample wells of the microplate to start the reaction.
8. Immediately start measuring fluorescence every minute for 60 min at 355/460 nm (excitation/emission) using a microplate reader with temperature set to 37 °C.
9. Correct fluorescence signals against corresponding samples lacking PTEN protein (blanks).

2.4. Comments

Note that recombinant PTEN may become partially inactivated over extended periods of time due to aggregation. It is therefore recommended to add the sulfide or sulfane species of interest to the samples soon after starting the assay and to complete the assay within 60 min.

We furthermore recommend verifying that all dephosphorylation activity originates solely from PTEN, by including the PTEN-C124S (or C124A) mutant as a catalytically inactive negative control.

While an assay buffer of pH 8.0 is commonly used for PTEN, in our experience, a pH of 7.4 works equally well in terms of PTEN activity (Greiner et al., 2013). It should be kept in mind, however, that any change in pH may influence the rate of spontaneous H_2S oxidation (Chen & Morris, 1972).

2.5. Expected results

Example response curves are shown in Fig. 1, which highlight the varying ability of different oxidants to inhibit PTEN.

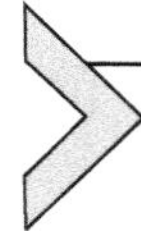

3. roGFP2 REDOX ASSAY

3.1. Principle

roGFP2 is a redox-sensitive variant of enhanced green fluorescent protein that contains two oxidizable cysteines of typical pK_a (~9) (Hanson et al.,

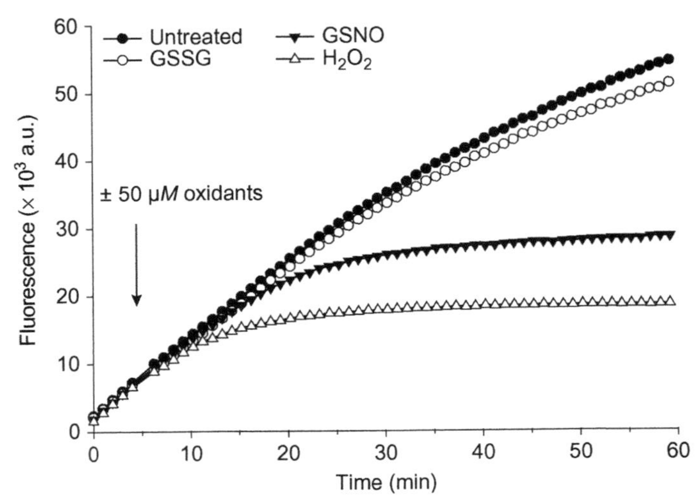

Figure 1 The PTEN activity assay allows to monitor the influence of oxidants on the active site cysteine in real time. The activity of PTEN-WT was measured under non-reducing conditions upon injection of buffer (untreated) or 50 μ*M* of GSSG, GSNO, or H_2O_2 (injection indicated by arrow). Curves represent means of triplicate wells.

2004). Ratiometric fluorescence measurements of roGFP2 can be used to monitor in real time the response of the thiol-disulfide state of these cysteines. It should be noted that ratiometric fluorescent changes strictly monitor the presence of the disulfide bond (for more information, see Meyer & Dick, 2010). Hence, this assay is ideally suited to assess which species can induce the formation or reduction of a disulfide bond.

In the protocol detailed below, reaction conditions are 1 μ*M* (2.7 μg) recombinant roGFP2 in a total volume of 100 μl.

3.2. Reagents and equipment

- Recombinant roGFP2-His

  We purify recombinant roGFP2-His under nonreducing conditions as described previously (Gutscher et al., 2008).
- roGFP2 assay buffer (100 m*M* potassium phosphate, 5 m*M* EDTA, pH 7.0)
- 220 m*M* DTT (freshly prepared in roGFP2 assay buffer)
- 11 m*M* diamide (1,1′-azobis(*N*,*N*-dimethylformamide)) (freshly prepared in roGFP2 assay buffer)

- 0.5 ml Zeba Desalt Spin Columns, 7 K MWCO (Thermo Scientific)
- Black flat-bottomed 96-well plate (e.g., BD Falcon #353219)
- Fluorescence microplate reader (such as FLUOstar Omega, BMG Labtech)

3.3. Procedure

1. Degas roGFP2 assay buffer for a minimum of 2 h before use by applying vacuum and stirring vigorously at 37 °C.
2. Reduce freshly thawed recombinant roGFP2-His with 20 m*M* DTT for 10 min at room temperature.
3. Pre-equilibrate a Zeba gel filtration column 3 × with degassed roGFP2 assay buffer (as detailed by manufacturer).
4. Pass prereduced roGFP2-His through the pre-equilibrated Zeba gel filtration column to remove DTT and adjust the concentration of the dialyzed roGFP2-His to 27 ng/μl with degassed roGFP2 assay buffer.
5. Dispense 100 μl of the diluted roGFP2-His solution per well into a black flat-bottomed 96-well plate (for 2.7 μg protein/well); use 100 μl degassed roGFP2 assay buffer as blank.
6. Immediately start measuring fluorescence emission at 520 nm with excitation at 390 nm and 480 nm every minute for 60 min using a microplate reader set to 37 °C.
7. After 5 min, pause the measurement and add 10 μl of 220 m*M* DTT to one well and 10 μl of 11 m*M* diamide to a second well (for 20 m*M* DTT/1 m*M* diamide final concentration); these samples serve as calibration controls to define the fully reduced and fully oxidized state of the protein.
8. Calculate the degree of oxidation of roGFP2 as described previously (Meyer & Dick, 2010).

3.4. Comments

In this protocol, roGFP2 is prereduced to allow monitoring of disulfide bond formation. To evaluate reductive effects on the pre-existing roGFP2 disulfide bond, skip the reduction step at the beginning of the protocol (step 2).

We prefer to perform this assay at 37 °C to reflect physiological conditions more closely; this assay is, however, equally functional at room temperature.

3.5. Expected results

Figure 2A shows the fluorescence emission traces for both excitation wavelengths (i.e., 390 nm and 480 nm) of the DTT and diamide controls as well as an H_2O_2-treated sample. Following the calculations detailed in Meyer and Dick (2010), the DTT and diamide measurements serve to define the fully reduced and fully oxidized state, respectively, of roGFP2, which then allows to determine the degree of probe oxidation as shown for the H_2O_2-treated sample (Fig. 2B).

4. APPLICATION OF "H_2S DONORS" AND POLYSULFIDES

4.1. General considerations

The preparation and handling of any source of H_2S is extremely challenging and poses a substantial risk of introducing artifacts. This topic has been expertly discussed in a number of recent publications that are all highly recommended (Hughes, Centelles, & Moore, 2009; Nagy et al., 2014; Olson, 2012; Wedmann et al., 2014). In particular, the use of sulfide salts such as NaHS and Na_2S require a number of considerations to be taken into account, two of which we would like to highlight. First, solvation of the sulfide salts leads to a significant alkalinization of the resulting solution (Olson, 2012); for instance, we measured a final pH of 8.5 for a 60 m*M* Na_2S solution that had been prepared in 200 m*M* Tris–HCl, pH 7.1. Sufficiently concentrated buffers thus need to be used to avoid an effect of sulfide salts on pH, and careful monitoring of the final pH of the solution is crucial. Second, sulfide salts generally contain numerous contaminants. For instance, on inquiry the Sigma-Aldrich technical service listed sodium sulfite, sodium sulfoxide, sodium carbonate, and iron as potential trace impurities for their $NaHS \cdot xH_2O$ product. Polysulfides, however, appear to be even more widespread contaminants of sulfide salts that are particularly problematic with regard to their oxidative effect on thiols (Greiner et al., 2013; Nagy et al., 2014). Regrettably, manufacturers typically do not test for polysulfides or elemental sulfur (that gives rise to polysulfides in the presence of HS^-) in their products; it is therefore good practice to employ the cold cyanolysis assay as described in Section 5.

The purest source of H_2S is the gas itself. In our experience $H_2S(aq)$ solutions, prepared by flushing ultrapure water with H_2S gas for 20 min and stored in brown glass bottles with minimum headspace and a double layer parafilm seal, are reasonably stable. This is due to the fact that $H_2S(aq)$ is

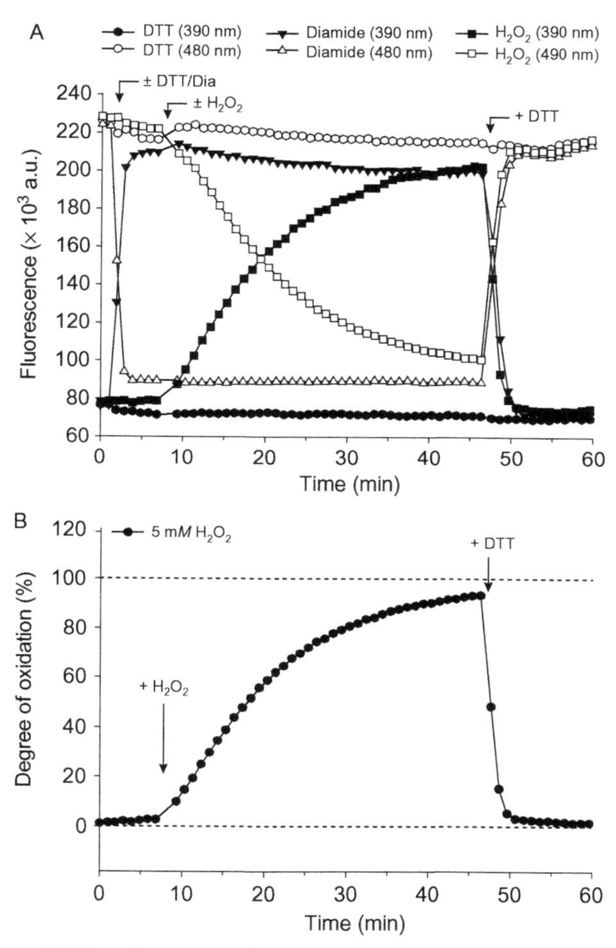

Figure 2 The roGFP2 redox assay allows the monitoring of a thiol/disulfide pair in real time. (A) Fluorescence emission at 520 nm was continuously measured for the excitation wavelengths 390 and 480 nm. Arrows indicate the injection of 20 m*M* DTT or 1 m*M* diamide to two samples of prereduced roGFP2 soon after start of the assay, and injection of 5 m*M* H_2O_2 to a third sample after 5 min. All samples received an injection of 20 m*M* DTT toward the end of the assay to show reversibility of the reaction. (B) The data in (A) were used to calculate the degree of oxidation for the H_2O_2-treated roGFP2.

relatively acidic (e.g., pH 5.0 at 80 m*M*), which precludes oxidation of H_2S (Chen & Morris, 1972). Nevertheless, we prefer to use H_2S(aq) solutions within 2 h of preparation. Other sufficiently pure alternative sources of H_2S are novel H_2S-donating compounds such as GYY4137 that continuously release low concentrations of H_2S. While the use of pure sources of H_2S may help avoid the introduction of pre-existing (polysulfide) contaminations into experimental systems, their use will not preclude the *de novo* generation of H_2S oxidation products (including polysulfides). Spontaneous oxidation of H_2S occurs in the presence of O_2 and transition metal catalysts, and is most rapid at pH 7–8 (Chen & Morris, 1972). Therefore, even the purest preparation of H_2S is at risk of forming polysulfides (and other oxidation products) during the course of an experiment that is performed under physiological conditions. While the use of chelators such as DTPA has been proposed to suppress air oxidation of H_2S (Hughes, Centelles, & Moore, 2009), in our hands even large concentrations of DTPA did not prevent the formation of polysulfides even in highly pure preparations of H_2S; this is in agreement with the observations of others (Nagy et al., 2014). Furthermore, frequently the addition of chelators is not an option, e.g., in experiments involving cells, animals, or enzymes harboring metal centers. Nevertheless, employing all possible measures to prevent spontaneous H_2S oxidation is good practice, and certainly recommended. For detailed advice, see Hughes et al. (2009), Nagy et al. (2014), Olson (2012), and Wedmann et al. (2014).

Given that in our experience the complete suppression of H_2S oxidation is practically impossible, we regard it as absolutely critical to perform control experiments that test for (a) the presence of polysulfides (and other sulfane sulfur species) in H_2S preparations and/or experimental samples (see Sections 5 and 6) and (b) for the ability of polysulfides to induce the experimental response in question. To control for the latter, reagents such as potassium (poly)sulfide (K_2S_x) or sodium (poly)sulfide (Na_2S_x) may be added to the experiment in place of H_2S. Should polysulfides efficiently induce the same experimental outcome as H_2S, polysulfides may well be the true inducers of the biological response in question. When preparing K_2S_x or Na_2S_x solution, note that polysulfides tend to precipitate as elemental sulfur (S_8) at pH <9.0. We therefore typically make K_2S_x stock solutions (e.g., 16.5 mg/ml) in 200 m*M* Tris–HCl, pH 9.2. Only immediately before use is the K_2S_x stock solution diluted into the required assay buffer (of lower pH). As K_2S_x is a mixture of K_2S and potassium polysulfides of different chain lengths, neither a molecular weight nor a molar concentration can be specified; to determine the concentration of polysulfides (or, more specifically, sulfane sulfur) in K_2S_x stock solutions, they need to be subjected to the cold cyanolysis protocol outlined in Section 5. In our hands, stock solutions of 16.5 mg/ml K_2S_x (in 200 m*M* Tris–HCl, pH 9.2) contain ~75 m*M* S^0 (and ~70 m*M* H_2S as determined by OD_{230} measurement).

As "H_2S donors" and polysulfides are intrinsically unstable in solution, we recommend to start the PTEN activity/roGFP2 redox assay *before* preparing the H_2S/polysulfide solutions and diluting them into the required assay buffer. Immediately after preparing and diluting the H_2S/polysulfide solution, pause the microplate reader, add the H_2S/polysulfide preparations to the sample wells, and restart the measurement. We prefer to add the H_2S/polysulfide species in volumes of 10 μl and prepare the dilutions accordingly; control samples should receive 10 μl of assay buffer. Ideally, the H_2S/polysulfide stock solutions should be subjected to the cold cyanolysis assay immediately after addition to the assay to accurately determine polysulfide levels (see Section 5).

4.2. Expected results

Figure 3 shows typical response curves for (A) PTEN and (B) roGFP2 upon treatment with sulfide salts and K_2S_x, respectively. Previous results have shown that polysulfide contaminations of sulfide salts are responsible for the observed oxidation of PTEN (Greiner et al., 2013). Varying levels of contaminating polysulfides explain the different degrees of PTEN oxidation in response to equimolar concentrations of sulfide salts (see figure legend).

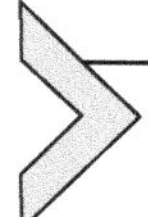

5. QUANTITATION OF SULFANE SULFUR BY COLD CYANOLYSIS

5.1. Principle

The reaction with cyanide (CN^-) at alkaline pH to form thiocyanate (SCN^-) is a defining characteristic of sulfane sulfur (Wood, 1987); it is thus often referred to as cyanolyzable sulfur. In many sulfane sulfur compounds, the sulfur is bonded only to other sulfur atoms, and thus exists in an oxidation state of zero (represented by S^0). These sulfane sulfur compounds are elemental sulfur (S_8), polysulfides (R–S–S_n–S–R, HS–S_n–SH), and polythionates (^-O_3S–S_n–SO_3^-). However, several other compounds contain a labile, highly reactive sulfur atom that can be transferred to CN^-. In these molecules, the sulfur atom may also exist in an oxidation state of -1, as it can additionally be attached to a hydrogen atom or an "activating group," which contains an unsaturated carbon adjacent to the sulfur-bonded carbon. Examples of these sulfane sulfur compounds include thiosulfate ($S_2O_3^{2-}$), thiosulfonates (R–SO_2–S–R), persulfides (R–S–SH), as well as disulfides of mercaptopyruvate, allyl mercaptan, mercaptoketimine, and others (Toohey, 1989). The chemical basis for the lability of sulfane sulfur has

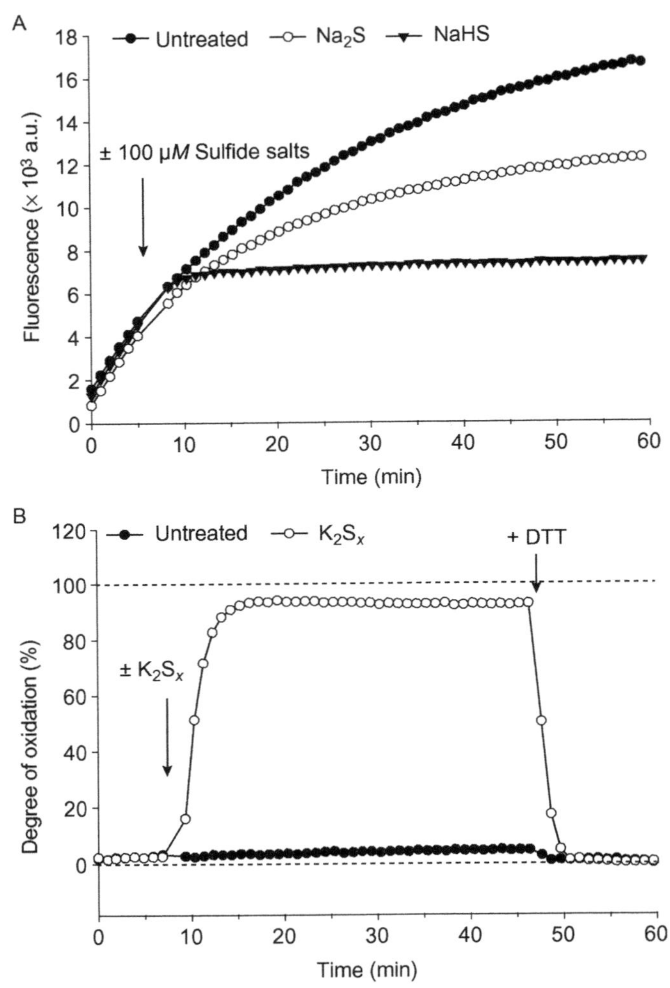

Figure 3 Sulfide salts and K_2S_x polysulfides lead to rapid oxidation of PTEN and roGFP2, respectively. (A) The activity of PTEN-WT was measured under nonreducing conditions upon injection of buffer (untreated) or 100 μ*M* of Na_2S (from Alfa Aesar) or NaHS (from Cayman). The applied concentrations of sulfide salts contained 30 n*M* (Na_2S) and 761 n*M* (NaHS) sulfane sulfur at the time of injection, as determined by a cyanolysis assay. Curves represent means of duplicate wells. (B) The degree of roGFP2 oxidation was recorded in response to injection of buffer (untreated) or 6.7 μg/ml K_2S_x (containing 24 μ*M* sulfane sulfur as determined by cyanolysis), followed by addition of 20 m*M* DTT to show reversibility of the reaction.

not been conclusively clarified, but is likely to be predicated on its presence in the thiosulfoxide form (−S=S), either by primary structure (thiosulfate, thiosulfonate) or by tautomerization (Kutney & Turnbull, 1982; Toohey, 1989). Sulfur in the thiosulfoxide form exists in an oxidation state of zero and is readily removed by nucleophilic acceptors such as thiolate anions ($R–S^-$), sulfite (SO_3^{2-}), or cyanide (CN^-) (Toohey, 1989).

The characteristic reaction of sulfane sulfur with CN^- at alkaline pH yields thiocyanate (SCN^-), which produces blood-red iron (III) thiocyanate ($[Fe(SCN)(H_2O)_5]^{2+}$) in the presence of excess iron chloride ($FeCl_3$). $[Fe(SCN)(H_2O)_5]^{2+}$ can be readily detected by measuring its absorbance at 460 nm. By use of a potassium thiocyanate (KSCN) standard curve it is possible to quantitate $[Fe(SCN)(H_2O)_5]^{2+}$ levels, which are a direct measure of the concentration of sulfane sulfur atoms in the solution.

While some sulfane sulfur species are relatively unreactive with CN^- and require high temperatures to facilitate the reaction, polysulfides (as well as persulfides, aryl thiosulfonates, and higher polythionates) are readily cyanolyzed at room temperature (cold cyanolysis) (Wood, 1987). The following protocol was modified and optimized for microplate analysis from the original protocol published by Wood (1987) and allows one to test for polysulfide contaminations in H_2S solutions freshly prepared for a PTEN or roGFP2 assay. Note that the cyanolysis assay is only sensitive enough to detect sulfane sulfur concentrations in the low micromolar range, which is why the H_2S *stock* solution needs to be subjected to cyanolysis, and not the dilutions.

5.2. Reagents and equipment

- H_2S stock solution to be tested (e.g., NaHS, Na_2S, H_2S(aq))
- 16.5 mg/ml K_2S_x (freshly prepared in 200 m*M* Tris–HCl, pH 9.2) as positive control
- 1 *M* DTT (freshly prepared in ultrapure H_2O)
- 0.5 *M* KCN (prepared in ultrapure H_2O)
- 37% formaldehyde
- 0.1 *M* potassium thiocyanate (KSCN; e.g., premade standard solution, Fluka #38140)
- Goldstein's Reagent

  To prepare 40 ml Goldstein's Reagent, dissolve 1 g ferric nitrate nonahydrate in 10 ml ultrapure water and add 10.5 ml of nitric acid (70% HNO_3) and 19.5 ml of ultrapure water

- Clear 96-well plate
- Microplate reader able to read absorbance at 460 nm
- Fume hood

5.3. Procedure

KCN and H_2S are highly toxic compounds! Please apply all necessary safety measures and perform this experiment in the fume hood.

1. Dilute freshly made 16.5 mg/ml K_2S_x 3× 1:5 in 200 m*M* Tris–HCl, pH 9.2.
2. Transfer 50 μl of all three K_2S_x dilutions and the H_2S stock solution (e.g., 100 m*M* NaHS in 200 m*M* Tris–HCl, pH 8.0) into microtubes containing 400 μl of ultrapure water. Use 50 μl of the buffer utilized for dissolution of the respective compound as blank. This dilution step helps to avoid interference by high buffer concentrations.
3. Add 50 μl of 0.5 *M* KCN (for 50 m*M* KCN final).
4. Allow cyanolysis to proceed for 45 min at room temperature.
5. Meanwhile, prepare KSCN standard solutions ranging from 3 m*M* down to 5.9 μ*M* by serial 1:2 dilution in ultrapure water. Transfer 500 μl of each dilution to fresh microtubes. Use 500 μl of ultrapure water as blank.
6. After completion of the cyanolysis reaction, add 10 μl of 37% formaldehyde to all samples, including the KSCN standard. Formaldehyde stabilizes the ferric thiocyanate complex by reacting with excess KCN.
7. To all samples add 100 μl of Goldstein's Reagent. The blood-red color of $[Fe(SCN)(H_2O)_5]^{2+}$ should develop immediately.
8. Transfer 200 μl into duplicate wells of a clear 96-well plate and measure OD_{460} using a microplate reader. If possible, read full absorbance spectra, as they reveal potentially elevated background that may be mistaken for a true OD_{460} signal.
9. Apply blank-correction to all sample readings and quantify sulfane sulfur levels by use of the KSCN standard curve. Be careful to take account of all dilution steps (e.g., 1:10 for the H_2S stock solution).

5.4. Comments

Cyanolysis occurs most rapidly at alkaline pH (i.e., pH 8.5–10). This pH range is usually achieved during the cyanolysis reaction if H_2S solutions are prepared in mildly alkaline buffers (pH 7–9), as KCN will further alkalinize the solution. If an acidic $H_2S(aq)$ stock solution is to be subjected to cyanolysis, however, adjustment of the pH may be necessary. This can be done using 1 *M* ammonium hydroxide (Wood, 1987).

Thiol reductants such as DTT may be used to degrade polysulfides (and other sulfane sulfur species), which is expected to nullify a sample's positive cyanolysis signal. To this end, the K_2S_x and/or H_2S stock solution may be split in two after step 1 and incubated with or without an excess of DTT for 10 min at RT before proceeding with step 2.

Note that in our experience GYY4137 interferes with the readout of the cyanolysis assay, thus precluding quantitation of polysulfides that may form from slowly released H_2S.

As an additional method to monitor the existence of polysulfides in H_2S solutions, their UV/vis-absorption spectra may be measured. Polysulfides show distinct absorption peaks at ~300 and ~372 nm (Debiemme-Chouvy, Wartelle, & Sauvage, 2004).

5.5. Expected results

In Fig. 4, exemplary cyanolysis results are depicted for NaHS and Na_2S from different suppliers, as well as K_2S_x.

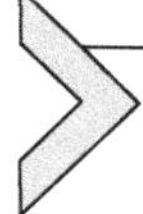

6. ELIMINATION OF SULFANE SULFUR BY COLD CYANOLYSIS

6.1. Principle

Cyanide can be added to assay samples in order to test if cyanolyzable sulfur (polysulfides or other sulfane sulfur species) plays a role in an observed thiol oxidation event. In the presence of substantial amounts of pre-existing polysulfides, the cyanolysis reaction may not be rapid enough to protect all free thiols from polysulfide-mediated oxidation, which is why in this case a preincubation of the H_2S solution with KCN is required. Also in case of *de novo* formation of polysulfides during the assay, an excess of KCN is necessary to outcompete the very facile reaction between polysulfides and free thiols. The protocols below detail the experimental steps to be taken for (a) preincubation of an H_2S solution with KCN (as exemplified by a roGFP2 redox assay) and (b) direct addition of KCN to samples before injection of H_2S (as exemplified by a PTEN activity assay).

It has been established previously that the assays *per se* are unaffected by the given concentrations of cyanide, that roGFP2 and PTEN disulfides are not cleaved by cyanide under the given conditions, and that H_2O_2-induced roGFP2 or PTEN oxidation is not affected by cyanide, as expected (Greiner

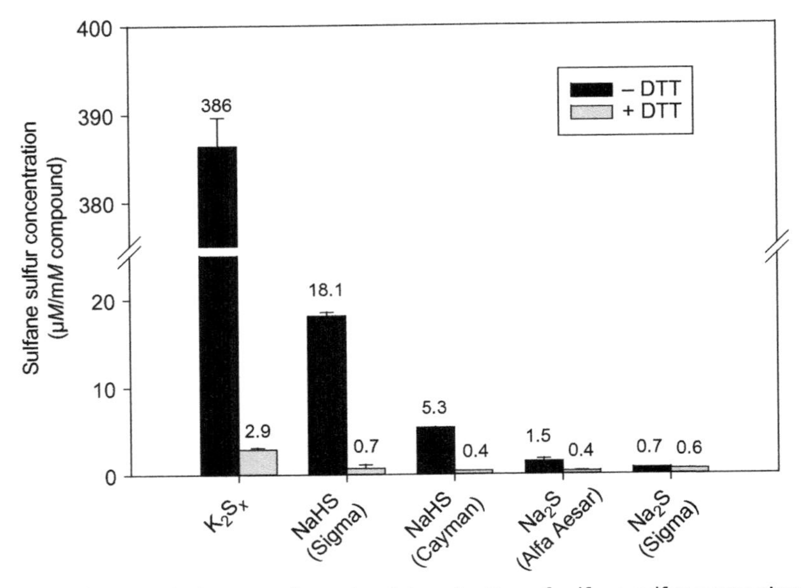

Figure 4 The cyanolysis assay allows the determination of sulfane sulfur contamination of sulfide salts. 150 m*M* stock solutions of sulfide salts were freshly prepared in degassed 200 m*M* Tris–HCl, pH 7.4, treated with or without 200 m*M* DTT for 10 min at room temperature and immediately subjected to the cyanolysis assay. To serve as positive control, a solution of 16.5 mg/ml K_2S_x was freshly prepared in 200 m*M* Tris–HCl, pH 9.2, treated $\pm$ DTT, serially diluted three times, and subjected to the cyanolysis assay. Data are given as µ*M* sulfane sulfur per m*M* sulfide salt, or µ*M* sulfane sulfur per 0.11 mg/ml K_2S_x. Bars denote mean $\pm$ range of duplicate wells.

et al., 2013 and Fig. 6). Thus, if any source of H_2S causes a different experimental outcome in the presence or absence of KCN, an involvement of polysulfides (or other forms of sulfane sulfur) is extremely likely.

6.2. Elimination of pre-existing sulfane sulfur

To degrade pre-existing polysulfides we recommend an incubation of the H_2S solution with at least a fivefold molar excess of KCN. In the protocol detailed below, prereduced roGFP2 is exposed to a 1 m*M* H_2S solution that has been preincubated with or without a fivefold molar excess of KCN for 1 h at RT.

6.2.1 Reagents and equipment

- All reagents required for a roGFP2 redox assay (see Section 3.2)
- H_2S stock solution to be tested (e.g., NaHS, Na_2S, H_2S(aq))
- 0.5 *M* KCN (prepared in ultrapure water)

- If necessary, 1 *M* ammonium hydroxide
- Fume hood

6.2.2 Procedure

Please apply all necessary safety measures when working with KCN and H_2S, and perform this experiment in the fume hood!

1. Degas roGFP2 assay buffer for a minimum of 2 h before use by applying vacuum and stirring vigorously at 37 °C.
2. Prepare a stock solution of your chosen H_2S source and dilute it to 22 m*M*.
3. Aliquot 2 × 300 μl of the 22 m*M* H_2S solution into fresh microtubes and add 66 μl of 0.5 *M* KCN to the first, and 66 μl of ultrapure water to the second aliquot. Depending on your chosen H_2S solution you may wish to adjust the pH to 8–10 using 1 *M* ammonium hydroxide. Then fill up these aliquots to 600 μl with ultrapure water (for 11 m*M* $H_2S \pm 55$ m*M* KCN final concentration). Include a KCN-only control (300 μl of the buffer used to prepare the H_2S solution + 66 μl of 0.5 *M* KCN ± 1 *M* ammonium hydroxide ± ultrapure water, to a total volume of 600 μl).
4. Allow cyanolysis to proceed for 1 h at room temperature.
5. Meanwhile, follow steps 2–7 of the roGFP2 redox assay protocol.
6. Five minutes after calibrating the redox sensor, pause the measurement and add 10 μl of your H_2S solution ± KCN as well as of the KCN-only control to as yet untreated samples (for 1 m*M* $H_2S \pm 5$ m*M* KCN final concentration), and restart the measurement. Do not add anything to the calibration controls.
7. Calculate the degree of oxidation of roGFP2 as described previously (Meyer & Dick, 2010).

6.2.3 Expected results

Figure 5 shows that preincubation with cyanide completely degrades all sulfane sulfur that is present in K_2S_x, thus preventing an oxidative effect of K_2S_x on roGFP2. Furthermore, cyanide alone is not able to cleave an existing disulfide bond in preoxidized roGFP2 (Fig. 5).

6.3. Prevention of the *de novo* formation of sulfane sulfur

For the cyanolysis reaction to be able to outcompete the very rapid reaction of polysulfides with free thiols, KCN will have to be present in very large excess. We recommend a 50- to 200-fold molar excess of KCN over the chosen H_2S concentration. In the protocol detailed below, prereduced PTEN is either exposed or not to 10 m*M* KCN, followed by addition of 50 μ*M* H_2S.

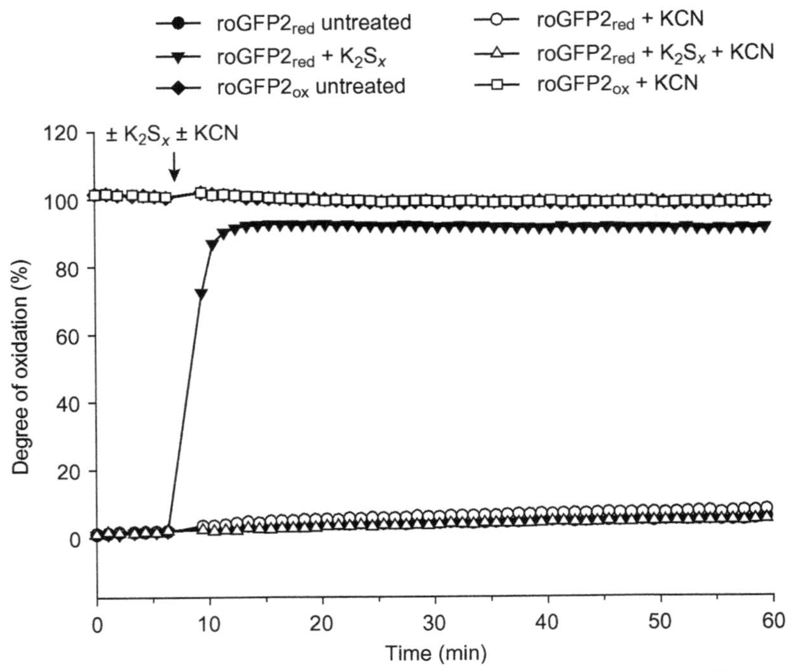

Figure 5 Preincubation with cyanide efficiently eliminates pre-existing sulfane sulfur. A solution of 1.2 mg/ml K_2S_x was exposed or not to 55 m*M* KCN for 1 h at room temperature, before it was added to prereduced roGFP2 ($roGFP2_{red}$) to a final concentration of 0.12 mg/ml $K_2S_x \pm 5$ m*M* KCN. Separately, 5 m*M* KCN or buffer were added to preoxidized roGFP2 ($roGFP2_{ox}$) to show that the disulfide bond is unaffected by KCN.

6.3.1 Reagents and equipment

- All reagents required for a PTEN activity assay (see Section 2.2)
- H_2S stock solution to be tested (e.g., NaHS, Na_2S, H_2S(aq), GYY4137)
- 0.5 *M* KCN (prepared in ultrapure water)
- Fume hood

6.3.2 Procedure

Again, please apply all necessary safety measures when working with KCN and H_2S, and perform this experiment in the fume hood!

1. Degas PTEN assay buffer for a minimum of 2 h before use by applying vacuum and stirring vigorously at 37 °C.
2. Dilute the 0.5 *M* KCN stock solution to 120 m*M* in PTEN assay buffer.
3. Follow steps 2–8 of the PTEN activity assay protocol.

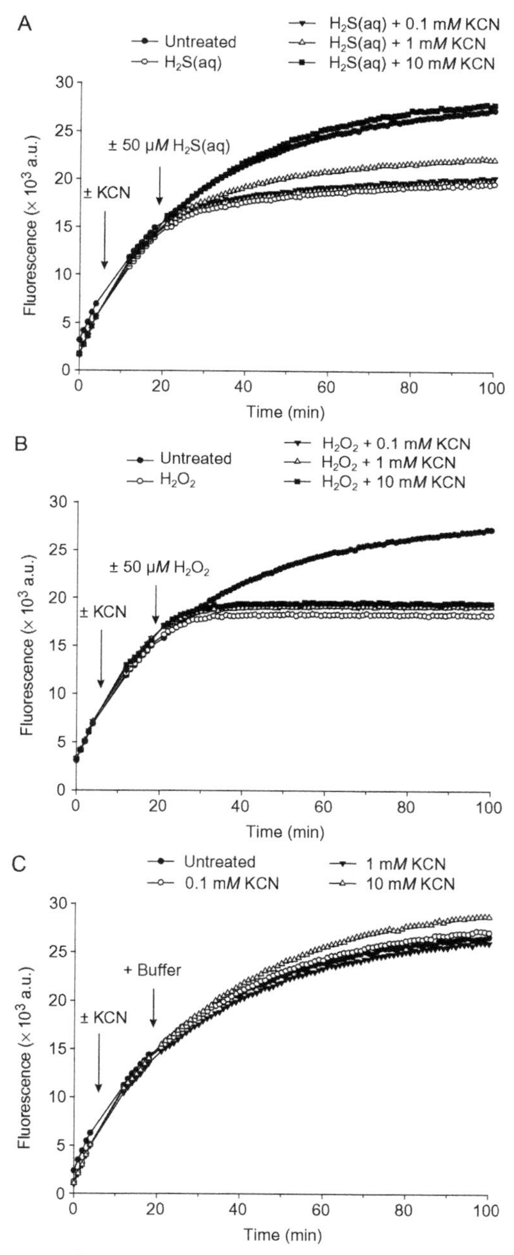

Figure 6 The presence of a high molar excess of cyanide prevents the *de novo* formation of polysulfides in solutions of H_2S. Activity of PTEN-WT was measured in response to injection of buffer (untreated) or different concentrations of KCN, followed by addition of 50 μ*M* H_2S(aq) (A) or H_2O_2 (B). Panel C depicts the KCN-only controls.

4. After 5 min, pause the fluorescence measurement and add 10 μl of 120 m*M* KCN (for a final concentration of 10 m*M* KCN upon addition of the H_2S solution in step 6). Restart measurement.
5. Prepare a stock solution of your chosen H_2S source and dilute it to 0.6 m*M* in PTEN assay buffer.
6. Pause the fluorescence measurement again and add 10 μl of the 0.6 m*M* H_2S solution (for a final concentration of 50 μ*M* H_2S). The KCN-only control receives 10 μl of the buffer that was used to prepare the H_2S solution. Restart measurement.
7. Correct fluorescence signals against corresponding samples lacking PTEN protein (blanks).

6.3.3 Expected results

Figure 6 shows that a solution of H_2S(aq) induces PTEN oxidation, unless in the presence of a high molar excess of KCN. Lower concentrations of KCN are not able to outcompete the rapid reaction of polysulfides with free thiols, and therefore cannot (fully) protect the PTEN active site cysteine from oxidation. Importantly, the PTEN assay *per se* as well as H_2O_2-induced oxidation of PTEN is unaffected by the applied concentrations of KCN (Fig. 6).

ACKNOWLEDGMENTS

Work on protein thiol oxidation mechanisms in our laboratory is supported by the German Research Foundation (SFB 1036, and SPP 1710). R. G. is a recipient of a PhD stipend from the German National Academic Foundation (Studienstiftung des deutschen Volkes). We thank Dr. Bruce Morgan for careful comments on the chapter.

REFERENCES

Chen, K. Y., & Morris, J. C. (1972). Kinetics of oxidation of aqueous sulfide by O_2. *Environmental Science & Technology*, *6*(6), 529–537. http://dx.doi.org/10.1021/Es60065a008.

Debiemme-Chouvy, C., Wartelle, C., & Sauvage, F. X. (2004). First evidence of the oxidation and regeneration of polysulfides at a GaAs electrode, under anodic conditions. A study by in situ UV-visible spectroelectrochemistry. *Journal of Physical Chemistry B*, *108*(47), 18291–18296. http://dx.doi.org/10.1025/Jp046977y.

Denu, J. M., & Dixon, J. E. (1998). Protein tyrosine phosphatases: Mechanisms of catalysis and regulation. *Current Opinion in Chemical Biology*, *2*(5), 633–641.

Greiner, R., Palinkas, Z., Basell, K., Becher, D., Antelmann, H., Nagy, P., et al. (2013). Polysulfides link H_2S to protein thiol oxidation. *Antioxidants & Redox Signaling*, *19*(15), 1749–1765. http://dx.doi.org/10.1089/ars.2012.5041.

Gutscher, M., Pauleau, A. L., Marty, L., Brach, T., Wabnitz, G. H., Samstag, Y., et al. (2008). Real-time imaging of the intracellular glutathione redox potential. *Nature Methods*, *5*(6), 553–559. http://dx.doi.org/10.1038/nmeth.1212.

Hanson, G. T., Aggeler, R., Oglesbee, D., Cannon, M., Capaldi, R. A., Tsien, R. Y., et al. (2004). Investigating mitochondrial redox potential with redox-sensitive green fluorescent protein indicators. *The Journal of Biological Chemistry*, *279*(13), 13044–13053. http://dx.doi.org/10.1074/jbc.M312846200.

Hughes, M. N., Centelles, M. N., & Moore, K. P. (2009). Making and working with hydrogen sulfide: The chemistry and generation of hydrogen sulfide in vitro and its measurement in vivo: A review. *Free Radical Biology & Medicine*, *47*(10), 1346–1353. http://dx.doi.org/10.1016/j.freeradbiomed.2009.09.018.
Kabil, O., Motl, N., & Banerjee, R. (2014). H_2S and its role in redox signaling. *Biochimica et Biophysica Acta*, *1844*(8), 1355–1366. http://dx.doi.org/10.1016/j.bbapap.2014.01.002.
Kutney, G. W., & Turnbull, K. (1982). Compounds containing the S=S bond. *Chemical Reviews*, *82*(4), 333–357. http://dx.doi.org/10.1021/Cr00050a001.
Lee, S. R., Yang, K. S., Kwon, J., Lee, C., Jeong, W., & Rhee, S. G. (2002). Reversible inactivation of the tumor suppressor PTEN by H_2O_2. *The Journal of Biological Chemistry*, *277*(23), 20336–20342. http://dx.doi.org/10.1074/jbc.M111899200.
Meyer, A. J., & Dick, T. P. (2010). Fluorescent protein-based redox probes. *Antioxidants & Redox Signaling*, *13*(5), 621–650. http://dx.doi.org/10.1089/ars.2009.2948.
Nagy, P., Palinkas, Z., Nagy, A., Budai, B., Toth, I., & Vasas, A. (2014). Chemical aspects of hydrogen sulfide measurements in physiological samples. *Biochimica et Biophysica Acta*, *1840*(2), 876–891. http://dx.doi.org/10.1016/j.bbagen.2013.05.037.
Olson, K. R. (2012). A practical look at the chemistry and biology of hydrogen sulfide. *Antioxidants & Redox Signaling*, *17*(1), 32–44. http://dx.doi.org/10.1089/ars.2011.4401.
Toohey, J. I. (1989). Sulphane sulphur in biological systems: A possible regulatory role. *The Biochemical Journal*, *264*(3), 625–632.
Wedmann, R., Bertlein, S., Macinkovic, I., Boltz, S., Miljkovic, J. L., Munoz, L. E., et al. (2014). Working with "HS": Facts and apparent artifacts. *Nitric Oxide*. http://dx.doi.org/10.1016/j.niox.2014.06.003.
Wood, J. L. (1987). Sulfane sulfur. *Methods in Enzymology*, *143*, 25–29.

CHAPTER FIVE

Protein Sulfhydration

Bindu D. Paul*, Solomon H. Snyder*,†,‡,1

*The Solomon H. Snyder Department of Neuroscience, Johns Hopkins University School of Medicine, Baltimore, Maryland, USA
†Department of Pharmacology, Johns Hopkins University School of Medicine, Baltimore, Maryland, USA
‡Department of Psychiatry, Johns Hopkins University School of Medicine, Baltimore, Maryland, USA
[1]Corresponding author: e-mail address: ssnyder@jhmi.edu

Contents

1. Introduction 80
   1.1 Protein S-sulfhydration 80
   1.2 Reciprocity of sulfhydration and nitrosylation 80
2. Detection of Sulfhydration Using the Modified Biotin Switch Assay 81
   2.1 Treatment of cell cultures to assess sulfhydration 83
   2.2 Reagents for the modified biotin switch assay 83
   2.3 Solutions and buffers for the modified biotin switch assay 83
   2.4 Protocol 84
3. Detection of Sulfhydration Using the Maleimide Assay 86
   3.1 Solutions and buffers for the maleimide assay 86
   3.2 Protocol 87
   3.3 Alternate protocol 88
   3.4 Considerations 88
4. Summary 88
Acknowledgments 89
References 89

Abstract

Hydrogen sulfide (H_2S) is one of the gasotransmitters that modulates various biological processes and participates in multiple signaling pathways. H_2S signals by a process termed sulfhydration. Sulfhydration has recently been recognized as a posttranslational modification similar to nitrosylation. Sulfhydration occurs at reactive cysteine residues in proteins and results in the conversion of an –SH group of cysteine to an –SSH or a persulfide group. Sulfhydration is highly prevalent *in vivo*, and aberrant sulfhydration patterns have been observed under several pathological conditions ranging from heart disease to neurodegenerative diseases such as Parkinson's disease. The biotin switch assay, originally developed to detect nitrosylation, has been modified to detect sulfhydration. In this chapter, we discuss the physiological roles of sulfhydration and the methodologies used to detect this modification.

Methods in Enzymology, Volume 555
ISSN 0076-6879
http://dx.doi.org/10.1016/bs.mie.2014.11.021

1. INTRODUCTION

Neurotransmitters, hormones, and growth factors have been long thought to signal exclusively via second messengers such as cyclic AMP/GMP or tyrosine phosphorylation of receptors. Work of Stamler et al. (1992) indicated that the gasotransmitter nitric oxide (NO) can also signal by covalent modification of cysteines in target proteins, a process designated S-nitrosylation. The biotin switch technique (Jaffrey, Erdjument-Bromage, Ferris, Tempst, & Snyder, 2001; Jaffrey & Snyder, 2001) permitted demonstration that S-nitrosylation occurs physiologically and can modify a substantial portion of endogenous proteins. Appreciation that hydrogen sulfide (H_2S) is also a physiologic gasotransmitter led to the discovery that it signals by an analogous method, sulfhydration transforming the SH of cysteines to SSH (Mustafa et al., 2009; Paul & Snyder, 2012, 2014).

H_2S is generated by cystathionine γ-lyase (CSE), cystathionine β-synthase (CBS), and 3-mercaptopyruvate sulfurtransferase (3-MST) *in vivo* (Paul & Snyder, 2012, 2014). CSE is predominantly expressed in peripheral tissues, whereas CBS and 3-MST are expressed in the central nervous system although these enzymes are also present in peripheral as well as in the central nervous system (Paul et al., 2014). H_2S produced by these enzymes has multifaceted roles in homeostatic pathways in the cell and in redox-regulated systems.

1.1. Protein S-sulfhydration

Sulfhydration typically occurs at cysteine residues with a low pK_a and converts the –SH group of cysteines to –SSH groups (Mustafa et al., 2009). Sulfhydration occurs for many proteins, modulating functions of the targeted protein (Table 1). Sulfhydration has emerged as a posttranslational modification comparable in importance to nitrosylation and phosphorylation. While nitrosylation results in capping of cysteine residues, sulfhydration enhances their nucleophilicity and reactivity. Sulfhydration, like nitrosylation, is reversible by the thioredoxin system (Benhar, Forrester, Hess, & Stamler, 2008; Krishnan et al., 2011). Sulfhydration is more abundant than nitrosylation. While only a small proportion of proteins are nitrosylated, 25–50% of hepatic proteins are sulfhydrated (Mustafa et al., 2009).

1.2. Reciprocity of sulfhydration and nitrosylation

Sulfhydration, like nitrosylation, occurs on reactive cysteine residues. In several instances, sulfhydration occurs on the same cysteine residue as

Table 1 Examples of sulfhydrated proteins

Protein modified	**References**
Glyceraldehyde 3-phosphate dehydrogenase	Mustafa et al. (2009)
β-Actin	Mustafa et al. (2009)
PTP1B 3	Krishnan, Fu, Pappin, and Tonks (2011)
Kir6.1 subunit of KATP	Mustafa et al. (2011)
IKCa	Mustafa et al. (2011)
p65 subunit of NF-κB 5	Sen et al. (2012)
Parkin	Vandiver et al. (2013)
Keap1	Yang et al. (2013)
MEK1	Zhao, Ju, et al. (2014) and Zhao, Li, et al. (2014)
Cu/Zn superoxide dismutase	de Beus, Chung, and Colon (2004)
p66Shc	Xie et al. (2014)
Androgen receptor	Zhao, Ju, et al. (2014) and Zhao, Li, et al. (2014)
TRPV6	Liu et al. (2014)

nitrosylation. The two modifications exert opposite effects on the protein being modified. For example, GAPDH can be sulfhydrated as well as nitrosylated on the same cysteine residue, Cys150 (Hara et al., 2005; Mustafa et al., 2009). Nitrosylation decreases, whereas sulfhydration increases the glycolytic activity of GAPDH. The E3 ubiquitin ligase parkin can be both nitrosylated and sulfhydrated at the same residue. Parkin acts under normal conditions to facilitate ubiquitination and clearance of toxic proteins, and sulfhydration enhances this catalytic activity. In Parkinson's disease, the active cysteine residue is nitrosylated, which suppresses its activity, leading to accumulation of aggregated proteins which mediate neurotoxicity.

2. DETECTION OF SULFHYDRATION USING THE MODIFIED BIOTIN SWITCH ASSAY

Sulfhydration can be detected by a modified biotin switch assay (Mustafa et al., 2009) in which unmodified cysteines in proteins are blocked by methyl methanethiosulfonate (MMTS) (Fig. 1). Sulfhydrated cysteines

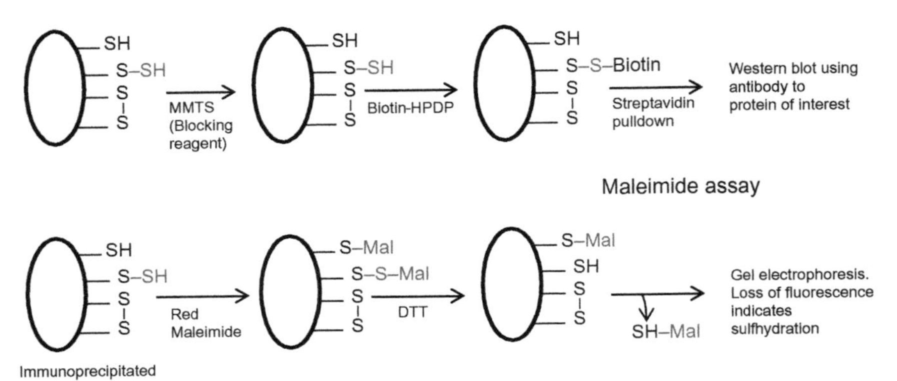

Figure 1 Detection of sulfhydration. (A) Modified biotin switch assay. Shown in the top panel is a protein, which has cysteines in different states: sulfhydrated (denoted as –SSH), involved in disulfide bonds (represented as S–S) and with free SH groups (denoted as SH). Free cysteines in the protein are blocked with methyl methanethiosulfonate (MMTS). Free MMTS is then removed by acetone precipitation or passage through desalting columns, and the reduced thiols are treated with biotin-HPDP. The biotinylated protein is pulled down using streptavidin beads and analyzed using Western blotting. (B) The maleimide assay. The protein of interest is immunoprecipitated and then treated with a fluorescent derivative of maleimide, which selectively labels thiols, both unmodified and those which are sulfhydrated. One portion of the reaction mix is treated with DTT, which would release the maleimide moiety, resulting in a decrease in fluorescence. The samples are then run on a polyacrylamide gel and scanned to detect the fluorescent proteins. The decrease in fluorescence is quantified using an image analysis software.

are then exposed by reduction utilizing dithiothreitol (DTT) followed by labeling with biotin which can be readily captured with streptavidin-conjugated sepharose and analyzed by immunoblotting. Alternatively, sulfhydration can be identified employing fluorescent maleimide to label both sulfhydrated and unmodified cysteines (Sen et al., 2012). DTT treatment then releases maleimide selectively from the sulfhydrated cysteines leading to a loss of signal. A modification of this approach, the tag-switch procedure (Zhang et al., 2013), employs methylsulfonyl benzothiazole to block thiols whereupon a reagent incorporating a nucleophilic component and a biotin reporter moiety selectively labels the sulfhydrated components. The modified protein is then captured with streptavidin and detected using Western blotting. A point of note is that the modified biotin switch can also detect other thiol modifications so that mass spectroscopy is often used to confirm the modification. Finally, the sulfhydration of the protein under study should be validated in a genetic background where the modification cannot occur. For example, sulfhydration of proteins in the peripheral tissues can be analyzed utilizing mice deleted for CSE, where sulfhydration is minimal.

2.1. Treatment of cell cultures to assess sulfhydration

Various donors of H_2S are commercially available. The most commonly used donors are NaHS and morpholin-4-ium 4 methoxyphenyl (morpholino) phosphinodithioate (GYY4137). NaHS generates H_2S rapidly. GYY4137 releases H_2S more slowly due to its hydrolysis in a more controlled manner. The concentration of donors used must be optimized for each cell line to ensure minimal cytotoxicity. Typical concentrations used for NaHS and GYY4137 for commonly used cell lines are 50–100 μ*M*. GYY4137 solutions are prepared in DMSO at a stock concentration of 100 m*M*, aliquoted, and stored at −80 °C in dark tubes. NaHS is freshly prepared in water and diluted to the concentrations required. Depending on the application and experiment being conducted, these compounds are added directly to the tissue culture medium, and incubation continued either overnight or for specific lengths of time, if a time course of sulfhydration is being studied. At the end of the incubation period, cells are harvested in HEN buffer (see below for recipe) and lysed as detailed below.

2.2. Reagents for the modified biotin switch assay

Acetone
Dimethylsulfoxide (DMSO)
Ethylenediaminetetraacetic acid (EDTA)
Glycerol
HEPES
DTT
MMTS (Sigma-Aldrich) *N*-[6-(biotinamido)hexyl]-3′-(2′-pyridyldithio) propionamide (biotin-HPDP) (Pierce)
Neocuproine (Sigma-Aldrich)
H_2S donor NaHS (Sigma-Aldrich) or GYY4137 (Cayman Chemicals
Streptavidin agarose (Sigma-Aldrich) or Neutravidin (Pierce)
Triton X-100

2.3. Solutions and buffers for the modified biotin switch assay

HEN buffer

HEPES–NaOH, pH 7.7, 250 m*M*
EDTA 1 m*M*
Neocuproine 0.1 m*M*

Blocking buffer

HEN buffer (Recipe 1):9 volumes

SDS (25% w/v in ddH_2O):1 volume

Adjust to 20 m*M* MMTS

HENS buffer

Adjust HEN to 1% SDS by addition of SDS solution.

Biotin-HPDP

Prepare biotin-HPDP as a 50-m*M* suspension in DMSO. Freeze in aliquots at −80 °C.

Biotinylation solution

Thaw the stock suspension of biotin-HPDP and vortex to resuspend the biotin-HPDP. Dilute with DMSO to a final concentration of 4 m*M*. Vortex to ensure that the biotin-HPDP goes into solution.

SDS-PAGE sample buffer (nonreducing)

50 m*M* Tris–HCl, pH 6.8, 7 mL

Glycerol 3 mL

SDS 1.028 g

Bromophenol blue 1.2 mg

Note: The sample buffer must be made without βME or DTT.

Neutralization buffer

HEPES–NaOH, pH 7.7, 20 m*M*

NaCl 100 m*M*

EDTA 1 m*M*

Triton X-100 0.5%

Neutralization buffer (high salt)

Prepare neutralization buffer (Recipe 10) with 600 m*M* NaCl.

Elution buffer

HEPES–NaOH, pH 7.7, 20 m*M*

NaCl 100 m*M*

EDTA 1 m*M*

β-ME 1% (v/v)

2.4. Protocol

1. Lyse frozen cells on plate in HEN buffer with the desired detergent (e.g., HEN + 1% Triton or + 0.1% SDS, and 0.1% sodium deoxycholate) containing the following components added fresh:
   a. 0.1 m*M* neocuproine
   b. Protease inhibitors
   c. Phosphatase inhibitors

2. Further lyse cells by passing crude lysate through a 1-mL syringe with 26-3/4 gauge needle 10 times.
3. Spin at 14,000 rpm at 4 °C for 5 min.
4. Estimate protein concentration, using 500 μg of overexpressed proteins or 1–2 mg for endogenous proteins. This will need to be determined empirically.
5. Dilute stock MMTS.
6. Block proteins in blocking buffer and vortex well. Block at 50 °C for 20 min at 1400 rpm on a thermomixer with rotation.
7. Precipitate proteins with 2.5 vol of cold acetone (precooled at −20 °C), centrifuging for 5 min at 3100 rpm using a tabletop centrifuge. Greater speeds make the pellet more difficult to resuspend later.
8. During the acetone precipitation step, thaw the frozen aliquots of biotin-HPDP. Dilute stock biotin-HPDP (50 m*M*) with DMSO to prepare 4 m*M* biotin-HPDP. Vortex well and incubate at room temperature for about 15 min while doing the next few steps.
9. Wash pellet gently two times with 70% acetone at room temperature. When complete, do not aspirate acetone until immediately before resuspension (i.e., process samples one at a time so that the pellet does not dry up).
10. Resuspend the pellets in HENS buffer and 0.8 m*M* biotin-HPDP (=200 μL/1 mL of the diluted biotin-HPDP from step 9), adjusted to approximately 0.5–1 mg/mL of protein (i.e., 500 μL or 1 mL total volume, determined empirically).
11. Label on a rotator in dark for 1 to 75–90 min at room temperature. The solution should be clear.

    If the solution becomes cloudy at the end of labeling, it indicates that the biotin-HPDP has precipitated out of solution. Labeling for less than 1 h 15 min helps avoid this. If precipitation occurs, remove as much of this precipitate as possible by spinning the tubes at 14,000 rpm.
12. Precipitate proteins and resuspend in HENS, followed by a second round of acetone precipitation.
13. Resuspend pellet in 250 μL of HENS buffer.
14. Add 750 μL of neutralization buffer, remove 15–30 μL for input, then add to washed streptavidin beads and rotate overnight at 4 °C.
15. Wash six to seven times in neutralization buffer + NaCl.
16. Aspirate to dryness with a syringe.

17. Add 35 μL elution buffer containing 1% βME, plus 100 m*M* NaCl and 0.1% sodium deoxycholate (salt and deoxycholate optional).
18. Elute in a thermomixer at 37 °C for 30 min, shaking at 1000 rpm.
19. Pellet down and collect the supernatant, making sure that the beads are undisturbed.
20. Add 4 × LDS and perform gel electrophoresis.

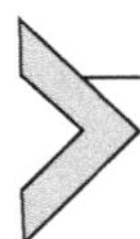

3. DETECTION OF SULFHYDRATION USING THE MALEIMIDE ASSAY

The maleimide assay was developed (Sen et al., 2012) to detect sulfhydration based on the principles elucidated earlier (Aracena-Parks et al., 2006; Wright, Christman, Snellinger, Johnston, & Mueller, 2006). Maleimide interacts selectively with sulfhydryl groups, labeling both sulfhydrated as well as unsulfhydrated groups. An advantage of this method is that oxidized and nitrosylated moieties are not detected. In the method, fluorescently labeled maleimide, namely Alexa Fluor 680-conjugated C2 maleimide (red maleimide), is reacted with the sample. This is followed by treatment with DTT, which cleaves the disulfide bond and releases the fluorescent "red" moiety from sulfhydrated but not unmodified cysteines. The samples are then analyzed using SDS-PAGE, and the decrease in fluorescence intensity quantitated using an image analyzer such as the LiCOR Odyssey system. The method has also been adapted to simultaneously measure sulfhydration and nitrosylation in the same sample, which can reveal important information on the reciprocity of the two modifications. To detect nitrosylation, the samples are first treated with ascorbate which selectively reduces the S–NO bonds and further labeled with Alexa Fluor 488-conjugated C5 maleimide or IRDye 800CW (green maleimide). This is followed by the procedure for detecting sulfhydration using DTT.

3.1. Solutions and buffers for the maleimide assay

Lysis buffer for cells

50 m*M* Tris–HCl, pH 7.5
150 m*M* NaCl
1 m*M* EDTA
0.1–1.5% Triton X-100 or 0–2% Tween 20 (v/v) (optimized for protein of interest)

Labeling reagents

Alexa Fluor 680-conjugated C2 maleimide or red maleimide (Life Technologies; Catalog number A-20344)

Alexa Fluor 680-conjugated C2 maleimide or green maleimide (Life Technologies; Catalog number A-10254). These reagents are to be dissolved in fresh anhydrous DMSO.

H_2S donors: NaHS or GYY4137

NO donors: GSNO

Other reagents: DTT, Sodium ascorbate, Zeba™ Spin Desalting Columns to remove excess NaHS or GSNO

3.2. Protocol

1. Lyse cells in lysis buffer with the desired detergent (Triton X-100 or Tween-20) containing freshly added protease inhibitors.
2. Further lyse cells by passing crude lysate through a 1-ml syringe with 26-3/4 gauge needle 10 times.
3. Spin at max speed at 4 °C for 5 min.
4. Estimate protein concentration, using 100–500 μg of protein for overexpressed, 1–2 mg for endogenous proteins. This will need to be determined empirically.
5. Immunoprecipitate the protein of interest using a highly specific antibody and standard procedures.
6. Add 25 μL Protein A/G agarose to capture the immune complexes. Wash the beads with 1 mL of lysis buffer and centrifuge the samples at 2500 rpm for 5 min. Wash three times. Resuspend beads in 500 μL lysis buffer with Alexa Fluor 680-conjugated C2 maleimide (2 μ*M* final concentration) and incubate for 2 h at 4 °C in the dark with occasional mixing. Pellet beads and wash four to five times.
7. Resuspend the beads after the final wash in 800 μL lysis buffer. Split into two parts of 380 μL each. Treat one set with 1 m*M* DTT in a final volume of 500 μL. Rotate at 4 °C for 1 h. Pellet beads, wash with lysis buffer three times, resuspend in 40 μL electrophoresis sample buffer, and boil at 95 °C for 3 min. Load 20 μL each on a 4–12% Bis-Tris NUPAGE gel (Life technologies) or its equivalent, then transfer to Immobilon-P membranes (Millipore). The membranes are then scanned with the Li-COR Odyssey sytem. After scanning, the membrane can also be analyzed by Western blotting using the antibody against the protein being studied.

3.3. Alternate protocol

1. Follow steps 1–4 of section 3.2. Incubate supernatant with Alexa Fluor 680-conjugated C2 maleimide (2 μ*M* final concentration) and incubate for 2 h at 4 °C in the dark with occasional mixing.
2. Add control IgG and 25 μL Protein A/G agarose and rotate at 4 °C for 30 min to preclear the lysate.
3. Pellet beads at 2500 rpm for 5 min at 4 °C and transfer supernatant to fresh tubes.
4. Use 100–500 μg protein and add primary antibody along with 25 μL Protein A/G agarose and rotate at 4 °C overnight.
5. Pellet down beads and wash four times.
6. Resuspend the beads after the final wash in 800 μL lysis buffer. Split into two parts of 380 μL each. Treat one set with 1 m*M* DTT in a final volume of 500 μL. Rotate at 4 °C for 1 h. Pellet beads, wash with lysis buffer three times, resuspend in 40 μL electrophoresis sample buffer, and boil at 95 °C for 3 min. Load 20 μL each on a 4–12% Bis-Tris NUPAGE gel (Life technologies) or its equivalent, then transfer to Immobilon-P membranes (Millipore). The membranes are then scanned with the Li-COR Odyssey sytem. After scanning, the membrane can also be analyzed by Western blotting using the antibody against the protein being studied.

3.4. Considerations

Several points should be considered while conducting the maleimide assay.

The antibody used to immunoprecipitate the protein should be very specific. The detergent used in the lysis buffer should not be too harsh. Labeling of proteins in their native folded configuration is ideal and hence conditions should be chosen such that the protein is not unfolded. This would minimize labeling of buried cysteine residues. Washing steps should be conducted rigorously to eliminate excess maleimide.

4. SUMMARY

Recent studies have demonstrated that S-sulfhydration is a widely prevalent posttranslational modification that regulates various physiological processes. Sulfhydration is important for optimal functioning of several proteins, and aberrant sulfhydration patterns have been observed in various disease conditions ranging from cardiovascular disorders to neurodegenerative diseases such as Parkinson's disease. The development of various techniques has permitted the study of the role of this newly discovered modification under physiological and pathological conditions, although some of these methods may lack sufficient specificity. Development of newer methods with advanced probes may advance the field and aid in the study of interplay between sulfhydration and other modifications that occur on reactive cysteine residues.

ACKNOWLEDGMENTS

This work was supported by USPHS Grants DA 000266 and MH18501 to S. H. S.

REFERENCES

Aracena-Parks, P., Goonasekera, S. A., Gilman, C. P., Dirksen, R. T., Hidalgo, C., & Hamilton, S. L. (2006). Identification of cysteines involved in S-nitrosylation, S-glutathionylation, and oxidation to disulfides in ryanodine receptor type 1. *The Journal of Biological Chemistry*, *281*, 40354–40368.

Benhar, M., Forrester, M. T., Hess, D. T., & Stamler, J. S. (2008). Regulated protein denitrosylation by cytosolic and mitochondrial thioredoxins. *Science*, *320*, 1050–1054.

de Beus, M. D., Chung, J., & Colon, W. (2004). Modification of cysteine 111 in Cu/Zn superoxide dismutase results in altered spectroscopic and biophysical properties. *Protein Science*, *13*, 1347–1355.

Hara, M. R., Agrawal, N., Kim, S. F., Cascio, M. B., Fujimuro, M., Ozeki, Y., et al. (2005). S-nitrosylated GAPDH initiates apoptotic cell death by nuclear translocation following Siah1 binding. *Nature Cell Biology*, *7*, 665–674.

Jaffrey, S. R., Erdjument-Bromage, H., Ferris, C. D., Tempst, P., & Snyder, S. H. (2001). Protein S-nitrosylation: A physiological signal for neuronal nitric oxide. *Nature Cell Biology*, *3*, 193–197.

Jaffrey, S. R., & Snyder, S. H. (2001). The biotin switch method for the detection of S-nitrosylated proteins. *Science's STKE*, *2001*, pl1.

Krishnan, N., Fu, C., Pappin, D. J., & Tonks, N. K. (2011). H_2S-induced sulfhydration of the phosphatase PTP1B and its role in the endoplasmic reticulum stress response. *Science Signaling*, *4*, ra86.

Liu, Y., Yang, R., Liu, X., Zhou, Y., Qu, C., Kikuiri, T., et al. (2014). Hydrogen sulfide maintains mesenchymal stem cell function and bone homeostasis via regulation of Ca(2+) channel sulfhydration. *Cell Stem Cell*, *15*, 66–78.

Mustafa, A. K., Gadalla, M. M., Sen, N., Kim, S., Mu, W., Gazi, S. K., et al. (2009). H_2S signals through protein S-sulfhydration. *Science Signaling*, *2*, ra72.

Mustafa, A. K., Sikka, G., Gazi, S. K., Steppan, J., Jung, S. M., Bhunia, A. K., et al. (2011). Hydrogen sulfide as endothelium-derived hyperpolarizing factor sulfhydrates potassium channels. *Circulation Research*, *109*, 1259–1268.

Paul, B. D., & Snyder, S. H. (2012). H_2S signalling through protein sulfhydration and beyond. *Nature Reviews. Molecular Cell Biology*, *13*, 499–507.

Paul, B. D., & Snyder, S. H. (2014). Modes of physiologic H_2S signaling in the brain and peripheral tissues. *Antioxidants & Redox Signaling*, Epub ahead of print.

Paul, B. D., Sbodio, J. I., Xu, R., Vandiver, M. S., Cha, J. Y., et al. (2014). Cystathionine γ-lyase deficiency mediates neurodegeneration in Huntington's disease. *Nature*, *509*, 96–100.

Sen, N., Paul, B. D., Gadalla, M. M., Mustafa, A. K., Sen, T., Xu, R., et al. (2012). Hydrogen sulfide-linked sulfhydration of NF-κB mediates its antiapoptotic actions. *Molecular Cell, 45*, 13–24.

Stamler, J. S., Simon, D. I., Osborne, J. A., Mullins, M. E., Jaraki, O., Michel, T., et al. (1992). S-nitrosylation of proteins with nitric oxide: Synthesis and characterization of biologically active compounds. *Proceedings of the National Academy of Sciences of the United States of America, 89*, 444–448.

Vandiver, M. S., Paul, B. D., Xu, R., Karuppagounder, S., Rao, F., Snowman, A. M., et al. (2013). Sulfhydration mediates neuroprotective actions of parkin. *Nature Communications, 4*, 1826.

Wright, C. M., Christman, G. D., Snellinger, A. M., Johnston, M. V., & Mueller, E. G. (2006). Direct evidence for enzyme persulfide and disulfide intermediates during 4-thiouridine biosynthesis. *Chemical Communications (Cambridge, England), 29*, 3104–3106.

Xie, Z. Z., Shi, M. M., Xie, L., Wu, Z. Y., Li, G., Hua, F., et al. (2014). Sulfhydration of p66Shc at cysteine59 mediates the antioxidant effect of hydrogen sulfide. *Antioxidants & Redox Signaling*, Epub ahead of print.

Yang, G., Zhao, K., Ju, Y., Mani, S., Cao, Q., Puukila, S., et al. (2013). Hydrogen sulfide protects against cellular senescence via S-sulfhydration of Keap1 and activation of Nrf2. *Antioxidants & Redox Signaling, 18*, 1906–1919.

Zhang, D., Macinkovic, I., Devarie-Baez, N. O., Pan, J., Park, C. M., Carroll, K. S., et al. (2013). Detection of protein S-sulfhydration by a tag-switch technique. *Angewandte Chemie (International Ed. in English), 53*, 575–581.

Zhao, K., Ju, Y., Li, S., Altaany, Z., Wang, R., & Yang, G. (2014). S-sulfhydration of MEK1 leads to PARP-1 activation and DNA damage repair. *EMBO Reports, 15*, 792–800.

Zhao, K., Li, S., Wu, L., Lai, C., & Yang, G. (2014). Hydrogen sulfide represses androgen receptor transactivation by targeting at the second zinc finger module. *The Journal of Biological Chemistry, 289*, 20824–20835.

H_2S in Cell Signaling in the Cardiovascular and Nervous System and Inflammatory Processes

CHAPTER SIX

Intravital Microscopic Methods to Evaluate Anti-inflammatory Effects and Signaling Mechanisms Evoked by Hydrogen Sulfide

Mozow Y. Zuidema*, **Ronald J. Korthuis**[†,‡,1]

*Harry S. Truman Veterans Administration Hospital, Cardiology, Columbia, Missouri, USA
[†]Department of Medical Pharmacology and Physiology, School of Medicine, One Hospital Drive, University of Missouri, Columbia, Missouri, USA
[‡]Dalton Cardiovascular Research Center, University of Missouri, Columbia, Missouri, USA
[1]Corresponding author: e-mail address: korthuisr@health.missouri.edu

Contents

Abstract

Hydrogen sulfide (H_2S) is an endogenous gaseous signaling molecule with potent anti-inflammatory properties. Exogenous application of H_2S donors, administered either acutely during an inflammatory response or as an antecedent preconditioning intervention that invokes the activation of anti-inflammatory cell survival programs, effectively limits leukocyte rolling, adhesion and emigration, generation of reactive oxygen species, chemokine and cell adhesion molecule expression, endothelial barrier disruption,

ISSN 0076-6879
http://dx.doi.org/10.1016/bs.mie.2014.11.022

capillary perfusion deficits, and parenchymal cell dysfunction and injury. This chapter focuses on intravital microscopic methods that can be used to assess the anti-inflammatory effects exerted by H_2S, as well as to explore the cellular signaling mechanisms by which this gaseous molecule limits the aforementioned inflammatory responses. Recent advances include use of intravital multiphoton microscopy and optical biosensor technology to explore signaling mechanisms *in vivo*.

1. INTRODUCTION

The cardinal signs of inflammation include vasodilation and increased blood flow, elaboration of proinflammatory mediators, upregulation of adhesion molecule expression on cell surfaces, leukocyte/endothelial cell adhesive interactions, endothelial barrier disruption and edema formation, and tissue injury/dysfunction. In many types of inflammation, generation of reactive oxygen species also occurs and plays important roles in producing many of the other sequelae to an inflammatory stimulus, including parenchymal cell injury. Ischemia and reperfusion (I/R) is now recognized as an inflammatory condition that provokes all of the aforementioned signs of inflammation, but can also be associated with development of capillary perfusion deficits termed the no-reflow phenomenon. Interestingly, capillary no-reflow develops by a neutrophil-dependent mechanism. This chapter focuses on application of intravital microscopic techniques that are used to assess the anti-inflammatory effects of both endogenous H_2S and those invoked by treatment with exogenous H_2S donors, with a particular emphasis on its effects to limit inflammatory responses to I/R.

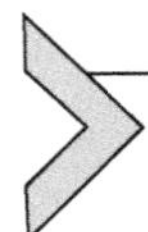

2. MOLECULAR DETERMINANTS OF NEUTROPHIL/ ENDOTHELIAL CELL ADHESIVE INTERACTIONS

The process of neutrophil recruitment to inflammatory sites occurs in 10 distinct phases: margination, capture (tethering) and rolling, slow rolling, arrest (firm adhesion), luminal crawling, transendothelial migration, abluminal crawling, penetration of the basement membrane at regions of low matrix protein deposition and through pericyte gaps, detachment from structures in the vessel wall, and emigration into interstitial tissue (reviewed in Voisin & Nourshargh, 2013; Fig. 1). The adhesion molecules mediating in these steps are different and can vary by leukocyte subset. For the discussion that follows, we focus on adhesive molecules involved in recruiting neutrophils to postischemic tissues, since these cells play the most prominent

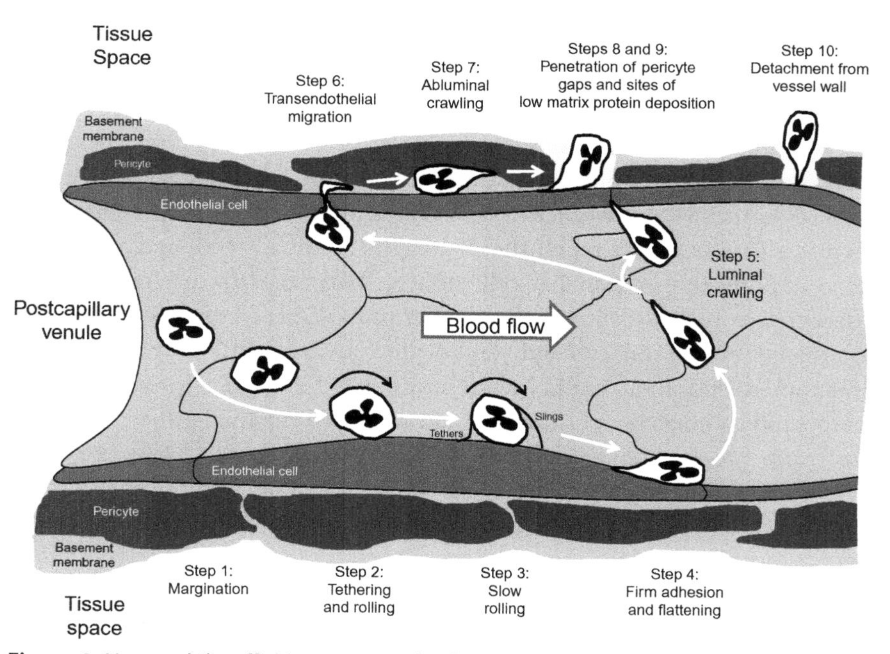

Figure 1 Neutrophil trafficking to sites of inflammation involves 10 steps. Step 1: *Margination.* As neutrophils exit capillaries and enter the larger diameter postcapillary venular segment of the microcirculation, hydrodynamic forces move granulocytes to the vessel margins (margination). Step 2: *Tethering and rolling.* If appropriate adhesion molecules are expressed on activated endothelial cells and the marginated neutrophil, granulocytes are capture by adhesive interactions that mediate rolling of the leukocyte along the vessel wall. Step 3: *Slow rolling.* The rolling neutrophil monitors its local environment for the presence of activating factors that promote adhesion molecule expression, thereby enabling the leukocyte to further slow its rolling behavior by forming more weak adhesive interactions that are mediated by the selectins. As the cells roll along and interact with P-selectin on the endothelial surface, excess membrane on the surface of neutrophils is pulled out into long nanotubes (microvilli) that form tethers at the rear of the rolling leukocyte. These tethers eventually detach as the cell continues to roll, but do not retract. Instead, they persist and are slung in front of the rolling leukocyte, to interact again with P-selectin. The membranous nanotube is now referred to as a sling, which the neutrophil rolls over to again be retarded in its movement down the vessel wall as the sling transitions to a tethering function. Step 4: *Firm or stationary adhesion.* By establishing strong adhesive interactions mediated by integrin-dependent interactions with endothelial ICAM-1 that are upregulated by chemokines, the slowly rolling leukocyte progresses to firm adhesion. Step 5: *Luminal crawling.* Integrin-ICAM-1-dependent adhesion activates intracellular signaling pathways that induce cytoskeletal changes and polarization of the cell that lead to luminal motility. The crawling neutrophils move along interendothelial junctions in search of preferred routes for diapedesis and are often observed moving against the direction of flow in this exploration. Step 6: *Transendothelial cell migration.* Neutrophils cross the endothelial cell barrier by traversing

(Continued)

role in the tissue injury invoked early during reperfusion and our work regarding the anti-adhesive effects of H_2S has focused on that particular question.

Complex and not well-understood hydrodynamic dispersal forces guide neutrophils exiting capillaries to the walls of postcapillary venules, in a process designated as margination. If appropriate adhesive receptors are expressed on both cell types, the neutrophil is initially tethered and begins to roll along the endothelial cell surface. This requires that the adhesive forces tethering leukocytes to the venular surface exceed the shear forces that normally sweep unactivated leukocytes away from quiescent (non-inflammed) endothelial cells. Ischemic vessels exhibit much lower shear rates compared to normally perfused vessels, naturally enhancing the probability for leukocyte–endothelial interactions (Gaboury, Johnston, Niu, & Kubes, 1995; Sanz, Johnston, Issekutz, & Kubes, 1999). Because reperfusion increases the number of rolling and adherent leukocytes and at the same time increases wall shear rate, it is clear that the strength of the adhesive interactions mediating rolling and adhesion are increased relative to nonischemic tissues.

The molecular determinants underlying leukocyte rolling, adhesion and crawling, and transmigration in I/R include specific integrins, members of the immunoglobulin superfamily, and the selectins, as well as a number of other molecular entities expressed on the surfaces of both the infiltrating leukocyte and the endothelium (Voisin & Nourshargh, 2013). These adhesion

Figure 1—Cont'd interendothelial junctions at preferential sites that overlie areas in the basement membrane that exhibit low matrix protein deposition. Occasionally, these inflammatory phagocytes cross the endothelial barrier by moving through cells in a transcellular route. Step 7: *Abluminal crawling*. Once through the endothelium, the diapedesing neutrophils crawl abluminally along pericyte processes, interacting with basement membrane structures at the same time. Steps 8 and 9: *Penetration of pericyte gaps and regions of low matrix protein expression in the basement membrane*. Abluminally crawling neutrophils breach the pericyte layer where gaps between these cells exist, which coincide with regions of low matrix protein deposition (depicted as a lighter shade of green) in the basement membrane. These steps occur more or less simultaneously and use different adhesion molecules to propel the leukocyte through these barriers. Step 10: *Detachment from the vessel Wall*. The last step in the process requires detachment of the migrating neutrophil from components of the vessel wall, which involves release of pseudopodial extensions from their sites of attachment to basement membrane components and adhesion receptors on pericytes and endothelial cells. Once this is achieved, the leukocyte follows directional cues provided by a chemotactic gradient and migrates through the tissue space towards inflammatory foci. *Modified from Voisin and Nourshargh (2013)*. (See the color plate.)

molecules interact in a highly coordinated and dynamic fashion to allow leukocyte diapedesis into postischemic tissues. E- and P-selectins are expressed on the surface of the endothelium after these cells are activated by proinflammatory mediators formed during I/R. These endothelial selectins interact with L-selectin and PSGL-1 expressed on neutrophils and mediate leukocyte rolling interactions. E-selectin is synthesized de novo, a process which requires 2–4 h, while P-selectin is initially mobilized to the cell surface from a stored pool contained in Weibel–Palade bodies. This is followed by expression of newly synthesized P-selectin 1–4 h later. Thus, endothelial P-selectin participates in leukocyte rolling during early reperfusion while both E- and P-selectin play a role in later periods. During this rolling phase, leukocytes are exposed to and become activated by chemoattractants and intracellular signals. Strengthening of the leukocyte-endothelial interactions occurs primarily through binding of neutrophilic integrin, CD11a/CD18, with its counter-receptor on endothelial cells, ICAM-1 (and perhaps ICAM-2), a member of the immunoglobulin superfamily. Once arrested, the neutrophil flattens out, which increases the contact area with the endothelium and decreases its cross-sectional profile, allowing it to better resist the effects of shear stress imposed by the flowing blood. The neutrophil then begins to crawl along the endothelial cell by a CD11b/CD18-ICAM-1/ICAM-2-dependent mechanism, often moving against the stream of flowing blood, in search of regions of this barrier that are more amenable to leukocyte diapedesis. Emigration occurs preferentially near tricellular endothelial junctions at sites where ICAM-1 expression is enriched and the underlying basement membrane exhibits low matrix protein deposition. Neutrophils may also exit the circulation by moving directly through endothelial cells, but this occurs much more rarely than egress via intercellular junctions. While mechanisms underlying the former are very unclear, diapedesis via the latter occurs by a process that involves the junctional proteins PECAM-1, ICAM-2, JAM-A, JAM-C, and CD99 interacting with PECAM-1 and CD177, CD11a and CD11b, CD11a, CD11b, CD99, respectively, on neutrophils. Two other endothelial ligands, ESAM and CD99L2, have been shown to play a role in neutrophil migration, but their binding partners on granulocytes are not known. Once the diapedesing neutrophil has penetrated the endothelial barrier, these phagocytic cells crawl along pericytes via CD11a- and CD11b-dependent binding with ICAM-1, interacting with the basement membrane at the same time via VLA-3- and VLA-6-mediated binding to laminins. The abluminally crawling neutrophil breaches these final two barriers by migrating through gaps between pericytes and regions of low matrix protein expression in the basement membrane. Once the neutrophil detaches from these structures in the vessel wall, the cell finally migrates into the interstitial matrix, where chemotactic gradients serve as guidance cues to direct neutrophils to inflammatory loci. For myocardial cells, extravasated neutrophils have been shown interact with ICAM-1, a process that results in increased myocyte ROS levels and cell injury. Oxidant-, platelet activating factor-, leukotriene B_4-, and chemokine-induced leukocyte/endothelial cell interactions are largely responsible for the microvascular dysfunction induced by reperfusion.

3. INTRAVITAL MICROSCOPIC APPROACHES TO STUDY LEUKOCYTE/ENDOTHELIAL CELL ADHESIVE INTERACTIONS

The ability of leukocytes to roll along, adhere (become stationary) to, and transmigrate across microvascular endothelium can be modified by a number of physiologic factors, including adhesion molecule expression on the surface of leukocytes or endothelium, electrostatic charge interactions between surface structures on both cell types, and hydrodynamic dispersal forces which tend to sweep leukocytes along the vessel wall. These adhesive interactions between leukocytes and endothelial cells lining postcapillary venules were first described using transillumination intravital microscopy. By this approach, anesthetized animals are placed on a custom-built animal board, the tissue of interest is surgically exposed and placed over an optically clear viewing pedestal or coverslip mounted over a hole in the animal board, and illuminated from one side using a mercury light source attached to the microscope, with the microscope objective collecting light passing through the tissue and focusing the image on a color camera, which projects the image to a television monitor for data collection (Fig. 2; Ley et al., 2008). However, other preparations that are less mobile because of tissue attachments, such as the rabbit tenuissimus muscle, require transillumination via a small diameter, curved fiber optic light pipe. The immersion objectives can help prevent motion artifacts while warm superfusate is dripping on the tissue examined. A reducing video coupler (range $0.3\times$ to $0.6\times$) placed between the camera and the microscope can improve resolution. Suitable tissues, which must be less than 100 μm thick to limit refractive light scattering that would otherwise blur the collected image of the microvessels, include the cremaster muscle or other thin muscles of the rat or mouse, the mesentery of cats, rats, mice, or rabbits, the hamster

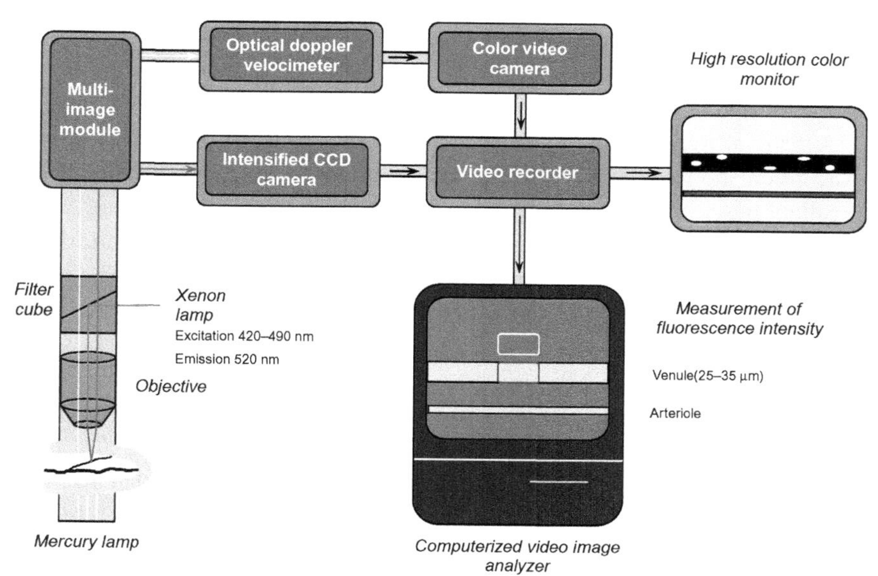

Figure 2 Graphical representation of components required for intravital microscopic observation of the living microcirculation by trans- or epi-illumination. For transillumination, light from a mercury light source is directed through exteriorized thin (<100 mm) tissues to a salt water immersion microscope objective that focuses the collected image onto a color video camera. Images from the camera are recorded by a DVD recorder and presented on a color television monitor. An optical Doppler velocimeter is incorporated into the light path between the microscope objective and video camera to monitor erythrocyte velocity. Computerized fluorescence intravital microscopy can also be incorporated into this system to quantify venular protein leakage, chemokine or adhesion molecule expression, or oxidant production. Using epifluorescence microscopy, light from a xenon source directed to the tissue under observation via a filter cube, which excites the fluorochrome of interest, causing it to emit light that is collected via the objective and directed to an intensified CCD camera, captured onto a computer (or to videotape or video disc by use of a video or digital video recorder, respectively) for computerized image analysis of fluorescence intensity. A multi-image module directs light to either the color video camera or the CCD camera. Vessel diameter is measured using a videocaliper (not shown) and the numbers of rolling and adherent leukocytes, leukocyte rolling velocity, and mean red cell velocity a quantified during off-line analysis of the collected images. Not shown for simplicity is the animal whose tissues are being observed.

cheek pouch, or the bat wing (Ley et al., 2008). The dorsal skin-fold chamber represents another approach that allows visualization of a preparation consisting primarily of striated muscle and subcutaneous tissue (Lehr, Vollmar, Vajkoczy, & Menger, 1999). Although surgery is required for chamber implantation, the microcirculation can be observed over extended periods in awake animals, once they recover from implantation. This can also be accomplished using the intact ear of the mouse or the bat wing.

Of the aforementioned models, murine models are now the most popular, given the wide availability of knockout and transgenic mice as well as excellent blocking antibodies that are directed against chemokines and adhesion molecules involved in the various stages of leukocyte–endothelial adhesive interactions. Except for the bat wing and mouse ear, all of the aforementioned preparations are accessible for intravital microscopy only after surgery, which requires anesthesia. It is important to note that choice of anesthetic is important because these agents can differentially influence a number of physiologic processes, including altering sympathetic tone, inducing hormone release, and reducing arterial blood pressure and PO_2. Exposing the bowel requires an abdominal midline incision so that an intestinal loop can exteriorized and draped over the viewing pedestal but no additional dissection to view the mesenteric microcirculation. Nevertheless, bowel exteriorization does some tissue manipulation which can invoke mild to severe inflammatory responses. For example, simply overstretching microvascular vessels or touching the tissue to be observed with surgical instruments, fingers, or moistened cotton swabs as the tissue is manipulated over the viewing pedestal can invoke significant leukocyte interactions with postcapillary venules and/or abolish arteriolar tone. Thus, special care must be taken to minimize tissue manipulation and to avoid touching the area to be viewed. Other preparations, such as the cremaster muscle and other thin muscles, require dissection prior to mounting over coverslip or viewing pedestal. Thus, care must be taken to grasp only the skin overlying the muscles with tweezers prior to making an incision to avoid damage to the underlying muscle tissue. For the cremaster, the connective tissue surrounding the muscle is carefully dissected free without touching the muscle. After pinning one end cremaster so that it lays flat over the viewing coverslip, an incision is made down the center with iris scissors, with care taken to avoid damage to the three main arterial feed vessels and veins. This opens of the saclike structure of the muscle, allowing it to be pinned out for viewing. Microvessels arising from the middle of the three arterial feed vessels are usually selected for study, as they are the farthest removed from the sites of surgical trauma.

In order to minimize inflammation to the tissues being evaluation, the surgical preparation of tissues should be performed with gentle manipulation by skilled surgeons and in a meticulous fashion. Additionally, a sterile environment is imperative to maintain an environment free of bioactive mediators that may interfere with the inflammatory cascade being studied. Other surgical considerations to maintain physiologic status while isolating the tissue selected for intravital microscopic viewing include intravenous access for anesthesia, intubation for breathing support, blood pressure monitoring, and maintenance of normal central core body temperature with a water-circulating or electrical heating mat and/or thermistor-controlled heat lamps.

Throughout the required surgery and during the remainder of the experimental protocols, the tissue under study must be kept warm (37 for most preparations, 35 °C for the cremaster) and moist in order to obtain physiologically meaningful data (Ley et al., 2008). To accomplish this, the tissue is occasionally rinsed in warmed, buffered physiologic salt solutions during surgery and then continuously superfused with similar solutions once placed over the viewing pedestal or coverslip. Superfusate pO_2 should approximate tissue pO_2 (30 mm Hg for muscle, 40 mm Hg for the mesentery) and pH (7.3–7.4) by bubbling the superfusion reservoir with 95% N_2, 5% CO_2. The tissue can be covered in plastic wrap, which helps in retaining heat and moisture. Because rodents have a large surface area to volume ratio, they lose heat readily, especially when anesthetized, which can dramatically modify physiologic responses. Thus, it is very important to maintain their core temperature 37 °C by use of thermistor-controlled heating pads and/or heating lamps.

When using transillumination intravital microscopy, salt water immersion objectives are used, typically with 20 × magnification and a numerical aperture of 0.5. This provides a wide field of view, encompassing about 300 μm × 400 μm tissue, which allows visualization leukocyte interactions in straight and unbranched portions of venules approximately 15–50 μm in diameter (Ley et al., 2008). Use of immersion objectives allows the tissue to be kept moist and warm by superfusion with physiologic salts solutions, as described above. To provide the greater level of resolution required for examination of the behavior of single adherent and transmigrating leukocytes, higher magnification objectives with larger numerical apertures are employed.

It is important to note that the particular leukocyte subset interacting with the blood vessel wall cannot be identified when using transilluminated intravital microscopy, unless the intravital observations are correlated with histochemical detection of a particular subtype in biopsies obtained at the end of the experiment. Using light microscopy and differential staining of mesenteric samples, nonischemic tissue is sparsely populated by mast cells and leukocytes, with 30–50% of the resident population being neutrophils. On the other hand, I/R produced a significant increase in the number of extravascular leukocytes, with neutrophils constituting greater than 80% of the total cell population. Thus, neutrophils are the predominant leukocyte subtype emigrating from postischemic venules over the time frame of these experiments.

For thicker tissues, such as the mouse intestinal wall, brain, lung, liver, pancreas, Peyer's patches and peripheral lymph nodes, epifluorescence intravital microscopy is used to visualize leukocyte interactions with the blood vessel wall. Leukocytes can be visualized following intravenous administration of carboxyfluorescein diacetate succinimidyl ester, rhodamine G, or acridine orange, with fluorescent images detected using a charge-coupled device camera. However, this approach again suffers from difficulties in identifying the leukocyte subset adhering to and emigrating across postcapillary venules. To more specifically visualize neutrophil interactions with the blood vessel walls, some investigators have isolated neutrophils, stained them with fluorescent dyes, and then injected into the experimental animal. This approach can be applied to other specific leukocyte subsets via use of density gradient centrifugation, flow cytometric cell sorting, or magnetic bead selection to isolate a specific type of cell. A problem with this approach is that the isolated, *ex vivo* labeled neutrophils become activated and behave differently when injected into mice. On the other hand, platelet interactions in the microcirculation can be visualized by this *ex vivo* isolation/reinjection approach because these thrombocytes are not activated by isolation and *ex vivo* labeling (Eichorn, Ney, Massberg, & Goetz, 2002; Massberg et al., 1998).

Rather than using nonselective stains, a low dose (3 μg) of neutrophil-specific Alexa-fluor 488 Gr1 antibody is administered intravenously (McDonald et al., 2008, 2010; Peters et al., 2008) to visualize neutrophils in intravital fluorescence microscopy experiments. Alternatively, neutrophils can be specifically visualized by expression of eGFP^{hi} (vs. eGFP^{low} monocytes and macrophages) in LysM-eGFP mice (McDonald et al., 2010; Peters et al., 2008). Several other transgenic and knock-in mice that express green fluorescent protein in only one leukocyte subset (e.g., NKT cells, monocytes, CD4+ or CD8+ T cells) under the control of a subset-specific promotor are also available, which allows adhesive interactions of each subset to be specifically detected and analyzed. Finally, labeling neutrophils with quantum dots conjugated to anti-RP-1 antibody has been employed to visualize these phagocytic cells and their adhesive behaviors in the microcirculation (Jayagopal, Russ, & Haselton, 2007). Care must be taken to illuminate tissues for only brief periods to prevent photobleaching and light toxicity secondary to generation of reactive oxygen species.

Imaging tissues deeper than 100 μm is difficult due to scattering of light, resulting in blurred images. Intravital confocal microscopy greatly improves the resolution and contrast of images in *in vivo* imaging (Jenne, Wong, Petri, & Kubes, 2011; Ley et al., 2008). Use of laser scanning confocal microscopy further enhances the imaging power of intravital microscopy by scanning the laser beam over every point of the focal plane, collecting a stack of sections, and using computer reconstruction algorithms to render a three-dimensional image that can be rotated in all planes. Laser scanning takes too long to apply and collect images and is thus not suitable for leukocyte trafficking studies. However, faster imaging is achieved by use of spinning disk confocal microscopy, which simultaneously detects fluorescence emitted from multiple points and utilizes a charge-coupled device instead of a photomultiplier tube, thereby increasing the rate at which images can be captured (Nakano, 2002). Imaging at greater tissue depths, thinner optical sections, and with decreased photobleaching or phototoxicity can be achieved by use of multiphoton confocal microscopy (Denk & Svoboda, 1997; Ley et al., 2008; Li et al., 2012). Other methods that have been developed to study leukocyte trafficking in deep tissues include single photon emission computed tomography, positron emission tomography, magnetic resonance imaging, and bioluminescence imaging. These methods can noninvasively track leukocytes over long periods of time, however, they lack the spatial and temporal resolution to visualize single-cell dynamics *in situ* (Mempel, Scimone, Mora, & von Andrian, 2004).

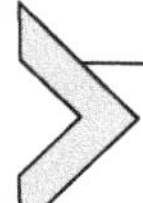

4. ASSESSING LEUKOCYTE ROLLING, ADHESION, AND EMIGRATION IN THE INTACT MICROCIRCULATION

Leukocyte rolling describes the low affinity adhesive interaction between leukocytes and the vascular endothelium whereby the force of blood flow induces a rotational motion (i.e., rolling) of the leukocyte along the vascular wall. Rolling behavior is most often quantified as the number (or flux) of rolling leukocytes, which is assessed by counting the number of leukocytes flowing past a fixed point (line drawn perpendicular to the long vessel axis on TV monitor) in the vessel of interest that are moving slower than the stream of erythrocytes.

Another way to characterize the behavior of rolling leukocytes is to quantify their velocity (leukocyte rolling velocity, V_{WBC}), determined from the time required for a leukocyte to traverse a given distance along the length of the venule. When expressed relative to the mean erythrocyte velocity (V_{RBC}) in the same vessel, the ratio of V_{WBC}/V_{RBC} provides information regarding the relative strength of the weak adhesive interactions between rolling leukocytes and the endothelium. In other words, V_{WBC}/V_{RBC} provides a measure of the fracture stress at the area of contact between the rolling leukocyte and the endothelium, such that a decrease in V_{WBC}/V_{RBC} indicates a strengthening of this weak adhesive interaction. If the fracture stress increases to a point where it is impossible for adhesive bonds to detach, the leukocyte becomes stationary ($V_{WBC}=0$), and is defined as a firmly adherent white cell. In practice, an adherent leukocyte is defined as a white cell that remains stationary for 30 s or more, because cells remaining adherent for this length of time then go on to emigrate. Adherent cells are expressed as the number per 100 μm length of venule. Some leukocytes roll along and stop intermittently along the vessel under view and are called saltating leukocytes. The significance of this phenomenon to inflammation is unclear. Emigrated leukocytes are quantified by counting the number of immunocytes outside the blood vessels and are presented as number/field number/mm^2.

Wall shear stress and wall shear rate influence leukocyte rolling and adhesion and thus must be estimated as part of experimental protocols examining leukocyte interactions in microvessels. Wall shear stress (τ) describes the mechanical energy required to peel a leukocyte away from an endothelial cell. Wall shear rate (γ) is defined as the rate at which adjacent layers of fluid move with respect to each other, usually expressed as reciprocal seconds. To obtain estimates of these variables, blood flow centerline velocity is measured. In our laboratory, we use a laser Doppler velocimeter system (Microcirculation Research Institute, Texas A&M University). Microparticle image velocimetry and off-line analysis tools can also be utilizated to assess leukocyte rolling *in vivo*, and can assist in calculating leukocyte rolling flux, leukocyte rolling flux fraction, leukocyte rolling velocity, wall shear rate, and stress and blood flow velocity (Sperandio, Pickard, Unnikrishnan, Acton, & Ley, 2006). Using these velocimetry approaches, venular diameter (D) and centerline red blood cell velocity are measured on-line with a video caliper and an optical Doppler velocimeter, respectively. The velocimeter is calibrated against a rotating glass disc coated with erythrocytes. Venular blood flow is calculated from the product of mean red cell velocity (V_{mean} = centerline velocity/1.6) (Davis, 1987) and microvascular cross-sectional area, assuming cylindrical geometry.

Wall shear rate (γ) is usually calculated based on the Newtonian definition: $\gamma = 8V_{mean}/D$, where V_{mean} = mean red cell velocity and D = diameter. Wall shear stress (τ) relates viscosity to mean flow velocity and wall shear rate, is calculated from the relation: $\tau = 4V_{mean}\eta/r$, where V = mean blood flow velocity, η = blood viscosity, and r = vessel radius, and is expressed as dyn/cm^2. The number of adherent and extravasated leukocytes and leukocyte rolling velocity are typically determined off-line during playback of videotaped images. Because blood is not a Newtonian fluid, these calculations of γ and τ are, at best, estimates (Ortiz, Briceno, & Cabrales, 2014; Sriram, Intaglietta, & Tartakovsky, 2014), but are useful in determining whether altered shear forces play a role in the changes in leukocyte rolling or adhesion that are invoked by an inflammatory state.

Leukocyte rolling occurs primarily in postcapillary venules and is rarely noted in normal arterioles or capillaries. The numbers of leukocytes rolling along postcapillary venules increases dramatically in inflammatory conditions and is usually associated with an increase in fracture stress (as assessed by decreased V_{WBC}/V_{RBC}). Leukocyte rolling also increases as wall shear rate is reduced below normal, as occurs in ischemic tissues. In normal arterioles, leukocyte adhesion does not occur but a relatively small number of adherent cells have been noted in certain inflammatory conditions, such as treatment with TNFα and in cigarette smoke-exposed animals. Reductions in blood flow and wall shear rate have very little effect on leukocyte adhesion in arterioles.

Recent advances in the use of multicolor fluorescence spinning disk and multiphoton confocal and intravital microscopic imaging has provided important insight regarding the mechanisms whereby neutrophils move abluminally to exit the vascular wall. This work shown that these inflammatory phagocytes migrate along pericytes to exit the vascular wall at regions of low matrix deposition in the basement membrane that are also characterized by gaps between pericytes (Proebstl et al., 2012; Stark et al., 2013; Voisin, Proebstl, & Nourshargh, 2010). Once in the tissue space, neutrophils move through the interstitium by a mechanism dependent on actin polymerization and on matrix metalloproteinase activity but without degradation of pericellular collagen (Lerchenberger et al., 2013).

Because the *in vivo* environment is very complex, unknown or unmeasured variables may influence leukocyte rolling and adhesion. For example, the walls of postcapillary venules are lined by endothelial cells that express a variety of adhesion receptors at different densities which influence leukocyte rolling and adhesion behaviors. As noted above, wall shear rates and the patterns of flow vary with venular diameter, curvature, and branching patterns, which can influence adhesive events in unrecognized ways. In addition, extrinsic factors such as nervous innervation, hormonal inputs, other formed elements in blood (e.g., platelets, other leukocytes, erythrocytes), inflammatory cells resident in the tissue space (e.g., lymphocytes, mast cells), and parenchymal cells all have the potential to influence leukocyte–endothelial interactions. Use of isolated, cannulated postcapillary venules perfused with neutrophils (or other cell types of interest) is an attractive *ex vivo* approach that obviates many of these limitations (Yuan, Mier, Chilian, Zawieja, & Granger, 1995; Fig. 3). Hence, this approach allows precise evaluation of the direct effects of adhesion molecules and inflammatory mediators in the absence of the complicating effects of these external factors. Moreover, vessels can be perfused with any type of cell that establishes adhesive interactions with postcapillary venular or arteriolar walls, including tumor cells (Kong, Dunn, Keefer, & Korthuis, 1996; Kong & Korthuis, 1997).

As another reductionist approach to study adhesive events under more precisely defined and reproducible conditions, several *in vitro* flow systems have been developed that mimic flow conditions in intact vessels, but can be constructed to eliminate variations in adhesion molecule site density and vascular caliber and orientation that occur *in vivo*. In these systems, microscopic examination of leukocyte interactions with endothelial cells grown on the bottom plate or with specific adhesion molecules or chemokines that are coated onto the plate can be observed and quantified (Ley et al., 2008).

Initial work beginning in the pate 1980s employed parallel plate flow chambers for examination of the effects of shear rate and flow patterns on leukocyte rolling and adhesion. These systems could be perfused with a particular type of leukocyte through the parallel plate with defined geometry, allowing a very accurate assessment of shear rate. This technology underwent several modifications over subsequent years to allow for examination of the role of specific adhesion molecules, expressed either on leukocytes or endothelial cells, in tethering, rolling and adhesion and for examining the molecular mechanisms whereby pro- and anti-adhesive molecules exerted their effects. For example, the plate could be coated with specific adhesion

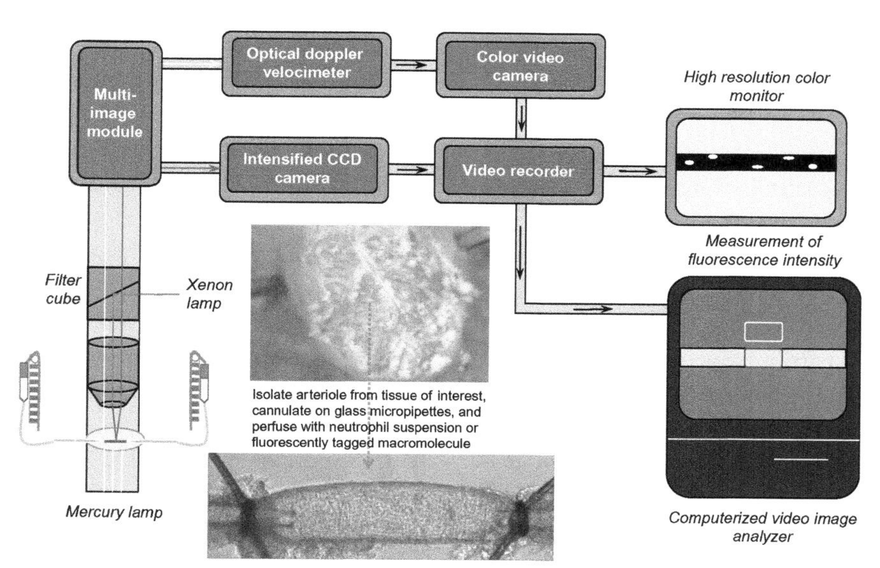

Figure 3 Schematic representation of essential components for assessment of neutrophil interactions and venular permeability isolated, perfused single postcapillary venules using *ex vivo* microscopy. This approach uses the same components as shown in Fig. 2 for intravital microscopy. The major difference is that an isolated vessel perfusion chamber takes the place of the viewing pedestal animal board. For this preparation, a postcapillary venule is isolated from the tissue of interest, cannulated on glass micropipettes and perfused with neutrophils or fluorescently tagged macromolecules. The height of the perfusion reservoirs can be adjusted in equal and opposite directions by use of a pulley system to alter perfusion pressure and flow without altering midpoint pressure in the vessel. Neutrophil–endothelial cell adhesive interactions and venular permeability are determined as described in the text. (See the color plate.)

molecules, alone or in various combinations and at varying densities on the plate and leukocyte subsets could be isolated and perfused over these surfaces at different shear rates and flow patterns. However, it soon became apparent that the procedures required to isolate leukocyte subsets altered their activation status, gene expression, and surface expression of adhesion molecules (e.g., L-selectin is shed and CD11b/CD18 is upregulated on neutrophils). To avoid the latter issues, autoperfused flow chamber systems were developed, in which cannula directs blood flowing from the animal through a parallel plate chamber (Ley et al., 2008).

More recently, this autoperfused parallel plate flow chamber approach has been adapted using sophisticated microfluidic technology to fabricate complex microchannel networks (Ley et al., 2008). Because of their small dimensions, these microchannel systems more close mimic the dimensions and flow characteristics in microvessels. In addition, the volume of blood required to perfuse such systems over the course of an experiment is very low, eliminating the requirement for recirculation of the perfusing blood to the donor animal. This obviates concerns related to return of blood containing activating factors that may accumulate during passage through the *in vitro* system. The fact that the microfluidic systems can contain several microchannels allows for high throughput analysis to be conducted. A particularly exciting application of technology is offered by coupling the microfluidic perfusion chamber to total internal reflection microscopy, which allows events at the area of contact between rolling leukocytes and the underyling substrate (Sundd et al., 2012; Sundd, Pospieszakska, & Ley, 2013). While these *in vitro* approaches have not been used to examine the anti-adhesive effects of H_2S, they hold particular promise for study the underlying molecular mechanisms in a highly controlled environment that more accurately mimics dynamic *in vivo* conditions than traditional static cell culture approaches, while still allowing evaluation of its effects on individual cell types.

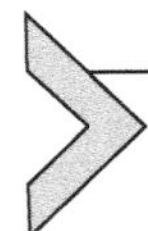

5. DETECTION OF CHEMOKINE AND ADHESION MOLECULE EXPRESSION USING INTRAVITAL MICROSCOPY

In addition to monitoring the effects of H_2S or other agents or perturbations on leukocyte rolling, adhesion and emigration in inflammatory states, intravital microscopy can also be applied to the detection of chemokine receptors and expression of adhesion molecules by endothelial cells (Fingar et al., 1997). For example, chemokine receptor expression has been detected by attaching a chemokine such as IL-8 to fluorescent microspheres, which are then injected into the blood stream (Fingar, Taber, et al., 1997). The fluorescent microspheres are 0.1–0.3 μm in diameter, which allows them to readily pass through capillaries, but are large enough to be retained in the circulation and to allow observation and resolution of individual microspheres, once the attached chemokine interacts with its receptor. This allows precise determination of the location of chemokine binding to their specific receptor in the microcirculation, while quantification of fluorescence intensity should provide an estimation of expression levels. A similar approach has been applied to detect ICAM-1 expression on pulmonary endothelial cells using intravital microscopy (Fingar, Guo, Lu, & Peiper, 1997). Sumagin and Sarelius (2010) locally infused an anti-ICAM-1 antibody conjugated to Alexa 388 to detect ICAM-1 expression in postcapillary venules using intravital fluorescence confocal microscopy. Others have employed near infrared light-emitting solid lipid nanoparticles conjugated to ligands specific for integrins to detect expression using intravital microscopy (Shuhendler et al., 2012). Jayagopal et al. (2007) created spectrally distinct quantum dot-antibody bioconjugates directed towards PECAM-1, ICAM-1, and VCAM-1, and following tail vein injection into control and diabetic rats, were able to demonstrate enhanced expression of ICAM-1 and VCAM-1 in large and microcirculatory vessels of the retina, while PECAM-1 fluorescence was similar.

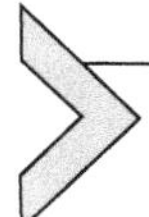

6. INTRAVITAL MICROSCOPIC METHODS TO ASSESS CHANGES IN MICROVASCULAR PERMEABILITY

Increased vascular permeability and edema formation are hallmarks of the inflammatory response. Disruption of endothelial barrier function facilitates blood–tissue exchange of large molecular weight components of the immune response, such as immunoglobulins and complement proteins, at sites of inflammation, thereby enhancing the ability of the immune system to fight infection. In the setting of sterile inflammation, as occurs in ischemic conditions, increased permeability can intensify the deleterious effects of ischemia by promoting edema formation. Excessive accumulation of fluid in the tissue space increases diffusion distance for oxygen and nutrients, exacerbating their already compromised delivery associated with reduced blood supply. In addition, edema raises interstitial fluid pressure, which compresses capillaries thereby contributing to the capillary perfusion deficit referred to as the no-reflow phenomenon. The narrowed capillaries also facilitate leukocyte-capillary plugging to further aggravate this perfusion impairment.

Intravital microscopic methods to assess changes in vascular leakage of large molecules involve administering a fluorescently labeled macromolecule and quantifying the appearance in the extravascular space. Vascular macromolecule leakage does not equate to permeability because it can be increased by enhanced convective transport secondary to elevated microvascular hydrostatic pressure and/or surface area. However, because surface area changes are negligible when observing leakage in single postcapillary venules, vascular macromolecule leakage provides a reasonable index of permeability changes if microvascular hydrostatic pressure changes are measured or if predicted to decrease. To quantify albumin leakage across mesenteric venules, 50 mg/kg of fluorescein isothiacyanate (FITC)-labeled bovine albumin is administered intravenously to the animals 15 min before the baseline recording. Fluorescence intensity is detected with a CCD camera, a CCD camera control unit, and an intensifier head attached to the camera. The fluorescence intensity of the venule under study (I_I), the fluorescence intensity of contiguous perivenular interstitium within 10–50 μm of the venular wall (I_O), and the background fluorescence (I_B) are measured at various times after the administration of FITC-albumin with a computer-assisted digital imaging processor. The windows to measure average fluorescence intensities within and along the venule are set at 20 μm length and 10 μm width. An index of vascular albumin leakage (VAL) is determined from the relation $(I_I - I_B)/(I_O - I_B)$. This method assumes that fluorescent-tagged albumin will move from the microvessel into the tissue space in a manner that reflects the behavior of native albumin.

A more sophisticated application of the same principle underlying the vascular leakage assay relates to assessment of dye accumulation in the interstitial space adjacent to a postcapillary venule that is cannulated with a double bore pipette, with one chamber containing the fluorescent macromolecule of interest (Scallan, Huxley, & Korthuis, 2011). The contents of each chamber in the pipette are under pressure that can control which pipette perfuses the vessel as well as controlling the hydrostatic pressure in the vessel segment. When pipette pressure is adjusted so the dye front is not moving, the pressure measured in the other pipette represents the pressure at the tip and the next downstream bifurcation in the microvascular network. To measure the flux of the labeled solute, pressure is elevated in the pipette to ensure rapid and complete filling of the vessel segment. Recording the step change in fluorescence intensity (I) from perfusion with no labeled solute to perfusion with the tagged macromolecule (I_0)and recording the slope of the tracing depicting the change in intensity as the solute diffuses into the interstitial space ($\Delta I/\Delta t$) allows calculation of permeability of the solute (Ps) from the relation: $(1/I_0)(\Delta I)(\Delta t)(r/2)$, where r = vascular radius and assuming a circular cross-sectional area for the vessel. Using a confocal microscope allows vascular cross-section to be more accurately quantified, providing a better estimate of Ps. The contribution of convective coupling to the flux of the macromolecule can be readily determined by repeating these measurements at a series of hydrostatic pressures. Analysis of the nonlinear relation between solute flux per unit surface area and pressure provides a true measure of diffusive permeability (Pd) that is unaffected by solvent drag (Scallan, Huxley, & Korthuis, 2010).

A measure of water permeability that can be obtained using intravital microscopy involves determination of hydraulic conductivity (Lp) in individually perfused rat mesenteric postcapillary venules, using the modified Landis technique (Scallan et al., 2010). By this technique, a single postcapillary venule is cannulated with a micropipette and perfused with albumin-ringer solution (control) containing 1% (vol/vol) hamster red blood cells as markers. A known hydrostatic pressure (40–60 cm H_2O), controlled by a water manometer, is applied through the micropipette to the vessel lumen, which allows the perfusate to continuously flow through the vessel. Water flux is measured when the downstream of the vessel is briefly occluded. If intravascular pressures exceed those outside the vessel, fluid filtration occurs and the marker cells will be observed to flow towards the occluder. Just the opposite occurs if intrascular pressures are less than those in the surrounding tissue. Volume flux per unit area (Jv/S) is then calculated from the relation: $(1/x_0)(\Delta x/\Delta t)(r/2)$; where Δx is the distance the marker cell moves over an interval of time (Δt), having started at distance x_0, and r = vessel radius. When these measures are repeated at several pressures, the slope of the line relating Jv/S to pressure is equal to Lp. The intercept of such a plot yields the effective osmotic pressure gradient.

As noted above, the inflammatory milieu is characterized by a large number of factors extrinsic to the vascular wall of postcapillary walls that influence leukocyte–endothelial cell interactions. These same factors also influence venular permeability responses, necessitating the development of approaches that allow precise control of these external factors, so that their individual contributions to barrier disruption can be assessed. Isolated, perfused postcapillary venules have proven useful in this regard (Breslin et al., 2006; Yuan, Chilian, Granger, & Zawieja, 1993; Fig. 3). Cell culture models, especially those involving endothelial cells grown on top of a three-dimensional extracellular matrix that can be seeded with inflammatory cells of interest should prove invaluable, especially in models that incorporate shear stress exposure.

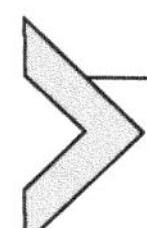

7. ASSESSMENT OF REACTIVE OXYGEN SPECIES GENERATION USING INTRAVITAL MICROSCOPY

The production of ROS in the microvasculature can be determined by monitoring the oxidation of dihydrorhodamine (DHR) to its fluorphore rhodamine 123, using intravital digital microfluorography. To make this measurement, long segments (300 μm) of postcapillary venules are identified and selected for viewing using intravital microscopy. Background fluorescence (I_{Bgrd}) of the first 100 μm of each 300-μm vessel length is recorded from the selected postcapillary venules with a xenon light source and a SIT camera system. Then, freshly prepared DHR-123 (1 mmol/L, a nonfluorescent dye that is oxidized by ROS to the fluorescent compound rhodamine-123) in bicarbonate-buffered saline is superfused over the tissue being observed for 15 min. The tissue is then washed with bicarbonate-buffered saline and the fluorescent image of each vessel section recorded (I_{DHR}). Images are captured onto a computer, and an area 100 μm long and 7 μm wide along the vessel wall is analyzed in each vascular segment using imaging software. The ratio of I_{DHR}:I_{Bgrd} is calculated for each vascular segment, and the average ratio for each animal determined. Although some have claimed that DHR oxidation may be specific detector for peroxynitrite, subsequent work indicated that it interacts with many radical species and peroxidases.

Dichlorofluorescein (DCFH) diacetate is another oxidant-sensitive probe that has been used in intravital microscopic studies to detect ROS production in the intact microcirculation (Suematsu et al., 1993). DCFH diacetate diffuses into cells in a nonfluorescent form that is cleaved by esterases to DCFH, which is trapped intracellularly. Exposure to hydroperoxides as they are formed in cells oxidizes DCFH to yield fluorescent dihydrofluorescein (DCF). After obtaining a background autofluorescence image, the fluorochrome is added to the superfusate bathing the tissue at a concentration of 5 μ*M* and perfused over the tissue under observation for 25 min, followed by 10 min rinse with DCFH diacetate-free superfusate to remove the dye precursor from the tissue. Fluorescence is detected using epifluorescence microscopy and quantified using image analysis software (Suematsu et al., 1993, 1994).

These techniques not only allowed for oxidant detection but also their topographic distribution in microcirculatory units. However, it is important to point out that data obtained using oxidant-sensitive probes *in vivo* are widely misinterpreted with regard to their ability to indicate specific types of reactive oxygen species that are being generated (Grisham, 2013; Winterbourn, 2014). Thus, while data obtained with DCFH diacetate or DHR dye detection approaches are often interpreted to be indicative of hydrogen peroxide production, these indicators react with a number of oxidizing species. Similarly, hydroethidine has been used as a superoxide-specific detector in intravital microscopic studies (Suzuki et al., 1998), but is also now recognized to be oxidized nonspecifically by a mechanism similar to DCFH (Winterbourn, 2014). Thus, oxidant-sensitive probes are useful for detecting ROS in tissues viewed by intravital microscopy, but conclusions based on this data should not be extrapolated to specific radical species. The recent development of genetically encoded ROS reporters and nanoparticle dye delivery systems may be useful in this regard, but have not yet been optimized for application in studies using intravital microscopy (Ezerina, Morgan, & Dick, 2014; Morgan, Sobotta, & Dick, 2011; Woolley, Stanicka, & Cotter, 2013).

8. FLUORESCENCE DETECTION OF CELL INJURY USING INTRAVITAL MICROSCOPY

Cell injury can be quantified in tissues using vital dyes. Propidium iodide (PI) and ethidium bromide have both been shown to be effective indicators of tissue injury and can be quantified using intravital microscopy (Dungey et al., 2006; Harris, Costa, Delano, Zweifach, & Schmid-Schonbein, 1998; Harris, Leiderer, Peer, & Messmer, 1996; Suematsu et al., 1994). PI is a positively charged indicator dye that is unable to cross cell membranes. However, when cells are injured an unable to maintain membrane potential difference across the plasmalemma, PI enters cells, binds to nuclear DNA, and becomes fluorescent. PI is either injected intravenously at a dose of 1 μM/kg body weight or can be superfused over the tissue at a concentration of 1 μM and the number of PI-positive cells in the field of view are counted as a measure of nonviable cells. As another approach, the fluorescent vital dyes bisbenzimide, which stains the nuclei of all cells, and the larger ethidium bromide molecule, which can only enter cells whose membranes have been injured, can be used. Ethidium bromide labels the nuclei of cells with a range of injury from minor (membrane permeable) to death. Ethidium bromide (5 μg/mL) and bisbenzimide (5 μg/mL) are added to the superfusate, and cell injury measured by recording fluorescent illumination with the appropriate filters for the two indicators, with injury expressed as a ratio cells stained with ethidium bromide to bisbenzimide-positive cells.

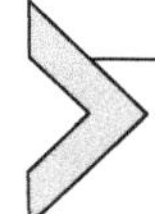

9. PERFUSED CAPILLARY DENSITY ASSESSMENT WITH INTRAVITAL MICROSCOPY

I/R causes the development of a capillary perfusion impairment called the no-reflow phenomenon. The extent of no-reflow can be readily assessed using intravital microscopy by counting the number of capillaries crossing perpendicular to a line drawn on the television monitor onto which the microvascular field is projected that exhibit flowing erythrocytes under control conditions or following I/R (Dungey et al., 2006). Others have quantified functional capillary density by measuring the length of capillaries perfused with red blood cells per observation area (Harris et al., 1996).

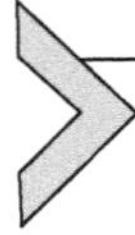

10. ACUTE AND PRECONDITIONING-INDUCED ANTI-INFLAMMATORY ACTIONS OF HYDROGEN SULFIDE: ASSESSMENT USING INTRAVITAL MICROSCOPY

Although once only regarded as a toxic gas, hydrogen sulfide (H_2S) was recently identified as an endogenously produced signaling molecule that exerts effects at the molecular, cellular, tissue, and system levels (Abe & Kimura, 1996; Kamoun, 2004; Kashiba, Kajimura, Godo, & Suematsu, 2002; Wang, 2003, 2012). Hydrogen sulfide and its major dissociation product, HS^-, are extremely reactive, most likely exerting biological effects through direct covalent modification of target molecules. On the one hand, HS^- is a potent chemical reductant and nucleophile that allows it to scavenge free radicals by single electron or hydrogen atom transfer, while on the other H_2S is a thiol and can interact with and "scavenge" free radicals (Wang, 2012).

H_2S is produced endogenously by the actions of cystathionine-β-synthase (CBS) or cystathionine-γ-lyase (CSE), which use cysteine or homocysteine as substrate. Another enzyme, 3-mercaptopyruvate sulfurtransferase, working in concert with cysteine aminotransferase, can metabolize cysteine to generate pyruvate and H_2S. These enzymes are differentially expressed across tissues and organs of the body, with CSE being the predominant H_2S-generating enzyme in both vascular smooth muscle and endothelial cells (Wang, 2012). Like the other gaseous signaling molecules (nitric oxide, carbon monoxide, ammonia, and perhaps methane). H_2S is a small molecule that easily traverses cell membranes, has a short half-life, and is synthesized on demand.

Use of intravital microscopy has proven to be an invaluable tool in evaluating the role of endogenously produced H_2S and hydrogen sulfide donors on leukocyte–endothelial cell adhesive interactions. Zanardo et al. (2006) were amongst the first to demonstrate that pharmacologic CSE inhibition produced a decrease in leukocyte rolling velocity and increased stationary adherence relative to control rats under baseline conditions. On the other hand, CSE inhibition enhanced carrageenan-induced leukocyte infiltration in an air pouch inflammatory model. These observations were interpreted to indicate that H_2S facilitates the presentation of a nonadhesive endothelial surface under baseline conditions and that endogenous production of this gaseous signaling molecule is an important modulator of acute inflammation. In this same study, pretreatment with the H_2S donors, NaHS, Na_2S, or Lawesson's reagent, markedly reduced the enhanced rolling and adhesive responses noted in mesenteric postcapillary venules that were induced by intragastric aspirin or superfusion with proinflammatory fMLP. Similar observations were made in mice heterozygous for the CBS gene expression, which should reduce the bioavailability of H_2S, and fed a hyperhomocysteinemic diet. The mutant mice demonstrated marked reductions in leukocyte rolling velocity and increased stationary adhesion relative to wild-type (WT) mice fed the same diet (Kamath et al., 2006). Intravital microscopy has also been used to quantify the effects of H_2S to limit the extent of capillary no-reflow in the kidney and hepatic sinusoidal leukocyte rolling and adhesion induced by renal ischemia/reperfusion (Zhu et al., 2012). Protective effects of H_2S donor treatment on neurovascular permeability induced by hyperhomocysteinemia and leukocyte trafficking in inflamed joints have also been reported (Andruski, McCafferty, Ignacy, Millen, & McDougall, 2008; Tyagi et al., 2010). H_2S also limits leukocyte adherence to vascular endothelium provoked by exposure to nonsteroidal anti-inflammatory drugs (Fiorucci et al., 2005; Zanardo et al., 2006).

The molecular signaling mechanisms invoked by H_2S to limit baseline or proinflammatory mediator-induced leukocyte rolling and adhesion are only now becoming apparent. It appears that ATP-sensitive potassium (K_{ATP}) channel activation by H_2S may be important, because its anti-adhesive and -inflammatory effects were blocked by coincident treatment with a pharmacologic K_{ATP} channel inhibitor (Zanardo et al., 2006). Preliminary work supports a role for neutrophilic annexin A1 (AnxA1) in the anti-adhesive effects of H_2S donor treatment (Brancaleone, Flower, Cirino, & Perretti, 2013). In this study, the ability of H_2S donor treatment to limit interleukin-1β-induced leukocyte adhesion noted in WT mice was not apparent in AnxA1 knockout mice. Moreover, treatment of human neutrophils with the same donor, produced a marked mobilization of AnxA1 to the cell surface. Interestingly, fluorescent beads coated with an antibody to P-selectin adhered in much greater numbers when infused into heterozygous CBS+/− mice compared to WT animals. The latter result supports the possibility that P-selectin expression could be an important end-effector target in the anti-adhesive effects of endogenous H_2S. From these isolated observations, it is not clear whether what signaling steps are invoked by H_2S to activate K_{ATP} channels or cause mobilization of AnxA1, or whether these two signaling elements are mechanistically linked in the same cell type or to endothelial P-selectin expression. These potential linkages could be readily addressed by intravital microscopic studies involving WT and chimeric mice.

In contrast to the aforementioned anti-adhesive effects of endogenous and exogenous H_2S, this signaling molecule appears to enhance leukocyte rolling and adhesion in mesenteric venules in a cecal ligation and puncture (CLP) sepsis model (Zhang, Zhi, Moochhala, Moore, & Bhatia, 2007). In these studies, treatment with an inhibitor of H_2S formation significantly reduced the numbers of rolling and firmly adherent leukocytes and increased leukocyte rolling velocity in septic mice. On the other hand, septic mice treated with an H_2S donor at the time of sepsis induction exhibited enhanced increases in leukocyte rolling and adherence. These changes were linked to an effect of H_2S to enhance adhesion molecule expression in septic mice. In stark contrast to the aforementioned results, Spiller et al. (2010) showed that H_2S donor treatment reduced mesenteric leukocyte rolling and adhesion in CLP sepsis. It is not clear what accounts for these divergent responses but could be due to differences in sepsis severity in the two studies.

In addition to these acute effects of H_2S treatment, a growing body of evidence indicates that antecedent preconditioning with H_2S donors elicits the development of an anti-inflammatory phenotype in tissues that are exposed to I/R 24 h later. Preconditioning refers to a phenomenon where antecedent exposure to a particular stimulus confers protection against the deleterious effects induced by subsequent exposure to prolonged I/R (Krenz, Baines, Kalogeris, & Korthuis, 2013). Preconditioning stimuli typically produce two phases of protection in I/R. Acute or early phase preconditioning develops rapidly (within minutes), involves activation of pre-existing effector molecules, and is short-lived, conferring protection for 2–6 h before disappearing. Delayed acquisition of tolerance to ischemia (late phase preconditioning) arises 12–24 h after the initial preconditioning stimulus is applied, is longer lived (24 h or longer), and requires expression of new gene products to mediate protection (Krenz et al., 2013).

The observations that H_2S donors produce vasodilation by activation of K_{ATP} channels and enhances the vasorelaxant effects of NO (Wang, 2003, 2012), coupled with the fact that K_{ATP} channel activators and NO donors elicit preconditioning (Krenz et al., 2013), led us to postulate that this gaseous signaling molecule could also serve as a preconditioning stimulus to limit the proinflammatory effects of I/R (Yusof, Kamada, Kalogeris, Gaskin, & Korthuis, 2009). To address this issue, we first sought to determine whether and H_2S donor (NaHS) preconditioning would elicit anti-inflammatory effects in tissues exposed to I/R one (acute preconditioning) or 24 h (delayed preconditioning) after NaHS treatment. Only delayed H_2S preconditioning caused postcapillary venules to shift to an anti-inflammatory phenotype in WT mice after I/R such that these vessels failed to support leukocyte rolling and adhesion during reperfusion. To our knowledge, this is the only pharmacologic preconditioning agent that produces late but not early phase preconditioning, suggesting that H_2S may induce preconditioning by mechanisms that have signaling elements distinct from other approaches.

The protective effect of delayed NaHS preconditioning to decrease post-ischemic leukocyte rolling was largely abolished by coincident NO synthase (NOS) inhibition in WT mice and was absent in eNOS knockout mice (Yusof et al., 2009). A similar pattern of response was noted for leukocyte rolling in WT mice treated with NaHS concomitant with p38 MAPK inhibitors. The effect of late phase NaHS preconditioning on stationary (firm) leukocyte adhesion was also attenuated by treatment with NOS and p38 MAPK inhibitors in WT mice, but the anti-adhesive effect of NaHS was still evident in eNOS-deficient mice. These studies indicate that delayed NaHS preconditioning prevents postischemic leukocyte rolling and adhesion by triggering activation of an eNOS- and p38 MAPK-dependent mechanism. However, the role of eNOS in the anti-adhesive effect of NaHS preconditioning was less prominent than its effect to reduce leukocyte rolling. In subsequent work, we demonstrated that large conductance, calcium-activated potassium (BKCa) channels (but not small (SKCa) or intermediate (IKCa) conductance, calcium-activated potassium channels) also play an important role in triggering or initiating the anti-adhesive effects of NaHS preconditioning that become manifest during I/R 24 h later (Zuidema et al., 2010). We have also obtained evidence implicating a role for activation of plasmalemmal (pK_{ATP}) and mitochondrial K_{ATP} ($mitoK_{ATP}$) channels as initiators of NaHS preconditioning (Zuidema, Gross, & Korthuis, unpublished observations). As we have also shown that NaHS preconditioning, activation of BKCa channels, or neutrophil depletion were also effective in preventing mucosal mitochondrial dysfunction induced by I/R (Liu et al., 2012), it is clear that the effect of antecedent H_2S donor treatment to limit postischemic leukocyte/endothelial interactions ultimately acts to attenuate tissue injury in I/R by preventing trafficking of neutrophils.

While the aforementioned studies clearly established roles for eNOS-derived NO, phosphorylation of p38 MAPK, and activation of pK_{ATP}, $mitoK_{ATP}$, and BKCa channels as essential triggers to inaugurate entrance into a preconditioned anti-inflammatory phenotype by antecedent NaHS (Fig. 4), the molecular entities responsible for mediating that ability of post-capillary venules to resist supporting leukocyte rolling and adhesion during I/R were unknown. Since the reaction products of heme oxygenase-1 (HO-1)-mediated heme degradation exhibit powerful anti-adhesive and antioxidant properties, we postulated that this stress protein might serve this

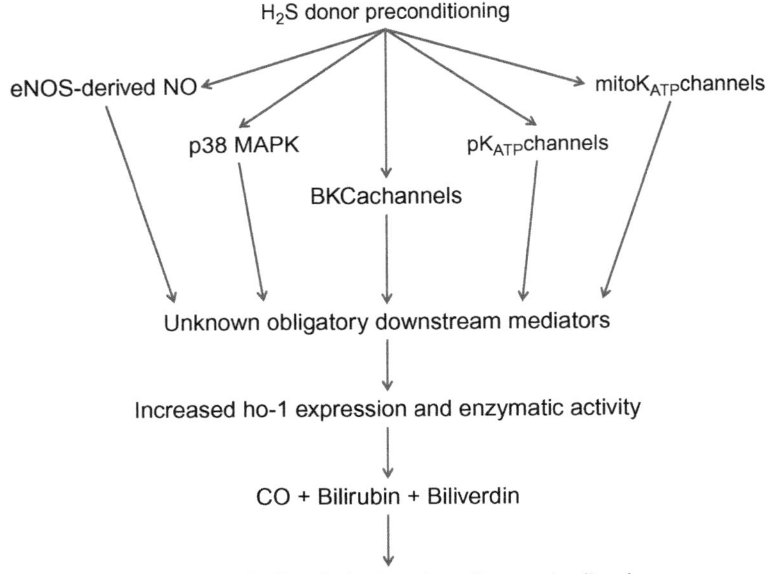

Figure 4 Signaling mechanisms invoked by antecedent preconditioning with H_2S donors 24 h prior to ischemia/reperfusion. Entrance into a protected state that limits postischemic leukocyte/endothelial cell adhesive interacts is triggered by an eNOS-derived NO-, p38 MAPK-, BKCa- and K_{ATP}-channel-dependent mechanism that is activated during the first hour after H_2S donor administration. These triggering events activate as yet unknown obligatory downstream signaling mediators to promote increased expression and activity of heme oxygenase-1 (HO-1) between 8 and 24 h later. The enzymatic activity of HO-1 generates the anti-adhesive gasotransmitter carbon monoxide (CO) as well as the powerful antioxidants bilirubin and secondarily derived biliverdin, which act in concert to prevent postischemic leukocyte trafficking and neutrophil-dependent parenchymal cell injury.

effector function in NaHS preconditioning (Zuidema, Peyton, Fay, Durante, & Korthuis, 2011). We showed that NaHS produced an increase in HO-1 expression at 8 and 24 h after administration, but not at earlier time points. HO activity in tissues samples was also elevated, when assessed 24 h after NaHS treatment. Moreover, pharmacologic HO inhibition during reperfusion prevented the ability of antecedent NaHS to abolish post-ischemic leukocyte rolling and adhesion in WT animals. NaHS preconditioning was also ineffective in mice genetically deficient in HO-1. These data indicate that HO-1 serves as an important end-effector of the anti-adhesive cytoprotective effects of antecedent H_2S treatment.

Taken together, the aforementioned late phase NaHS preconditioning studies support the view that H_2S initiates entrance into a preconditioned state by an eNOS-derived NO-, p38 MAPK-, BKCa channel-, and K_{ATP} channel-dependent mechanism to invoke increased expression and activity of HO-1 24 h later (Fig. 4). HO-1-catalyzed formation of carbon monoxide, bilirubin, and secondarily derived biliverdin act to limit postischemic leukocyte rolling and adhesion by preventing adhesion molecule expression and scavenging ROS as they are produced.

11. CONCLUSION AND PERSPECTIVES

Use of intravital microscopic approaches to study the effects of endogenously produced H_2S on postcapillary venular function have clearly established that baseline production of the gaseous signaling molecule, most likely by endothelial cells but perhaps also by the underlying pericyte layer, downregulates endothelial adhesion molecule expression to maintain a non-adhesive surface. Moreover, it is clear that use of exogenous H_2S donors exert powerful and direct anti-adhesive effects in inflamed tissues that target the expression of adhesion molecules to limit leukocyte–endothelial cell adhesive interactions. Although surprisingly little research has been directed at uncovering the anti-inflammatory mechanisms that are invoked by the acute administration of H_2S donors during inflammation, we have begun to uncover the signaling pathways invoked by preconditioning tissues with H_2S donors 24 h prior to I/R. Because of potential off-target or even deleterious effects of H_2S (e.g., poisoning mitochondrial respiratory chain by binding to the iron of cytochrome c oxygenase), uncovering the signaling mechanisms that underlie the anti-inflammatory effects of H_2S donors may lead to the development of novel therapies to treat conditions characterized by increased leukocyte–endothelial interactions. Promising new directions in this regard relate to use of BKCa activators or inducers of HO-1. Other promising developments with regard to pharmacologic development relate to development of analogues that modify the rate and amount of H_2S release from donor molecules or allow their administration via routes (inhalation, injection, skin patch, oral intake) that enhance patient compliance and/or treatment efficacy when these compounds enter human trials. Other strategies that may enhance translational therapeutics include formulation of H_2S-releasing nonsteroidal anti-inflammatory drugs (NSAIDs) as an approach to reduce gastrointestinal bleeding and ulceration that often accompanies chronic use of NSAIDS in arthritic patients (Fiorucci et al., 2005; Wallace et al., 2014; Zanardo et al., 2006). H_2S-releasing derivatives of other drugs, such as mesalamine, have also been developed to improve their anti-inflammatory potency and effectiveness (Fiorucci et al., 2007).

The power of intravital microscopy for the assessment of inflammatory responses and their modulation by H_2S donors in the intact living microcirculation has not been fully utilized. Nearly all studies have concentrated on effects of these agents on leukocyte rolling and adhesion, with very few or no studies focusing on platelet adhesion, chemokine and adhesion molecule expression, venular permeability, generation of ROS and RNOS, perfused capillary density, or cellular injury via use of techniques described above. As a consequence, the field has relied on *in vitro* and or *ex vivo* approaches to obtain much of what is known about the effect of H_2S donors on changes in chemokine and adhesion molecule expression, venular permeability, ROS/RNOS generation, and cellular injury induced by proinflammatory stimuli. Certainly, important new information has arisen from these *in vitro* approaches and immunohistochemical data obtained from tissue sections regarding the mechanisms underlying the anti-inflammatory effects of H_2S. However, these results do not necessarily inform us about the relative contributions of these mechanisms in the living microcirculation. Indeed, the unanticipated complexity of the *in vivo* milieu is often cited as a major contributing factor to the very high percentage (>93%) of failed clinical trials related to new cardiovascular (and other) drugs. Comparing the effects of H_2S on these variables in the more complicated *in vivo* milieu to data obtained from *in vitro* systems allows assessment of which of the many possibilities defined *in vitro* actually occur in the living microcirculation.

One new area for application of intravital microscopy that may significantly enhance the power of *in vivo* approaches to study mechanisms at the cell and molecular level relates to the development of novel approaches to visualize and quantify signal transduction events in intact vessels and microvascular networks of the living animal. As an example, Tallini et al. (2007) genetically engineered transgenic mice to study endothelial cell-specific calcium signaling *in vivo*. In these mice, the Ca^{2+} sensor GCaMP2 was placed under the control connexin 40 transcription regulatory elements within a bacterial artificial chromosome, a transgenesis approach that produces a mouse with endothelium lineage-specific calcium signaling. Using these mice, these investigators showed that bidirectional conduction of vasodilation mediated by relaxation of vascular smooth muscle cells along arteriolar networks involves a wave of Ca^{2+} that travels along the endothelium. Mauban, Fairfax, Rizzo, Zhang, and Wier (2014) used the GCaMP2 approach and a FRET-based genetically encoded Ca^{2+}/calmodulin sensor biomolecule (exMLCK) indicator, combined with two-photon fluorescence microscopy, to noninvasively quantify vascular smooth muscle Ca^{2+} in specific vessels in the ear of unanesthetized mice over extended periods of time as hypertension developed (Mauban et al., 2014). Continued development of new FRET-based optical biosensors that target other signaling molecules, coupled with spinning disk confocal or multiphoton imaging, would allow us to learn exactly what cells, and where in those cells, the molecule of interest was being activated in the living microcirculation. In addition to these approaches, use of immunocytes derived from GFP and YFP mice that are adoptively immunotransferred into wild type mice will facilitate the study of cell–cell interactions in signaling inflammatory responses and how this intercellular communication is influenced by H_2S donor treatment.

ACKNOWLEDGMENT

This work was supported by a grant from the National Institutes of Health (PO1 HL095486).

REFERENCES

Abe, K., & Kimura, H. (1996). The possible role of hydrogen sulfide as an endogenous neuromodulator. *Journal of Neuroscience, 16*(3), 1066–1071.

Andruski, B., McCafferty, D. M., Ignacy, T., Millen, B., & McDougall, J. J. (2008). Leukocyte trafficking and pain behavioral responses to a hydrogen sulfide donor in acute monoarthritis. *American Journal of Physiology. Regulatory, Integrative and Comparative Physiology, 295*(3), R814–R820.

Brancaleone, V., Flower, R. J., Cirino, G., & Perretti, M. (2013). OP06 annexin A1 (AnxA1) mediates hydrogen sulfide (H2S) effects in the control of inflammation. *Nitric Oxide, S1–31*(Suppl. 2), S21–S22.

Breslin, J. W., Sun, H., Xu, W., Rodarte, C., Moy, A. B., Wu, M. H., et al. (2006). Involvement of ROCK-mediated endothelial tension development in neutrophil-stimulated microvascular leakage. *American Journal of Physiology. Heart and Circulatory Physiology, 290*(2), H741–H750.

Davis, M. J. (1987). Determination of volumetric flow in capillary tubes using an optical Doppler velocimeter. *Microvascular Research, 34*, 223–230.

Denk, W., & Svoboda, K. (1997). Photon upmanship: Why multiphoton imaging is more than a gimmick. *Neuron, 18*(3), 351–357.

Dungey, A. A., Badhwar, A., Bihari, A., Kvietys, P. R., Harris, K. A., Forbes, T. L., et al. (2006). Role of heme oxygenase in the protection afforded skeletal muscle during ischemic tolerance. *Microcirculation, 13*(2), 71–79.

Eichorn, M. E., Ney, L., Massberg, S., & Goetz, A. E. (2002). Platelet kinetics in the pulmonary microcirculation in vivo assessed by intravital microscopy. *Journal of Vascular Research, 39*, 330–339.

Ezerina, D., Morgan, B., & Dick, T. P. (2014). Imaging dynamic redox processes with genetically encoded probes. *Journal of Molecular and Cellular Cardiology, 73*, 43–49.

Fingar, V. H., Guo, H. H., Lu, Z. H., & Peiper, S. C. (1997). Expression of chemokine receptors by endothelial cells: Detection by intravital microscopy using chemokine-coated fluorescent microspheres. *Methods in Enzymology, 288*, 148–158.

Fingar, V. H., Taber, S. W., Buschemeyer, W. C., ten Tije, A., Cerrito, P. B., Tseng, M., et al. (1997). Constitutive and stimulated expression of ICAM-1 protein on pulmonary endothelial cells in vivo. *Microvascular Research, 54*(2), 135–144.

Fiorucci, S., Antonelli, E., Distrutti, E., Rizzo, G., Mencarelli, A., Orlandi, S., et al. (2005). Inhibition of hydrogen sulfide generation contributes to gastric injury caused by anti-inflammatory nonsteroidal drugs. *Gastroenterology, 129*(4), 1210–1224.

Fiorucci, S., Orlandi, S., Mencarelli, A., Caliendo, G., Santagada, V., Distrutti, E., et al. (2007). Enhanced activity of a hydrogen sulphide-releasing derivative of mesalamine (ATB-429) in a mouse model of colitis. *British Journal of Pharmacology, 150*(8), 996–1002.

Gaboury, J. P., Johnston, B., Niu, X. F., & Kubes, P. (1995). Mechanisms underlying acute mast cell-induced leukocyte rolling and adhesion in vivo. *Journal of Immunology, 154*(2), 804–813.

Grisham, M. B. (2013). Methods to detect hydrogen peroxide in living cells: Possibilities and pitfalls. *Comparative Biochemistry and Physiology. Part A, Molecular & Integrative Physiology, 165*(4), 429–438.

Harris, A. G., Costa, J. J., Delano, F. A., Zweifach, B. W., & Schmid-Schonbein, G. W. (1998). Mechanisms of cell injury in rat mesentery and cremaster muscle. *American Journal of Physiology, 274*(3 Pt 2), H1009–H1015.

Harris, A. G., Leiderer, R., Peer, F., & Messmer, K. (1996). Skeletal muscle microvascular and tissue injury after varying durations of ischemia. *American Journal of Physiology, 271*(6 Pt 2), H2388–H2398.

Jayagopal, A., Russ, P. K., & Haselton, F. R. (2007). Surface engineering of quantum dots for in vivo vascular imaging. *Bioconjugate Chemistry, 18*(5), 1424–1433.

Jenne, C. N., Wong, C. H., Petri, B., & Kubes, P. (2011). The use of spinning-disk confocal microscopy for the intravital analysis of platelet dynamics in response to systemic and local inflammation. *PLoS One, 6*(9), e25109.

Kamath, A. F., Chauhan, A. K., Kisucka, J., Dole, V. S., Loscalzo, J., Handy, D. E., et al. (2006). Elevated levels of homocysteine compromise blood-brain barrier integrity in mice. *Blood, 107*(2), 591–593.

Kamoun, P. (2004). Endogenous production of hydrogen sulfide in mammals. *Amino Acids, 26*(3), 243–254.

Kashiba, M., Kajimura, M., Godo, N., & Suematsu, M. (2002). From O_2 to H_2S: A landscape view of gas biology. *Keio Journal of Medicine, 51*(1), 1–10.

Kong, L., Dunn, G. D., Keefer, L. K., & Korthuis, R. J. (1996). Nitric oxide reduces tumor cell adhesion to isolated rat postcapillary venules. *Clinical & Experimental Metastasis, 14*(4), 335–343.

Kong, L., & Korthuis, R. J. (1997). Melanoma cell adhesion to injured arterioles: Mechanisms of stabilized tethering. *Clinical & Experimental Metastasis, 15*(4), 426–431.

Krenz, M., Baines, C., Kalogeris, T., & Korthuis, R. J. (2013). Cell survival programs and ischemia/reperfusion: Hormesis, preconditioning, and cardioprotection. In D. N. Granger & J. Granger (Eds.), *Colloquium series on integrated systems physiology: From molecule to function to disease*. San Rafael, CA: Morgan and Claypool.

Lehr, H. A., Vollmar, B., Vajkoczy, P., & Menger, M. D. (1999). Intravital fluorescence microscopy for the study of leukocyte interaction with platelets and endothelial cells. *Methods in Enzymology, 300*, 462–481.

Lerchenberger, M., Uhl, B., Stark, K., Zuchtriegel, G., Eckart, A., Miller, J., et al. (2013). Matrix metalloproteinases modulate ameoboid-like migration of neutrophils through inflamed interstitial tissue. *Blood, 122*, 770–780.

Ley, K., Mestas, J., Pospieszalska, M. K., Sundd, P., Groisman, A., & Zarbock, A. (2008). Chapter 11. Intravital microscopic investigation of leukocyte interactions with the blood vessel wall. *Methods in Enzymology, 445*, 255–279. http://dx.doi.org/10.1016/S0076-6879(08)03011-5.

Li, W., Nava, R. G., Bribriesco, A. C., Zinselmeyer, B. H., Spahn, J. H., Gelman, A. E., et al. (2012). Intravital 2-photon imaging of leukocyte trafficking in beating heart. *Journal of Clinical Investigation, 122*(7), 2499–2508.

Liu, Y., Kalogeris, T., Wang, M., Zuidema, M. Y., Wang, Q., Dai, H., et al. (2012). Hydrogen sulfide preconditioning or neutrophil depletion attenuates ischemia-reperfusion-induced mitochondrial dysfunction in rat small intestine. *American Journal of Physiology. Gastrointestinal and Liver Physiology, 302*(1), G44–G54.

Massberg, S., Enders, G., Leiderer, R., Eisenmenger, S., Vestweber, D., Krombach, F., et al. (1998). Platelet-endothelial cell interactions during ischemia/reperfusion: The role of P-selectin. *Blood, 92*, 507–515.

Mauban, J. R., Fairfax, S. T., Rizzo, M. A., Zhang, J., & Wier, W. G. (2014). A method for noninvasive longitudinal measurements of [Ca2+] in arterioles of hypertensive optical biosensor mice. *American Journal of Physiology. Heart and Circulatory Physiology, 307*(2), H173–H181.

McDonald, B., McAvoy, E. F., Lam, F., Gill, V., de la Motte, C., Savani, R. C., et al. (2008). Interaction of CD44 and hyaluronan is the dominant mechanism for neutrophil sequestration in inflamed liver sinusoids. *Journal of Experimental Medicine, 205*(4), 915–927.

McDonald, B., Pittman, K., Menezes, G. B., Hirota, S. A., Slaba, I., Waterhouse, C. C., et al. (2010). Intravascular danger signals guide neutrophils to sites of sterile inflammation. *Science, 330*(6002), 362–366.

Mempel, T. R., Scimone, M. L., Mora, J. R., & von Andrian, U. H. (2004). In vivo imaging of leukocyte trafficking in blood vessels and tissues. *Current Opinion in Immunology, 16*(4), 406–417. http://dx.doi.org/10.1016/j.coi.2004.05.018.

Morgan, B., Sobotta, M. C., & Dick, T. P. (2011). Measuring E(GSH) and H2O2 with roGFP2-based redox probes. *Free Radical Biology and Medicine, 51*(11), 1943–1951.

Nakano, A. (2002). Spinning-disk confocal microscopy—A cutting-edge tool for imaging of membrane traffic. *Cell Structure and Function, 27*(5), 349–355.

Ortiz, D., Briceno, J. C., & Cabrales, P. (2014). Microhemodynamic parameters quantification from intravital microscopy videos. *Physiological Measurement, 35*(3), 351–367.

Peters, N. C., Egen, J. G., Secundino, N., Debrabant, A., Kimblin, N., Kamhawi, S., et al. (2008). In vivo imaging reveals an essential role for neutrophils in leishmaniasis transmitted by sand flies. *Science, 321*(5891), 970–974.

Proebstl, D., Voisin, M. B., Woodfin, A., Whiteford, J., D'Acquisto, F., Jones, G. E., et al. (2012). Pericytes support neutrophil subendothelial cell crawling and breaching of venular walls in vivo. *Journal of Experimental Medicine, 209*, 1219–1234.

Sanz, M. J., Johnston, B., Issekutz, A., & Kubes, P. (1999). Endothelin-1 causes P-selectin-dependent leukocyte rolling and adhesion within rat mesenteric microvessels. *American Journal of Physiology*, *277*(5 Pt 2), H1823–H1830.

Scallan, J., Huxley, V. H., & Korthuis, R. J. (2010). *In capillary fluid exchange: Regulation, functions, and pathology*. San Rafael, CA: Morgan Claypool.

Shuhendler, A. J., Prasad, P., Leung, M., Rauth, A. M., Dacosta, R. S., & Wu, X. Y. (2012). A novel solid lipid nanoparticle formulation for active targeting to tumor alpha(v) beta(3) integrin receptors reveals cyclic RGD as a double-edged sword. *Advanced Healthcare Materials*, *1*(5), 600–608.

Sperandio, M., Pickard, J., Unnikrishnan, S., Acton, S. T., & Ley, K. (2006). Analysis of leukocyte rolling in vivo and in vitro. *Methods in Enzymology*, *416*, 346–371. http://dx.doi.org/10.1016/S0076-6879(06)16023-1.

Spiller, F., Orrico, M. I., Nascimento, D. C., Czaikoski, P. G., Souto, F. O., Alves-Filho, J. C., et al. (2010). Hydrogen sulfide improves neutrophil migration and survival in sepsis via K+ATP channel activation. *American Journal of Respiratory and Critical Care Medicine*, *182*(3), 360–368.

Sriram, K., Intaglietta, M., & Tartakovsky, D. M. (2014). Non-Newtonian flow of blood in arterioles: Consequences for wall shear stress measurements. *Microcirculation*, *21*(7), 628–639.

Stark, K., Eckart, A., Haidari, S., Tirniceriu, A., Lorenz, M., von Bruhl, M.-L., et al. (2013). Capillary and arteriolar pericytes attract innate leukocytes exiting through venules and 'instruct' them with pattern-recognition and motility patterns. *Nature Immunology*, *14*, 41–51.

Suematsu, M., DeLano, F. A., Poole, D., Engler, R. L., Miyasaka, M., Zweifach, B. W., et al. (1994). Spatial and temporal correlation between leukocyte behavior and cell injury in postischemic rat skeletal muscle microcirculation. *Laboratory Investigation*, *70*(5), 684–695.

Suematsu, M., Schmid-Schonbein, G. W., Chavez-Chavez, R. H., Yee, T. T., Tamatani, T., Miyasaka, M., et al. (1993). In vivo visualization of oxidative changes in microvessels during neutrophil activation. *American Journal of Physiology*, *264*(3 Pt 2), H881–H891.

Sumagin, R., & Sarelius, I. H. (2010). Intercellular adhesion molecule-1 enrichment near tricellular endothelial junctions is preferentially associated with leukocyte transmigration and signals for reorganization of these junctions to accommodate leukocyte passage. *Journal of Immunology*, *184*(9), 5242–5252.

Sundd, P., Gutierrez, E., Koltsova, E. K., Kuwano, Y., Fukuda, S., Pospieszalska, M. K., et al. (2012). 'Slings' enable neutrophil rolling at high shear. *Nature*, *488*(7411), 399–403.

Sundd, P., Pospieszakska, M. K., & Ley, K. (2013). Neutrophil rolling at high shear: Flattening, catch bond behavior, tethers and slings. *Molecular Immunology*, *53*, 59–69.

Suzuki, H., DeLano, F. A., Parks, D. A., Jamshidi, N., Granger, D. N., Ishii, H., et al. (1998). Xanthine oxidase activity associated with arterial blood pressure in spontaneously hypertensive rats. *Proceedings of the National Academy of Sciences of the United States of America*, *95*(8), 4754–4759.

Tallini, Y. N., Brekke, J. F., Shui, B., Doran, R., Hwang, S. M., Nakai, J., et al. (2007). Propagated endothelial Ca2+ waves and arteriolar dilation in vivo: Measurements in Cx40BAC GCaMP2 transgenic mice. *Circulation Research*, *101*(12), 1300–1309.

Tyagi, N., Givvimani, S., Qipshidze, N., Kundu, S., Kapoor, S., Vacek, J. C., et al. (2010). Hydrogen sulfide mitigates matrix metalloproteinase-9 activity and neurovascular permeability in hyperhomocysteinemic mice. *Neurochemistry International*, *56*(2), 301–307.

Voisin, M. B., & Nourshargh, S. (2013). Neutrophil transmigration: Emergence of an adhesive cascade within venular walls. *Journal of Innate Immunity*, *5*(4), 336–347.

Voisin, M. B., Proebstl, D., & Nourshargh, S. (2010). Venular basement membranes ubiquitously express matrix low-expression regions: Characterization in multiple tissues and remodeling during inflammation. *American Journal of Pathology, 176*, 482–495.
Wallace, J. L., Blackler, R. W., Chan, M. V., Da Silva, G. J., Elsheikh, W., Flannigan, K. L., et al. (2014). Anti-inflammatory and cytoprotective actions of hydrogen sulfide: Translation to therapeutics. *Antioxidants & Redox Signaling*.
Wang, R. (2003). The gasotransmitter role of hydrogen sulfide. *Antioxidants and Redox Signaling, 5*(4), 493–501.
Wang, R. (2012). Physiological implications of hydrogen sulfide: A whiff exploration that blossomed. *Physiological Reviews, 92*(2), 791–896.
Winterbourn, C. C. (2014). The challenges of using fluorescent probes to detect and quantify specific reactive oxygen species in living cells. *Biochimica et Biophysica Acta, 1840*(2), 730–738.
Woolley, J. F., Stanicka, J., & Cotter, T. G. (2013). Recent advances in reactive oxygen species measurement in biological systems. *Trends in Biochemical Sciences, 38*(11), 556–565.
Yuan, Y., Chilian, W. M., Granger, H. J., & Zawieja, D. C. (1993). Permeability to albumin in isolated coronary venules. *American Journal of Physiology, 265*(2 Pt 2), H543–H552.
Yuan, Y., Mier, R. A., Chilian, W. M., Zawieja, D. C., & Granger, H. J. (1995). Interaction of neutrophils and endothelium in isolated coronary venules and arterioles. *American Journal of Physiology, 268*(1 Pt 2), H490–H498.
Yusof, M., Kamada, K., Kalogeris, T., Gaskin, F. S., & Korthuis, R. J. (2009). Hydrogen sulfide triggers late-phase preconditioning in postischemic small intestine by an NO- and p38 MAPK-dependent mechanism. *American Journal of Physiology. Heart and Circulatory Physiology, 296*(3), H868–H876.
Zanardo, R. C., Brancaleone, V., Distrutti, E., Fiorucci, S., Cirino, G., & Wallace, J. L. (2006). Hydrogen sulfide is an endogenous modulator of leukocyte-mediated inflammation. *FASEB Journal, 20*(12), 2118–2120.
Zhang, H., Zhi, L., Moochhala, S. M., Moore, P. K., & Bhatia, M. (2007). Endogenous hydrogen sulfide regulates leukocyte trafficking in cecal ligation and puncture-induced sepsis. *Journal of Leukocyte Biology, 82*(4), 894–905.
Zhu, J. X., Kalbfleisch, M., Yang, Y. X., Bihari, R., Lobb, I., Davison, M., et al. (2012). Detrimental effects of prolonged warm renal ischaemia-reperfusion injury are abrogated by supplemental hydrogen sulphide: An analysis using real-time intravital microscopy and polymerase chain reaction. *BJU International, 110*(11 Pt C), E1218–E1227.
Zuidema, M. Y., Peyton, K. J., Fay, W. P., Durante, W., & Korthuis, R. J. (2011). Antecedent hydrogen sulfide elicits an anti-inflammatory phenotype in postischemic murine small intestine: Role of heme oxygenase-1. *American Journal of Physiology. Heart and Circulatory Physiology, 301*(3), H888–H894.
Zuidema, M. Y., Yang, Y., Wang, M., Kalogeris, T., Liu, Y., Meininger, C. J., et al. (2010). Antecedent hydrogen sulfide elicits an anti-inflammatory phenotype in postischemic murine small intestine: Role of BK channels. *American Journal of Physiology. Heart and Circulatory Physiology, 299*(5), H1554–H1567.

CHAPTER SEVEN

Attenuation of Inflammatory Responses by Hydrogen Sulfide (H_2S) in Ischemia/Reperfusion Injury

Neel R. Sodha, Frank W. Sellke[1]
Division of Cardiothoracic Surgery, Department of Surgery, Alpert Medical School of Brown University, Providence, Rhode Island
[1]Corresponding author: e-mail address: fsellke@lifespan.org

Contents

1. Introduction 128
2. Ischemia–Reperfusion Injury 128
3. Central Nervous System 130
4. Respiratory System 131
5. Cardiovascular System 133
6. Gastrointestinal System 135
   6.1 Stomach 135
   6.2 Small intestine 136
7. Hepatobiliary System 138
8. Renal System 139
9. Musculoskeletal 141
10. Summary 142
References 143

Abstract

Cellular and tissue injury induced by ischemia is often exacerbated by restoring perfusion to the affected organ system. The injury, termed ischemia–reperfusion injury, is mediated in large part by the inflammatory response generated in the setting of reperfusion. Recent research has demonstrated that the administration of hydrogen sulfide as a therapeutic agent in the setting of ischemia–reperfusion can markedly attenuate the inflammatory response with subsequent mitigation of tissue injury and improved function. This beneficial anti-inflammatory effect has been observed in multiple organ systems, subject to ischemia–reperfusion injury, the details of which are the subject of this chapter.

Methods in Enzymology, Volume 555
ISSN 0076-6879
http://dx.doi.org/10.1016/bs.mie.2014.11.041

1. INTRODUCTION

Ischemia–reperfusion injury (I/R injury) refers to a series of pathologic events, which occur after restoration of perfusion to ischemic tissue. While restoration of perfusion is necessary to salvage tissue from the inevitable necrosis, which would otherwise occur, reperfusion can trigger additional tissue injury via free radical injury, programmed cell death, and inflammation. Inflammation has been found to be central to reperfusion injury and thus a major focus of investigation for therapeutic targets. After initially demonstrating protection against ischemia by reducing metabolic demand, hydrogen sulfide has also been shown to function as an effective agent in limiting tissue inflammation after ischemia–reperfusion. This robust anti-inflammatory effect has been demonstrated in multiple organ systems, but the underlying mechanism and involved signaling pathways remain poorly elucidated. Herein, we will discuss the available research investigating the role hydrogen sulfide plays in attenuating ischemia–reperfusion injury, specifically addressing studies which have examined inflammation.

2. ISCHEMIA–REPERFUSION INJURY

Tissue ischemia can occur in any organ system under a number of conditions. Broadly speaking, the impairment in perfusion may result from the interruption of arterial inflow or venous outflow, or from global hypoperfusion secondary to impaired blood pressure as can be seen in the setting of shock. Interruption from arterial inflow may be secondary to acute mechanical factors such as thrombosis, embolism, and arterial dissection, or via iatrogenic arterial occlusion. Venous obstruction from thrombosis or mechanical compression leads to impaired outflow and increased pressure gradients across capillary beds which subsequently can compromise arterial perfusion. Global hypoperfusion results from hypotension, which can occur from any of the common etiologies of shock, such as neurogenic (spinal cord injury with subsequent vasodilation), cardiogenic (heart failure with impaired cardiac output), septic (systemic vasodilation), hypovolemic (hemorrhage, gastrointestinal losses), and anaphylactic (systemic vasodilation). The resultant impairment in tissue perfusion interrupts aerobic metabolism allowing toxic metabolites to accumulate. This process invariably leads to cellular necrosis unless perfusion is restored. Unfortunately, the restoration of perfusion can impart additional tissue injury, with damage occurring beyond areas of ischemia alone. The resulting injury, termed I/R injury, can be seen in any organ system and has tremendous clinical relevance as detailed in the sections below.

I/R injury occurs secondary to several processes, all of which are interrelated. These include inflammation, free radical formation, endothelial and microvascular dysfunction, and cell death via apoptosis and autophagy. Elucidating the role each plays individually is especially difficult, as one component of the process may itself be triggered by or trigger another process (e.g., the interrelationship between inflammation and endothelial cell dysfunction). Cell death can occur via necrosis, apoptosis, or autophagy. Necrosis results in cellular swelling, followed by rupture leading to the release of antigens capable of eliciting a robust inflammatory response. Apoptosis is triggered by a highly regulated signaling cascade which results in cellular shrinkage and death, but the release of ATP during this process can also trigger a strong inflammatory response. The proinflammatory cytokines liberated during the inflammatory response can also serve to trigger cell death-signaling cascades, highlighting the complex interrelationship between the components of I/R injury. The capillary and microvascular endothelium are injured during periods of ischemia, with consequent impairment in nitric oxide synthase function and enhanced expression of cell adhesion molecules, which facilitate the migration of inflammatory cells. Disruption of cell to cell barriers develops, and with the restoration of reperfusion, inflammatory cells are able to adhere to the area of injury and migrate between cell junctions adding to tissue edema and inflammation. Depletion of endogenous antioxidant mechanisms as well as invasion of free radical producing inflammatory cells results in the presence of increased free radical injury. Impaired nitric oxide production by endothelial cells can lead to decreases in vasorelaxation and result in a "no reflow" phenomenon at the microvascular level despite restoration of blood flow in larger vessels. Upon cellular injury resulting in necrosis or apoptosis, ligands can be released which can induce the generation of proinflammatory cytokines via Toll-like receptor activation (which can also be triggered by oxidative stressors) and subsequent induction of NF-κB and mitogen-activated protein kinase signaling pathways. The subsequent release of cytokines and chemokines can serve to attract inflammatory cells, propagating the proinflammatory milieu.

3. CENTRAL NERVOUS SYSTEM

Cerebral ischemia may result from multiple causes. Thrombosis and embolism (air, fat, clot, calcium, endocarditic vegetation) occur most commonly and result in ischemic cerebrovascular accidents (stroke). Global hypoperfusion can occur due to hypotension in the setting of shock or cardiac arrest. Interruption of cerebral blood flow also occurs during iatrogenic interventions as seen with carotid arterial surgery and during deep hypothermic circulatory arrest for complex cardiovascular procedures. Flow can be restored via pharmacologic interventions such as the administration of thrombolytic agents or anticoagulants, mechanical interventions such as thromboembolectomy or stenting, and restoration of cardiac function after cardiac arrest. The resulting edema during reperfusion can have catastrophic consequences due to increased intracranial pressure. Two studies have examined the effects of sulfide on inflammation in the setting of cerebral I/R injury.

Yin et al. (2013) examined the role of sulfide in a rat model of global brain ischemia followed by reperfusion. In this model, ischemia was induced by electrocauterization of the bilateral vertebral arteries, followed by vascular clipping of the right and left common carotid arteries. Ischemia was maintained for 30 min, followed by the removal of the carotid clips and reperfusion for 24 h. Sulfide was administered as NaHS as a low-dose (0.2 μM/kg) or high-dose (0.4 μM/kg) solution 30 min after the onset of ischemia via intraperitoneal injection. Apoplexy and neurologic scoring were performed at 6, 12, and 24 h after the surgical protocol and rats were sacrificed at 24 h. Both apoplexy index and neurological symptom scoring were significantly improved in sulfide-treated groups in a dose-dependent manner. Tetrazolium chloride (TTC) staining was performed to assess brain infarct size and demonstrated marked reductions with sulfide therapy—control animals receiving placebo infarcted 55% of the area at risk, whereas low-dose sulfide decreased this to 23% and high-dose sulfide to 9%. Tissue homogenates were assayed for superoxide dismutase (SOD) activity, and while reaching statistical significance, SOD activity was marginally higher with sulfide, in a dose-dependent manner. TNF-α and IL-10 were assessed with ELISA studies and decreased by roughly 25% or increased by 25%, respectively, with sulfide treatment.

Wang et al. (2014) studied the blood–brain barrier after cerebral ischemia and as part of their study investigated the effects of a sulfide donor on inflammation. Using a mouse model, ischemia was achieved with middle cerebral artery (MCA) occlusion. Occlusion was performed with a monofilament catheter introduced into the MCA via the internal carotid with flow measured by laser Doppler. The contralateral side served as control. Ischemia was maintained for 60 min followed by 3 h of reperfusion. Based on prior reports from their group demonstrating that ADT [5-(4-methoxyphenyl)-3H-1,2-dithiole-3-thione] can act as a sulfide donor, ADT was used as the interventional agent administered at 50 mg/kg. Mice were sacrificed at 24 or 48 h after ischemia. No mortality differences were observed between sulfide-treated animals or controls. Neurologic scoring based on the motor function was improved in animals receiving ADT. As with the previously discussed study, infarct size as assessed by TTC staining decreased by about 1/3. Modest reductions in cerebral edema (as assessed by wet-to-dry weight ratio) occurred in ADT-treated groups. Quantitative PCR to assess IL-1β demonstrated that IL-1β levels in ADT-treated animals were about 20% of control levels. ELISA was performed to assess IL-10 and demonstrated a 40% increase relative to controls with ADT treatment. Tissue myeloperoxidase (MPO) assays, used to assess neutrophil infiltration and activity, demonstrated a reduction of near 50% in MPO activity when mice were administered ADT. Both matrix metalloproteinase-9 (MMP-9) expression and activity, as assessed with gelatin zymography, were reduced in ADT-treated mice, as was the degree of NF-κB nuclear translocation.

4. RESPIRATORY SYSTEM

The lung is relatively resistant to ischemic injury given that the majority of cardiac output will traverse the pulmonary vasculature. Three clinical scenarios in which I/R injury may be encountered include massive pulmonary embolism, trauma, and lung transplantation. In the setting of massive pulmonary embolism, flow may be acutely restored via the administration of thrombolytic agents, catheter-based thromboembolectomy, or surgical thromboembolectomy. In the trauma setting, prolonged lung ischemia may occur when arterial inflow is occluded for control of hemorrhage. In the setting of transplantation, prolonged ischemia followed by reperfusion can lead to graft dysfunction after transplantation requiring prolonged oxygen or ventilator dependence.

Wu et al. (2013) studied the role of endogenously produced hydrogen sulfide in relation to I/R injury in the setting of lung transplantation. Utilizing a rat model of orthotopic left lung transplantation, they harvested donor lung preserved it with Perfadex solution at 4 °C in a low potassium dextran solution for 3 h prior to implantation, creating 3 h of total ischemia. The experimental intervention involved administration of NaHS (14 μ*M*/kg) as a sulfide donor, or the CSE inhibitor propargylglycine (PAG; 37.5 mg/kg) to inhibit endogenous sulfide production. Agents were administered intraperitoneally 15 min before the start of lung implantation to the recipient. Animals were sacrificed at 24 h after transplantation. Using the alveolar-arterial (A-a) gradient of oxygen as a functional outcome, the studies revealed that administration of NaHS resulted in a 20% reduction in the A-a gradient and 20% decrease in edema (as assessed by wet-to-dry weight ratio) relative to untreated mice undergoing lung transplantation. In mice receiving the CSE inhibitor PAG, A-a gradients and edema were 20% higher than in mice undergoing lung transplant alone and nearly 40% higher than in mice receiving exogenous sulfide. Lung inflammation scoring on histology was performed based on changes in architecture, leukocyte infiltration, edema, presence of pneumonitis, or lung field obliteration with scores ranging from 0 to 5. Control mice undergoing lung transplant had a mean histopathologic score of 3.33, which was reduced to 1.83 in NaHS-treated animals and increased to 4.33 in PAG-treated animals. To assess neutrophil infiltration and inflammation, tissue MPO assay was performed which demonstrated a decrease in MPO activity by 1/3 in NaHS-treated animals. IL-1β and IL-10 levels assessed using ELISA kits demonstrated a 40% decrease in IL-1β levels and doubling of IL-10 levels with NaHS treatment. These anti-inflammatory effects were not observed in animals treated with PAG.

The lung is particularly susceptible to inflammatory injury secondary to remote ischemia–reperfusion injury, as clinically seen with acute respiratory distress syndrome occurring after limb ischemia and reperfusion. Based on this clinical scenario, Qi, Chen, Li, Wang, and Xie (2014) examined the role of sulfide in lung injury after limb ischemia. Using a rat model, limb ischemia was induced for 4 h by bilateral lower limb band ligation. This was followed by 4 h of reperfusion after band removal. Animals were administered NaHS (0.78 mg/kg) or DL-propargylglycine (PPG)—an inhibitor of CSE (60 mg/kg) intraperitoneally before I/R injury. Based on the H&E staining, alveolar congestion, disruption of alveolar architecture, alveolar wall thickness, and inflammatory cell infiltration were all reduced with NaHS administration and increased with PPG relative to controls. Lung edema as assessed by wet-to-dry ratio was reduced with NaHS treatment, but increased with PPG administration relative to untreated controls. When compared with animals undergoing the I/R protocol without any other treatment, NaHS administration was associated with a 20% decrease in protein levels of TLR-4 and NF-κB, whereas PPG administration was associated with a 20% increase in these levels relative to untreated animals.

5. CARDIOVASCULAR SYSTEM

Myocardial ischemia can occur secondary to multiple causes, including thrombosis as seen with acute myocardial infarction, embolization from tumor, calcium, or infection, coronary arterial trauma or vasospasm, demand ischemia in the setting of increased work, or globally during cardioplegic arrest for open heart surgery and transplantation. Restoration of perfusion generally occurs via pharmacological means (thrombolytics), catheter-based interventions (angioplasty or coronary artery stenting), or surgical intervention. Reperfusion injury is particularly difficult in the setting of myocardial ischemia, as it can trigger lethal arrhythmia and depressed cardiac function which can further exacerbate hypotension and hypoperfusion.

Elrod et al. (2007), utilizing a mouse model of myocardial I/R injury, were among the first to demonstrate the beneficial effects of sulfide in the setting of cardiac ischemia. Left coronary artery occlusion was performed to induce ischemia for time periods between 30 and 45 min, followed by 24–72 h of reperfusion. Sulfide was administered as Na_2S using H_2S gas as a starting material, with sulfide doses ranging from 10 to 500 μg/kg into the left ventricular lumen. The optimal dosing to reduce infarct size was 50 μg/kg and subsequently used in further studies. TTC staining demonstrated marked reductions in infarct size with sulfide administration from 48% of the area at risk in controls to 13% in sulfide-treated mice, correlating with a 40% decrease in Troponin I. Echocardiography was utilized to assess cardiac function and demonstrated preservation of left ventricular end diastolic and systolic dimensions after myocardial infarction (MI) in sulfide treated animals. Fractional shortening in the left ventricle was improved by 50% relative to controls. Overall left ventricular ejection fraction decreased 39% in control mice after I/R injury, but only 18% in sulfide-treated mice. H&E staining showed reduced myocardial neutrophilic infiltrate, necrosis, and hemorrhage compared to controls. MPO assays demonstrated a 26% reduction in MPO activity in sulfide-treated mice. ELISA studies for inflammatory cytokines demonstrated no difference in TNF-α or IL-10 levels between treated and untreated groups, but IL-1β levels were twofold higher in animals given placebo. Using intravital microscopy of leukocyte rolling, the authors demonstrated that sulfide completely inhibited thrombin-induced leukocyte–endothelial cell interactions.

Sivarajah et al. (2009) used a rat model of myocardial I/R injury, inducing 25 min of left anterior descending (LAD) coronary artery ischemia via snare occlusion followed by 2 h of reperfusion. Sulfide was administered as an NaHS (3 mg/kg) bolus intravenously 15 min before reperfusion. TTC staining demonstrated a decrease in infarct size from 58% to 45% with sulfide treatment. NF-κB translocation was decreased with sulfide therapy, and MPO activity decreased by 50%. Most interestingly, these effects were abolished by blocking K-ATP channels with 5-HD, providing mechanistic insight into the protection provided by sulfide.

In order to determine if the cardioprotective benefits of sulfide therapy observed in murine models would translate to clinically relevant large animal models, Sodha et al. (2009) utilized sulfide as a therapeutic agent in a swine model of I/R injury. In this model, the LAD was occluded distal to the second diagonal artery for 60 min followed by 120 min reperfusion. Sulfide was administered as NaHS in a 100 μg/kg bolus, followed by a 1 mg/kg/h infusion 10 min before the onset of reperfusion. From a functional perspective, global left ventricular function as determined by LV + dP/dt was significantly better in sulfide-treated groups by 40%, and regional left ventricular function was nearly two times better. No differences in the incidence of pulseless ventricular tachycardia or ventricular fibrillation were observed between groups. Infarct size, as assessed by TTC staining, was reduced 1.94-fold. The degree of free radical stress was assessed via nitrotyrosine staining of myocardial sections and demonstrated significantly lower scores with sulfide therapy. MPO activity decreased 15-fold in sulfide-treated swine, and tissue levels of IL-6, IL-8, and TNF-α lower by 1/3 as assessed using a microarray panel.

While the previous studies had administered sulfide in an intravenous fashion, Snijder et al. (2013) studied the gaseous administration of sulfide specifically because inhaled sulfide may induce hypometabolism which the intravenous or intraperitoneal routes do not. Using a mouse model, inhaled sulfide was administered as O_2/10 ppm H_2S/N_2 mixture in a 49:1 ratio. The gas was given 30 min prior to ischemia and 5 min prior to reperfusion. Coronary ischemia was induced via LAD suture ligation for 30 min. Mice were sacrificed 1 and 7 days after surgery. Gaseous H_2S induced a state of hypometabolism as evidenced by reduction in CO_2 production by 72% relative to basal levels. Sulfide administered at 10 ppm did not reduce infarct size, but increasing the dose to 100 ppm reduced infarct size by 62% based on histology sections. Troponin T levels decreased by 47% on day 1 with the 100 ppm dosing regimen. To assess inflammation and inflammatory cell infiltration, Ly-6G staining for granulocytes was used. Ly-6G-positive granulocyte staining increased 12-fold in controls, relative to sham animals, and the administration of sulfide at a 100 ppm concentration decreased this by 60%.

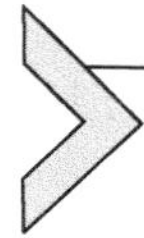

6. GASTROINTESTINAL SYSTEM

6.1. Stomach

The stomach is relatively resistant to ischemia secondary to a rich network of collateral arterial and venous flow, as well as a blood supply from multiple arterial branches. Despite this extensive vascularization, the stomach may become ischemic in the setting of mechanical obstruction to blood flow (intrathoracic herniation, volvulus) or profound hypotension.

Based on the prior studies which demonstrated exogenous administration of H_2S resulted in gastric mucosal protection from I/R injury, Guo, Liang, Shah Masood, and Yan (2014) have investigated the effects of hydrogen sulfide administration on inflammation in gastric I/R injury. Utilizing an *in vitro* model with human gastric epithelial cells (cell line GES-1), cells were subjected to simulated ischemia utilizing an ischemia solution (glucose free Krebs buffer containing 5 m*M* sodium lactate, 20 m*M* 2-DOG, 20 m*M* sodium dithionite at a pH of 6.6). The cells were maintained in this solution for 15 min, followed by replenishment with Dulbecco's modified Eagle's medium supplemented with 10% fetal bovine serum. The treatment groups were administered NaHS in concentrations ranging from 30 to 300 μ*M* for 30 min prior to the induction of simulated ischemia. About 100 μ*M* was found to be the optimal concentration and used for studies. Cell viability was examined at different time points after reperfusion. After 4 h of "reperfusion," MTT was utilized to assay cell viability. Cell viability as per MTT staining demonstrated an approximately 20–25% increase in viable cells relative to controls with 100 μ*M* of sulfide treatment. TNF-α and IL-6 levels in the cell culture supernatant were assayed after 6 h of reperfusion utilizing a commercial ELISA kit and were reduced by 40% with administration of sulfide. NaHS also inhibited the nuclear translocation of the proinflammatory NF-κB p65 subunit indicating a block to inflammation.

Using a rat model of gastric I/R injury, Mard et al. (2012) investigated the role of endogenously produced sulfide and exogenously administered sulfide in protection from I/R injury. Ischemia was induced in rats via clamping of the celiac artery for 30 min, followed by clamp removal and reperfusion for 3 h. To assess the protection provided by endogenous H_2S protection, one experimental group was given PAG (a CSE inhibitor) 5 min before the end of ischemia. Another experimental group was administered exogenous NaHS as an intravenous bolus in concentrations of 160, 320, or 640 ng/kg 5 min prior to the end of ischemia. Histologic assessment of gastric mucosal lesion size demonstrated that lesions in control animals were approximately 4 mm in size, 1 mm or less in size with NaHS treatment (decreasing in a dose-dependent manner), and 5.5 mm in size in animals given PAG. Serum levels of IL-1β, TNF-α, IL-10, and TGF-β assayed using ELISA kits and demonstrated that in sulfide-treated animals, serum levels of IL-1β were approximately 1/3 of control, TNF-α 2/3 of control, and no different for IL-10 or TGF-β. With sulfide treatment, quantitative real-time PCR demonstrated that gastric mucosal mRNA levels of IL-1β decreased over 50% relative to controls, TNF-α decreased over 50% relative to controls, and no significant changes in TGF-β were seen.

6.2. Small intestine

The small intestine is at risk for ischemia in the setting of acute thrombosis or from embolization from a variety of sources. Aortic pathology such as acute aortic dissection can result in visceral arterial occlusion. Additionally, dehydration and hypovolemia may induce mesenteric ischemia in patients with underlying chronic atherosclerotic disease. Without timely reperfusion, intestinal ischemia can be rapidly fatal secondary to subsequent sepsis which can occur.

Zuidema et al. (2010) investigated whether pretreatment with sulfide would induce an "anti-inflammatory" phenotype in the small intestine and whether this was in part related to BK channels. BK channels are a class of voltage- and calcium-activated potassium channels characterized by their large unitary conductance and sensitivity to blockade by iberiotoxin (IBTX). Their activation hyperpolarizes the endothelial cell membrane and enhances calcium influx through nonvoltage gate calcium channels, and thereby thought to contribute to nitric oxide production via calcium sensitive eNOS. IBTX and paxilline are BK channel blockers. Using a mouse model, intestinal ischemia was induced by clamping of the superior mesenteric artery for 45 min. Carboxyfluorescein diacetate succinimidyl ester was dissolved in DMSO and administered intravenously to label leukocytes for intravital microscopy studies at 30–40 min postischemia and

60–70 min postischemia. Intravital fluorescence microscopy was performed by exteriorizing a segment of small intestine for imaging. Venules ranging from 20 to 50 μm in diameter were imaged over a 100 μm length for periods of 30 s, counting the number of attached leukocytes and those with slowed velocities (indicating rolling). Experimental groups were administered NaHS at a dose of 14 μg/kg intraperitoneally 24 h before induction of ischemia, or the selective BK channel inhibitor paxilline 25 mg/kg intraperitoneally, or NS 1619 the BK channel opener at 100 μ*M*. I/R injury induced a near sevenfold increase in the number of adherent and rolling leukocytes, an effect which was completely blocked by NaHS, but remained the same if NaHS and paxilline were administered together. Administering NS 1619 induced a similar affect as observed with giving sulfide. Investigation into the role of small and intermediate conductance K channels was performed as well, but found to be noncontributory. The presence of BKα channels in endothelial and vascular smooth muscle cells of small intestinal venules in the mouse was confirmed with immunostaining.

The same group (Liu et al., 2012) utilized a rat model to investigate if leukocytes affected mitochondrial function in the setting of small intestinal I/R injury. Superior mesenteric artery ischemia was induced with a vascular clamp for 45 min followed by 60 min of reperfusion. Sulfide was administered as NaHS (14 μ*M*/kg) intraperitoneally 24 h prior to surgery. Another group was administered paxilline (2.5 mg/kg) or penitrem A (0.4 μg/kg)—both of which are selective BKca channel inhibitors—10 min prior to sulfide treatment. NS 1619, a selective BKca channel opener, was administered at a dose of 1 mg/kg intraperitoneally 24 h prior to surgery. After completion of 60 min of reperfusion, 20 cm of ileum was harvested. Ileal segments were opened longitudinally and mucosal cells were removed. Harvested cells were checked for mitochondrial function, MPO content, and TNF-α levels. Mitochondrial membrane potential was assessed using 5,5′,6,6′-tetrachloro-1,1′,3,3′-tetraethyl-imidacarbocyanine iodide (JC-1), which exists as a green-fluorescent monomer at low membrane potential, but forms red fluorescent aggregates at higher potentials. Membrane integrity was assessed by translocation of mitochondrial cytochrome *c* to the cytosol. Mitochondrial respiration of intestinal enterocytes (scraped mucosal cells) was assayed by measuring mitochondrial dehydrogenase-dependent reduction of MTT to its formazan derivative. I/R injury was associated with a significant decrease in mitochondrial membrane potential, respiratory activity, and an increase in the cytosolic fraction of cytochrome *c*. All of these effects were mitigated with sulfide treatment, but administration of paxilline or penitrem A abolished this protection. To determine what role neutrophils play in the mitochondrial dysfunction, antineutrophil serum was administered and largely reversed the effects seen with I/R injury on mitochondrial dysfunction. Interestingly, no changes in neutrophil infiltration, MPO activity, or TNF-α levels were seen with sulfide therapy, indicating that mitochondrial protection may be independent of neutrophil infiltration.

7. HEPATOBILIARY SYSTEM

The liver may be subject to ischemia followed by reperfusion injury in the setting of shock, liver resection, or trauma during which time bleeding is controlled with vascular inflow occlusion, and in the setting of liver transplantation.

Kang et al. (2009) examined the effects of sulfide on liver I/R injury after reports of the protective effect of sulfide in the heart. Using a rat model, ischemia was induced by portal vein and hepatic artery occlusion using a vascular clamp for 30 min prior to reperfusion. Sulfide was administered as NaHS intraperitoneally 30 min before ischemia at a dose of 14 μ*M*/kg. PAG was administered as an intraperitoneal injection (50 mg/kg) 30 min before ischemia and 30 min after reperfusion. Rats were sacrificed at 1, 3, and 6 h after the I/R protocol. Liver injury histopathologic scores based on H&E staining were significantly lower with NaHS (by nearly 1/3 compare to controls), but when PAG was administered, histopathologic scores were worse than I/R alone by 20%. ELISA measurements of TNF-α 20% lower in sulfide-treated animals than in controls, whereas PAG treatment resulted in an increase of TNF-α by 20%. Interestingly, levels of the anti-inflammatory cytokine IL-10 were 20% lower in sulfide-treated animals than in controls, and 20% higher in PAG-treated animals.

Based on their previous work, Wang et al. (2012) sought to identify the mechanistic protection afforded by sulfide. As the PI3K–Akt1 pathway controls a variety of cellular processes, including cell survival and proliferation, Wang's group hypothesized this pathway may be involved in the hepatoprotection afforded by sulfide in the setting of liver I/R injury. Using a mouse model and isolated hepatocytes from the mice, models of anoxia/reperfusion or I/R were used. For *in vitro* cellular studies, hepatocytes were pretreated with NaHS in concentrations ranging from 5 to 100 μ*M*. For hepatic I/R injury, the left lateral and median lobes of the liver were clipped with an atraumatic vascular clip for 90 min, followed by reperfusion. NaHS was administered as an intraperitoneal solution of 1.5 mg/kg

1 h prior to the onset of ischemia. LY294002, an Akt blocker, was administered via the tail vein at a dose of 1.5 mg/kg 30 min before NaHS was given. Liver harvest occurred 6 h after reperfusion. Histopathologic scoring of cellular injury on H&E stains demonstrated that NaHS administration decreased injury scores by over 50%, an effect which was partially negated by the addition of LY294002. TNF-α and IL-6 levels as assessed with an ELISA kit were decreased by NaHS, an effect negated by LY294002, demonstrating that the activation of akt1 is key in protection provided by H_2S.

Using a mouse model, Cheng et al. (2014) induced hepatic ischemia by clamp occlusion of the portal vein, bile duct, and hepatic artery to the left and median lobes of the liver for 60 min followed by reperfusion. NaHS was administered intraperitoneally at 14 μM/kg 30 min before I/R injury. Selected mice were killed at 4, 12, and 24 h. TNF-α and IL-6 mRNA were assessed by RT-PCR and increased threefold after I/R, but sulfide treatment reduced this to a 1- to 1.5-fold increase.

Instead of intravenous or intraperitoneal dosing of hydrogen sulfide, Bos et al. (2012) delivered sulfide as a gas to induce hypometabolism in their mouse model of hepatic I/R injury. In this model, the left hepatic artery and portal vein were clamped for 60 min causing ischemia to median and left lateral lobes. This was followed by reperfusion and tissue harvest at 1, 6, and 24 h later. Sulfide was administered as 100 ppm H_2S/17% O_2 mixture. Mice were pretreated for 25 min prior to I/R injury. Core body temperature maintained with a heat lamp. Treatment with inhaled sulfide prevented the normally seen increases in inflammatory cytokines after I/R injury as assessed by RT-PCR for mRNA levels of TNF-α, IL-6, and IL-1β. Staining for Ly-6G-positive granulocytes demonstrated a large influx of these granulocytes in the liver after I/R injury, an influx which was prevented by sulfide. Ly-6G granulocyte levels were over 10-fold higher in controls at 24 h.

8. RENAL SYSTEM

The kidney is exquisitely sensitive to hypoperfusion and ischemia, with even short periods able to induce kidney injury in the form of acute tubular necrosis. Prolonged end-organ ischemia can be seen in the setting of thromboembolic disease, renal arterial involvement in aortic dissection, trauma, and kidney transplantation.

Similar to studies examining the protective effect of gaseous sulfide on hepatic I/R injury, Bos et al. (2009) investigated the protective effect of

H_2S gas on renal I/R injury. Using a mouse model, bilateral renal ischemia was achieved with renal artery vascular clamping for 30 min followed by reperfusion for 1–3 days. To separate the effects of H_2S from the known protective effects of hypothermia, they maintained core body temperature at 37 °C during and after the procedure. Hypometabolism was confirmed by recording a lower respiratory rate and measuring CO_2 production using a closed ventilator loop. Pretreatment consisted of giving H_2S gas at 100 ppm/17% O_2 for 30 min prior to ischemia, followed by 30 min of reperfusion. Posttreatment consisted of administering H_2S gas 5 min before the onset of reperfusion and continuing it for 30 more minutes for 35 min total of treatment. When pretreated, CO_2 production decreased by 40%, respiratory rate by 60%, and renal O_2 consumption by over 50%. When sulfide was given prior to ischemia, whether pre-, or pre- and post-, 100% animal survival occurred at 72 h, whereas the other groups, including post sulfide alone and controls, had a 25% survival at the 72-h timepoint. Serum creatinine 24 h after surgery increased 3-fold with sulfide pretreatment, 15-fold in control animals, and 13-fold in posttreatment animals. Immunohistochemical staining for Mac-1 (CD11b marker on macrophages, monocytes, granulocytes, and natural killer cells) and Ly-6G (expressed on mature granulocytes to assess inflammation) was markedly reduced if pretreatment with sulfide was given.

Hunter et al. (2012) utilized a swine in a large animal model of renal I/R injury. The left renal pedicle was clamped for 60 min of warm ischemia, followed by reperfusion and right nephrectomy. Sodium sulfide was administered as a 100 μg/kg bolus 10 min before reperfusion, followed by a 1 mg/kg continuously for 30 min after reperfusion. ELISA assays for TNF-α demonstrated a reduction in serum TNF-α by 40% at the 6-h time point, but no other time points. Neutrophilic infiltration based on histology demonstrated a 2/3 decrease on renal biopsies obtained at the 30 min timepoint after reperfusion, but no other timepoints. Serum creatinine was lower on days 1–5 in sulfide-treated swine, but was equivalent between sulfide-treated and -untreated swine by days 6 and 7.

To investigate a role for sulfide in transplant preservative solutions, Lobb et al. (2012) utilized a rat model of kidney transplant. Syngenic Lewis rats were transplanted to eliminate confounding effects of immunosuppression. Donor kidneys were flushed with 25 mL of cold, 4 °C University of Wisconsin (UW) preservation solution, or UW + 150 μ*M* NaHS until effluent was clear, and then placed in 50 cc of same solution and stored at 4 °C for 24 h. Recipients underwent bilateral nephrectomy followed by implant. Survival was markedly different between groups. There were no survivors after 5 days in UW group, but up to 80% survival at 14 days in sulfide group. MPO staining demonstrated a 50% decrease in neutrophilic inflammation on days 3–5. There were no significant differences in TNF-α levels based on quantitative PCR.

In a subsequent study by the same group, Lobb et al. (2014) used a rat model to investigate warm I/R injury. Rats underwent right nephrectomy, followed by atraumatic clamping of the left renal pedicle for 60 min, followed by reperfusion. Sulfide was administered as a 150 μM NaHS solution intraperitoneally, after which the abdomen was closed. Rats were monitored for 7 days with blood sampling at days 3 and 7. There were no significant differences in serum creatinine between treatment and control groups until day 7, at which time Cr was about 30% lower in the sulfide-treated group. No differences in necrosis or apoptosis were visualized on H&E staining or TUNEL staining, respectively. Renal expression of proinflammatory and apoptotic markers via qRT-PCR on days 1 and 7 examining TLR-4, TNF-α, and ICAM-1, and IL-2 did not demonstrate significant differences at 7 days.

Simon et al. (2011) utilized a porcine model of aortic occlusion-induced kidney I/R injury to investigate the effects of sulfide on inflammation. An intra-aortic balloon was utilized to induce ischemia for 90 min, followed by 8 h of reperfusion. Na_2S. Sulfide was administered intravenously as an Na_2S bolus of 0.2 mg/kg followed by a continuous infusion of 2 mg/kg/h during 2 h before aortic occlusion, 0.5 mg/kg/h during 90 min of aortic occlusion, and 1 mg/kg/h during an 8-h reperfusion period. Inflammatory cytokines were assessed using an ELISA kit and demonstrated decreased levels of IL-6 and IL-1β at 2 and 8 h of reperfusion. No changes in TNF-α levels were observed.

9. MUSCULOSKELETAL

Skeletal muscle is subject to ischemia in both pathologic conditions and iatrogenically in the setting of tissue transfer. The most common clinical scenario encountered involves limb ischemia in the setting of thrombotic or embolic occlusion in the vascular tree resulting in impaired arterial perfusion of the limb. Extensive deep venous thrombosis may result in impaired tissue perfusion from impairing outflow. In the setting of reconstructive surgery, free muscle flap tissue transfer subjects the muscle flap to a period of ischemia as well. As the nascent field of limb transplantation expands, I/R injury will no doubt play some role here as well.

Two studies have examined the effects of hydrogen sulfide on inflammation in the setting of I/R injury in skeletal muscle. As L-selectin-mediated adhesion and cellular signaling are responsible for leukocyte localization after injury, Ball et al. (2013) examined the effects of hydrogen sulfide on L-selectin expression on *in vitro* neutrophils and neutrophil infiltration into tissues in a murine model of I/R injury. Utilizing 8-week-old mice, hind-limb ischemia was achieved via tourniquet application proximal to the greater trochanter of the right hind-limb, with the left limb serving as a control. Ischemia was maintained for 3 h, followed by 3 h of reperfusion. The bilateral gastrocnemius muscles were then harvested. Sulfide was administered as NaHS (20 m*M* solution) via the tail vein with a goal bloodstream concentration of 20 μ*M*. Controls were administered PBS. H&E staining was performed on gastrocnemius sections to assess for neutrophil infiltration. No differences were seen if NaHS was given 20 min before ischemia or 20 min before reperfusion. NaHS resulted in a fivefold decrease in fraction of PMN/total cells observed on histologic analysis.

As the field of reconstructive surgery expands, muscle-free flaps, essentially auto-transplanting a muscle group for the purposes of reconstruction with subsequent revascularization is increasing. These free flaps are necessarily subjected to a period of ischemia followed by reperfusion after vascular anastomoses take place. Villamaria and colleagues (Villamaria et al., 2014) have investigated the ability of H_2S to mitigate I/R injury in a porcine model of free flaps. In this model, gracilis free muscle flaps weighing between 200 and 400 g were harvested and then subjected to 3 h of cold ischemia at 4 °C. In the treatment arm, flaps were infused via the artery with H_2S solution (1 cc of 11.6 mg/mL solution per 400 g of flap). Controls were given heparinized saline. After 3 h of ischemia, flaps were anastomosed in the neck to the right external carotid artery and right internal jugular vein. Serial assessments were made for TNF-α and IL-6 on POD 1, 2, 7, and 14. IL-6 levels remains similar between groups (trended toward but did not reach significance), whereas TNF-α levels were 68 pg/mL in control, 30 pg/mL in the treatment group on POD 1, and 93 pg/mL versus 39 pg/mL on POD 7. No difference in flap failure rates was observed.

10. SUMMARY

Sulfide has emerged as a promising therapeutic agent for lessening the degree of injury after ischemia and reperfusion. The majority of studies published to date have consistently demonstrated that administration of sulfide prior to ischemia or prior to reperfusion results in reduced inflammation, as evidenced by decreased inflammatory cell infiltration and reduced levels of proinflammatory cytokines. These anti-inflammatory effects are associated with less necrotic and apoptotic cell death, and improved function. While the data supporting the therapeutic efficacy of sulfide in mitigating I/R injury in small and large animal models are robust, there remains a limited mechanistic understanding of why these protective effects occur. To date, it seems that a significant degree of protection provided by sulfide is related to BK channel signaling as agents which block these channels have abolished the protective effects of sulfide. Beyond investigation into mechanistic signaling, additional studies into dosing are needed as routes of administration and doses vary widely between publications. Despite the current limitations, sulfide offers promise as a solution to a commonly encountered clinical problem affecting multiple organ systems.

REFERENCES

Ball, C. J., Reiffel, A. J., Chintalapani, S., Kim, M., Spector, J. A., & King, M. R. (2013). Hydrogen sulfide reduces neutrophil recruitment in hind-limb ischemia–reperfusion injury in an L-selectin and ADAM-17-dependent manner. *Plastic and Reconstructive Surgery*, *131*(3), 487–497.

Bos, E. M., Leuvenink, H. G., Snijder, P. M., Kloosterhuis, N. J., Hillebrands, J. L., Leemans, J. C., et al. (2009). Hydrogen sulfide-induced hypometabolism prevents renal ischemia/reperfusion injury. *Journal of the American Society of Nephrology*, *20*(9), 1901–1905.

Bos, E. M., Snijder, P. M., Jekel, H., Weij, M., Leemans, J. C., van Dijk, M. C., et al. (2012). Beneficial effects of gaseous hydrogen sulfide in hepatic ischemia/reperfusion injury. *Transplant International*, *25*(8), 897–908.

Cheng, P., Wang, F., Chen, K., Shen, M., Dai, W., Xu, L., et al. (2014). Hydrogen sulfide ameliorates ischemia/reperfusion-induced hepatitis by inhibiting apoptosis and autophagy pathways. *Mediators of Inflammation*, *2014*, 935251.

Elrod, J. W., Calvert, J. W., Morrison, J., Doeller, J. E., Kraus, D. W., Tao, L., et al. (2007). Hydrogen sulfide attenuates myocardial ischemia–reperfusion injury by preservation of mitochondrial function. *Proceedings of the National Academy of Sciences of the United States of America*, *104*(39), 15560–15565.

Guo, C., Liang, F., Shah Masood, W., & Yan, X. (2014). Hydrogen sulfide protected gastric epithelial cell from ischemia/reperfusion injury by Keap1 s-sulfhydration, MAPK dependent anti-apoptosis and NF-kappaB dependent anti-inflammation pathway. *European Journal of Pharmacology*, *725*, 70–78.

Hunter, J. P., Hosgood, S. A., Patel, M., Rose, R., Read, K., & Nicholson, M. L. (2012). Effects of hydrogen sulphide in an experimental model of renal ischaemia–reperfusion injury. *British Journal of Surgery*, *99*(12), 1665–1671.

Kang, K., Zhao, M., Jiang, H., Tan, G., Pan, S., & Sun, X. (2009). Role of hydrogen sulfide in hepatic ischemia–reperfusion-induced injury in rats. *Liver Transplantation*, *15*(10), 1306–1314.

Liu, Y., Kalogeris, T., Wang, M., Zuidema, M. Y., Wang, Q., Dai, H., et al. (2012). Hydrogen sulfide preconditioning or neutrophil depletion attenuates ischemia–reperfusion-induced mitochondrial dysfunction in rat small intestine. *American Journal of Physiology. Gastrointestinal and Liver Physiology*, *302*(1), G44–G54.

Lobb, I., Mok, A., Lan, Z., Liu, W., Garcia, B., & Sener, A. (2012). Supplemental hydrogen sulphide protects transplant kidney function and prolongs recipient survival after prolonged cold ischaemia–reperfusion injury by mitigating renal graft apoptosis and inflammation. *BJU International*, *110*(11 Pt C), E1187–E1195.

Lobb, I., Zhu, J., Liu, W., Haig, A., Lan, Z., & Sener, A. (2014). Hydrogen sulfide treatment ameliorates long-term renal dysfunction resulting from prolonged warm renal ischemia–reperfusion injury. *Canadian Urological Association Journal*, *8*(5–6), E413–E418.

Mard, S. A., Neisi, N., Solgi, G., Hassanpour, M., Darbor, M., & Maleki, M. (2012). Gastroprotective effect of NaHS against mucosal lesions induced by ischemia–reperfusion injury in rat. *Digestive Diseases and Sciences*, *57*(6), 1496–1503.

Qi, Q. Y., Chen, W., Li, X. L., Wang, Y. W., & Xie, X. H. (2014). H2S protecting against lung injury following limb ischemia–reperfusion by alleviating inflammation and water transport abnormality in rats. *Biomedical and Environmental Sciences*, *27*(6), 410–418.

Simon, F., Scheuerle, A., Groger, M., Stahl, B., Wachter, U., Vogt, J., et al. (2011). Effects of intravenous sulfide during porcine aortic occlusion-induced kidney ischemia/reperfusion injury. *Shock*, *35*(2), 156–163.

Sivarajah, A., Collino, M., Yasin, M., Benetti, E., Gallicchio, M., Mazzon, E., et al. (2009). Anti-apoptotic and anti-inflammatory effects of hydrogen sulfide in a rat model of regional myocardial I/R. *Shock*, *31*(3), 267–274.

Snijder, P. M., de Boer, R. A., Bos, E. M., van den Born, J. C., Ruifrok, W. P., Vreeswijk-Baudoin, I., et al. (2013). Gaseous hydrogen sulfide protects against myocardial ischemia–reperfusion injury in mice partially independent from hypometabolism. *PLoS One*, *8*(5), e63291.

Sodha, N. R., Clements, R. T., Feng, J., Liu, Y., Bianchi, C., Horvath, E. M., et al. (2009). Hydrogen sulfide therapy attenuates the inflammatory response in a porcine model of myocardial ischemia/reperfusion injury. *Journal of Thoracic and Cardiovascular Surgery*, *138*(4), 977–984.

Villamaria, C. Y., Fries, C. A., Spencer, J. R., Roth, M., & Davis, M. R. (2014). Hydrogen Sulfide mitigates reperfusion injury in a porcine model of vascularized composite autotransplantation. *Annals of Plastic Surgery*, *72*(5), 594–599.

Wang, Y., Jia, J., Ao, G., Hu, L., Liu, H., Xiao, Y., et al. (2014). Hydrogen sulfide protects blood–brain barrier integrity following cerebral ischemia. *Journal of Neurochemistry*, *129*(5), 827–838.

Wang, D., Ma, Y., Li, Z., Kang, K., Sun, X., Pan, S., et al. (2012). The role of AKT1 and autophagy in the protective effect of hydrogen sulphide against hepatic ischemia/reperfusion injury in mice. *Autophagy*, *8*(6), 954–962.

Wu, J., Wei, J., You, X., Chen, X., Zhu, H., Zhu, X., et al. (2013). Inhibition of hydrogen sulfide generation contributes to lung injury after experimental orthotopic lung transplantation. *Journal of Surgical Research*, *182*(1), e25–e33.

Yin, J., Tu, C., Zhao, J., Ou, D., Chen, G., Liu, Y., et al. (2013). Exogenous hydrogen sulfide protects against global cerebral ischemia/reperfusion injury via its anti-oxidative, anti-inflammatory and anti-apoptotic effects in rats. *Brain Research*, *1491*, 188–196.

Zuidema, M. Y., Yang, Y., Wang, M., Kalogeris, T., Liu, Y., Meininger, C. J., et al. (2010). Antecedent hydrogen sulfide elicits an anti-inflammatory phenotype in postischemic murine small intestine: Role of BK channels. *American Journal of Physiology. Heart and Circulatory Physiology*, *299*(5), H1554–H1567.

CHAPTER EIGHT

CD47-Dependent Regulation of H_2S Biosynthesis and Signaling in T Cells

Sukhbir Kaur, Anthony L. Schwartz, Thomas W. Miller[1], David D. Roberts[2]
Laboratory of Pathology, Center for Cancer Research, National Cancer Institute, National Institutes of Health, Bethesda, Maryland, USA
[2]Corresponding author: e-mail address: droberts@helix.nih.gov

Contents

[1] Current address: Paradigm Shift Therapeutics, Rockville, Maryland, USA

Methods in Enzymology, Volume 555
ISSN 0076-6879
http://dx.doi.org/10.1016/bs.mie.2014.11.023

Abstract

Pharmacological concentrations of H_2S donors inhibit some T cell functions by inhibiting mitochondrial function, but evidence is also emerging that H_2S at physiological concentrations produced via chemical sources and endogenously is a positive physiological mediator of T cell function. Expression of the H_2S biosynthetic enzymes cystathionine γ-lyase (CSE) and cystathionine β-synthase (CBS) is induced in response to T cell receptor signaling. Inhibiting the induction of these enzymes limits T cell activation and proliferation, which can be overcome by exposure to exogenous H_2S at submicromolar concentrations. Exogenous H_2S at physiological concentrations increases the ability of T cells to form an immunological synapse by altering cytoskeletal actin dynamics and increasing the reorientation of the microtubule-organizing center. Downstream, H_2S enhances T cell receptor-dependent induction of *CD69*, *CD25*, and *Interleukin-2 (IL-2)* gene expression. The T cell stimulatory activity of H_2S is enhanced under hypoxic conditions that limit its oxidative metabolism by mitochondrial and nonenzymatic processes. Studies of the receptor CD47 have revealed the first endogenous inhibitory signaling pathway that regulates H_2S signaling in T cells. Binding of the secreted protein thrombospondin-1 to CD47 elicits signals that block the stimulatory activity of exogenous H_2S on T cell activation and limit the induction of *CSE* and *CBS* gene expression. CD47 signaling thereby inhibits T cell receptor-mediated T cell activation.

1. INTRODUCTION

Before one can address how physiological levels of H_2S regulate the function of circulating T cells and T cells at sites of inflammation, the concentrations of H_2S that are produced by T cells and encountered in their local microenvironment while in circulation need to be determined. While we will survey the relevant recent literature, the reader is referred to an excellent earlier review on this subject (Olson, 2009). Early estimates of the physiological concentrations of H_2S in circulation ranged from 10 to 100 μM in healthy animals and humans (Hongfang et al., 2006; Hyspler et al., 2002; Richardson, Magee, & Cummings, 2000). Consistent with this range, the H_2S levels produced by human peripheral blood lymphocytes have been determined to be 11.6 ± 6.4 μmol/min/mg protein (Barathi, Vadhana, Angayarkanni, & Ramakrishnan, 2007). In contrast, Furne et al. found only low nanomolar concentrations of H_2S in mouse brain and liver tissue by analysis of the gas head space using chromatographic separation with electrochemical detection (Furne, Saeed, & Levitt, 2008), which is consistent with HPLC analysis of derivatized H_2S showing that plasma sulfide is below 0.55 μ*M* (Olson, 2009; Sparatore et al., 2009). More recent studies have confirmed low or submicromolar levels of H_2S, and it is clear that the rapid catabolism of H_2S and the limited availability of sulfur-containing amino acids in the diet make >10 μ*M* levels of H_2S unsustainable physiologically. More recent measurements based on derivatization with monobromobimane showed approximately 1.7 μ*M* H_2S circulating in plasma of wild-type (WT) C57Bl/6 mice versus 0.3 μ*M* in the plasma of cystathionine-γ-lyase (CSE)-null mice (Shen et al., 2011). In the plasma of healthy humans, the mean free H_2S concentration was determined to be 370 n*M*, although elevations to 440–510 n*M* were found in some vascular disease states (Peter et al., 2013).

In addition to the pH-dependent equilibrium between free H_2S and HS^-, it is becoming clear that H_2S can interconvert with other sulfur metabolites that can both buffer the circulating H_2S levels and mediate its physiological functions in tissues. Methods have recently been developed to assess free, acid-labile, and bound sulfane sulfur (Shen, Peter, Bir, Wang, & Kevil, 2012). Acid-labile sulfur includes that which is contained in iron-sulfur clusters, and sulfane sulfur includes H_2S oxidized to thiosulfate and that which is bound to protein and glutathione (GSH) thiols in the form of persulfides and polysulfides. Using these methods, the levels of free H_2S and bound sulfane sulfur circulating in human and mouse plasma were consistently less than 1 μ*M*, whereas acid-labile sulfur was in the low micromolar range. The acid-labile pool is thought to provide a buffer for free H_2S levels. Recent studies are indicating that persulfides and polysulfides in the bound sulfane pool play major roles in H_2S signaling (Greiner et al., 2013; Ida et al., 2014).

Although studies of the direct effects of H_2S on T cells were only recently initiated, several early observations suggested that H_2S regulates T cell function. Addition of L-cysteine, 2-mercaptoethanol, α-thioglycerol, and other thiol compounds to cell culture media enhanced the proliferation of peripheral blood lymphocytes (Fanger, Hart, Wells, & Nisonoff, 1970; Iwata et al., 1994; Noelle, 1980) and murine lymphoma cells, as well as increased the activation of mouse splenic lymphocytes (Broome & Jeng, 1973). Irrigation of inflamed tissues with sulfurous water, containing dissolved H_2S, has long history of use in traditional medicine and was recently confirmed in a prospective trial to be beneficial for treating sinusitis (Ottaviano et al., 2011). Sulfurous water was reported not to affect peripheral blood mononuclear cells but specifically inhibited T cell activation and proliferation of memory T cells. Reduction in secretion of interleukin-2 suggested that H_2S inhibits T cell activation (Valitutti, Castellino, & Musiani, 1990).

2. REGULATION OF H_2S BIOSYNTHESIS IN T CELLS

2.1. Biosynthesis of H_2S in normal T cells

H_2S is synthesized in mammalian cells by three enzymes: cystathionine-β-synthase (CBS), CSE, and 3-mercaptopyruvate sulfurtransferase (MST) (Chen, Jhee, & Kruger, 2004; Eto & Kimura, 2002; Hughes, Centelles, & Moore, 2009; Julian, Statile, Wohlgemuth, & Arp, 2002; Kabil & Banerjee, 2014; Levonen, Lapatto, Saksela, & Raivio, 2000; Yusuf et al., 2005). Because the behavioral phenotype of *Mst*-null mice only supports a central nervous system function of this enzyme (Nagahara et al., 2013), the present discussion of T cell biology will only consider CBS and CSE. In addition to their role in H_2S biosynthesis, CBS and CSE are required for GSH biosynthesis and have been studied in T cells mostly in the latter context. Early studies found that rat splenic lymphocytes do not produce detectable levels of CSE or CBS (Brodie, Potter, & Reed, 1982; Glode, Epstein, & Smith, 1981). Depletion of GSH in these cells resulted in elevated utilization of cysteine sulfur rather than cystine for GSH synthesis (Brodie et al., 1982). Human naïve T cells were also shown to lack detectable levels of CBS and CSE enzymes; however, this pathway was intact in naïve and malignant B-cell lines (Kamatani & Carson, 1982).

Activation of $CD4^+$ T cells using anti-CD3 plus anti-CD28 antibodies resulted in detectable CBS mRNA expression at 48 h, but mRNA expression was not affected by Th1- or Th2-polarizing cytokines (Lund, Aittokallio, Nevalainen, & Lahesmaa, 2003). This suggested that H_2S biosynthesis might be induced late in T cell activation. As discussed below, other cells that interact closely with T cells express H_2S biosynthetic enzymes, suggesting that paracrine exposure to H_2S also plays a significant regulatory role in T cells (Fig. 1).

2.2. Biosynthesis of H_2S by lymphoid malignancies

Increased cerebrospinal fluid levels of H_2S were reported in children with central nervous system acute lymphocytic leukemia (Du et al., 2011). Isolated leukemic cells from these patients expressed CBS and CSE mRNAs, but expression levels in control leukocytes were not reported. Polymorphisms in CBS have been associated with risk for developing non-Hodgkins lymphoma (Li et al., 2013; Metayer et al., 2011; Weiner et al., 2011). However, it remains unclear whether this risk relates to altered H_2S biosynthesis versus other

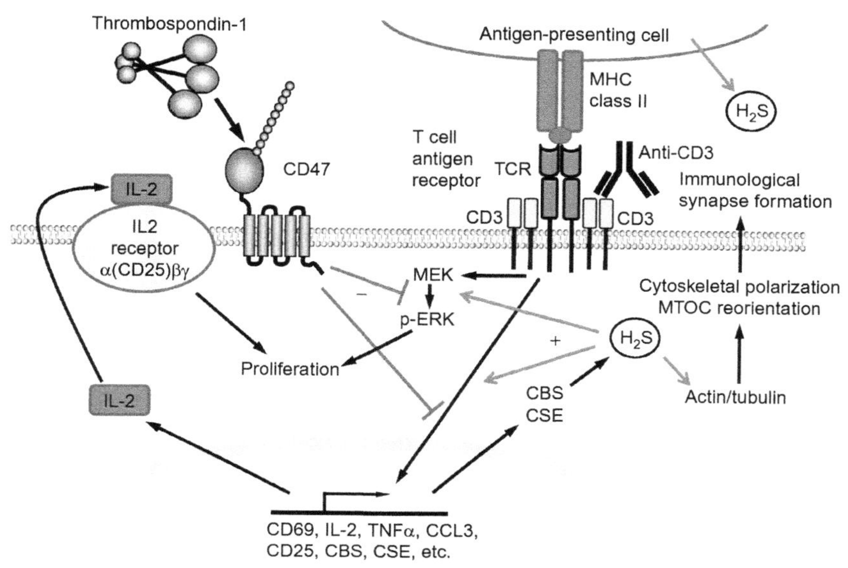

Figure 1 Enhancement of T cell activation by H_2S and inhibition by thrombospondin-1/CD47 signaling. T cell antigen receptor signaling is initiated by recognition of a specific peptide antigen in the context of MHC on an antigen-presenting cell. Optimal signaling to induce expression of T cell activation genes and proliferation requires expression of the H_2S biosynthetic enzymes CBS and CSE, but exogenous H_2S, probably in part from the APC, also enhances TCR signaling to activate the ERK MAP kinase pathway, modify actin and tubulin to induce reorientation of the microtubule-organizing center (MTOC) to facilitate formation of an immunological synapse with the antigen-presenting cell and induction of T cell activation genes. The latter include IL-2 that drives T cell proliferation and CD25 that encodes the α-subunit of the IL-2 receptor. Thrombospondin-1 signaling through its receptor CD47 inhibits these responses of T cells to exogenous H_2S and inhibits the induction of CBS and CSE during T cell activation. (See the color plate.)

metabolic functions of CBS. Although specific studies in T cell malignancies are lacking, increased expression of H_2S biosynthetic enzymes, usually CBS, has been proposed to be a general characteristic of cancers and has been well documented in colon and ovarian cancers (Hellmich, Coletta, Chao, & Szabo, 2014). Knockdown of CBS expression slowed the growth of those tumors.

3. CATABOLISM OF H_2S

The primary catabolic pathway for H_2S in mammals is in mitochondria, where H_2S is oxidized to thiosulfate and sulfate (Kabil & Banerjee, 2014). Oxidation can be coupled to the electron transport chain via ubiquinone and involves initial oxidation mediated by the FAD-dependent sulfide quinone oxidoreductase (SQRDL) followed by oxygen-dependent conversion to sulfite by persulfide dioxygenase (ETHE1). In the presence of adequate levels of the ETHE1 substrate oxygen, this pathway rapidly depletes H_2S and sulfane H_2S in the form of GSH persulfide. Consequently, *Ethe1*-null mice exhibit a 20-fold increase in H_2S levels (Tiranti et al., 2009), and inhibition of this pathway under hypoxic conditions correspondingly stabilizes H_2S in the presence of cells and enhances the amplitude and duration of T cell signaling responses (Miller et al., 2012).

Limited data exist concerning the expression of H_2S catabolic enzymes in T cells and B cells. Activation of $CD4^+$ T cells using anti-CD3 plus anti-CD28 antibodies resulted in late induction of ETHE1 mRNA expression at 48 h, but expression was not affected by Th1- or Th2-polarizing cytokines (Lund et al., 2003). Expression of ETHE1 mRNA was induced in Jurkat T cells treated with the toxin azaspiracid-1, but SQRDL was unchanged (Twiner et al., 2008). ETHE1 and SQRDL mRNA expression was elevated in a variety of lymphomas relative to expression in naïve lymphocytes and plasma cells (Brune et al., 2008). Altered expression of the biosynthetic enzyme CBS was not associated with malignancy in this data set.

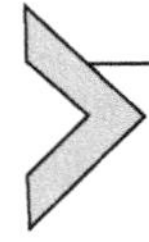

4. REGULATION OF T CELL ACTIVATION BY H_2S SIGNALING

4.1. Biphasic effects of H_2S on T cells

Different studies have reported H_2S to be an activator or inhibitor of T cell viability and function, which leads to some confusion concerning whether H_2S is a pro- or anti-inflammatory mediator in T cells or in the broader immune system. Some studies have showed that H_2S is proinflammatory using different proinflammatory models such as acute pancreatitis and associated lung injury (Bhatia et al., 2005), caerulein-induced pancreatitis (Ang, Rivers-Auty, Hegde, Ishii, & Bhatia, 2013), and burn injury (Zhang, Sio, Moochhala, & Bhatia, 2010), while others have reported anti-inflammatory effects of H_2S on caerulein-induced acute pancreatitis (Sidhapuriwala, Ng, & Bhatia, 2009), endotoxin-induced systemic inflammation (Tokuda et al., 2012), inflammation and ulceration of the colon (Wallace, Vong, McKnight, Dicay, & Martin, 2009), and endotoxic shock in the rat (Li, Salto-Tellez, Tan, Whiteman, & Moore, 2009). This apparent contradiction may be resolved in part by considering the concentrations of H_2S employed in each study. Like nitric oxide, H_2S can act on different primary cellular targets with different dose-dependencies. Unlike the well-studied nitric oxide signaling mediators, the identities of the physiological sensors that mediate H_2S signaling at each concentration range are less clear, and it is unlikely that a single high-affinity H_2S sensor will be identified that would be equivalent to the soluble guanylate cyclase sensor for nitric oxide in mammalian cells. Sulfhydration of specific reactive thiols in proteins is currently the leading candidate to mediate H_2S signaling at nanomolar to low micromolar concentrations (Paul & Snyder, 2012, 2014). Identification of physiologically relevant sulfhydration targets is an emerging field, but the specific targets that sense H_2S in T cells remain to be defined. At hundred micromolar to low millimolar concentrations, inhibition of mitochondrial metabolism becomes the dominant effect of H_2S (Wedmann et al., 2014). However, current assessments of H_2S concentrations indicate that physiological signaling occurs at submicromolar levels. Thus, our current understanding is that stimulatory and inhibitory signaling mediated by specific sulfhydration targets can occur at physiological concentrations of H_2S, but metabolic inhibition becomes the dominant mechanism of action at higher pharmacological concentrations.

H_2S can diffuse across the cell membrane and induces cell proliferation and apoptosis (Baskar & Bian, 2011; Li, Rose, & Moore, 2011). The inflammation and stress signaling response involves the activation of phospho-ERK (pERK), which phosphorylates Eukaryotic Initiation Factor 2 (eIF2α) and inhibits global translation in eukaryotes. Recently, H_2S was implicated in regulation of the endoplasmic reticulum stress response via sulfydration (Krishnan, Fu, Pappin, & Tonks, 2011). pERK is activated by tyrosine phosphorylation, and this ERK activation is downregulated via removal of the phosphate group by PTP1B. The sulfhydration of PTP1B leads to stimulation of pERK activity and suppression of protein translation. Thus, H_2S permits stressed cells to turn off protein synthesis and enhance damage repair (Dickhout et al., 2012; Harding et al., 2003). Another *in vitro* study using a bactericidal activity assay showed that H_2S alters the cell cycle of granulocytes and prevents apoptosis by inhibiting p38 MAP kinase and caspase-3 (Rinaldi et al., 2006). However, lymphocytes and eosinophils were not affected.

We reported that 300 n*M* H_2S enhances ERK 1 and ERK 2 phosphorylation in a time-dependent manner in Jurkat T cells (Miller, Kaur, Ivins-O'Keefe, & Roberts, 2013). H_2S-induced ERK activation and T cell activation were prevented in the presence of the MEK inhibitor PD184161. This suggests that H_2S induces T cell activation in part via the MEK-ERK MAP kinase pathway. Thrombospondin-1 (TSP1) also inhibited H_2S-induced ERK phosphorylation. This was consistent with prior studies showing that TSP1 inhibited ERK phosphorylation induced by nitric oxide in endothelial cells (Ridnour et al., 2005). We also found that H_2S-induced ERK activation was perturbed by TSP1 in T cells derived from WT mice but not T cells from mice lacking CD47. Therefore, CD47 is necessary for this inhibitory TSP1 signal.

In contrast to the stimulatory activity of 300 nM H_2S on Jurkat cells, a H_2S-releasing (dithiolethione) aspirin derivative inhibited the growth of Jurkat T lymphoma cells with an $IC_{50}=1.9\ \mu M$ (Chattopadhyay et al., 2013). β-Catenin protein levels were decreased, and downstream expression of cyclin D1 and cMyc were decreased. Mechanistic studies in other cell types indicated that dithiolethiones regulate Hsp27 phosphorylation, activate protein phosphatase 2A (PP2A), and covalently modify the p50 and p65 subunits of NF-κB to inhibit DNA binding (Isenberg et al., 2007; Switzer et al., 2012, 2009). Activation of PP2A was shown to mediate downstream inhibition of Akt signaling in breast and lung cancer cell lines. However, it is unclear that the activity of the above dithiolethione in T lymphoma cells is mediated by H_2S or is predictive of the physiological regulation of the same processes by H_2S.

The Akt/GSK3β pathway is an important mechanism for regulation of cell growth, proliferation, and viability of lymphocytes. Akt activation is controlled by phosphorylation of Ser^{473}, regulating its downstream target GSK3β, which induces apoptosis of lymphocytes. Systemic lupus erythematosus (SLE) and SLE with renal disease is an autonomous immune disease associated with increased lymphocyte activation and cell proliferation. Phosphorylation of Akt at Ser^{473} and the Akt target GSK3β at Ser^{9} was enhanced when lymphocytes from SLE patients were treated with the mitogenic lectin phytohemagglutinin (PHA). Pretreatment with H_2S attenuated Akt phosphorylation at Ser^{473} and increased phosphatase and tensin homolog (PTEN). This is consistent with the above-reported effects of dithiolethiones on Akt activation and suggests that H_2S induces stabilization of SLE lymphocytes. NaHS treatment in the absence of PHA attenuated Akt (Ser^{473}) phosphorylation but was unable to alter GSK3β (Ser^{9}) phosphorylation (Han et al., 2013).

H_2S at 25–200 μM dose-dependently overcame the inhibitory effects of cobalt chloride ($CoCl_2$), a hypoxia mimic that stabilizes HIFα (Yuan, Hilliard, Ferguson, & Millhorn, 2003), on the viability of peripheral blood lymphocytes or CEM, an acute lymphoblastic leukemia cell line with $CD4^+$ T lymphoblast features (Bruzzese et al., 2014). At 25–50 μM, H_2S inhibited the induction by $CoCl_2$ of intranuclear NF-κB p65 subunit and cellular HIF1α levels, implicating these transcriptional regulators in the anti-inflammatory activity of H_2S. The same concentrations of H_2S enhanced adenosine production in CEM cells but decreased adenosine deaminase activity, adenosine receptor $A_{2A}R$ expression, and cAMP production. Because cAMP is an inhibitor of T cell activation (Bjorgo, Moltu, & Tasken, 2011), this result predicts that H_2S would enhance T cell activation under these pseudohypoxic conditions, but that hypothesis was not tested.

Several studies have identified inhibitory effects on T cells associated with higher H_2S concentrations. A clinical case report found decreased numbers of blood cells, lymphocytes, and neutrophils associated with chronic exposure to toxic levels of H_2S as compared to healthy controls (Saadat & Bahaoddini, 2004). H_2S at 0.20–4 m*M* induced T cell death and reduction in IL-2 production via loss of mitochondrial potential within 4 h. The toxic effects of H_2S were primarily observed in $CD8^+$ T cells, $CD4^+$ T cells, and $CD16^+$ NK cells (Mirandola et al., 2007). Treatment of Jurkat T cells with 5 m*M* NaHS at pH 6.0 induced cell blebbing and apoptosis (Kanno et al., 2013). The cell blebbing of NaHS-treated cells was suppressed by cytochalasin D but not by Nocodazole, suggesting that actin filaments but not microtubule structures are involved in the process of cell blebbing. Treatment with 5 m*M* of NaHS also resulted in cleavage of Rock-1 and -2 to generate truncated forms during apoptosis. However, Jurkat cells treated at 1 m*M* NaHS did not result in cell toxicity.

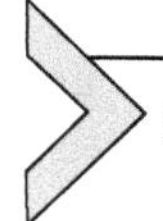

5. AUTOCRINE AND PARACRINE ROLES OF H_2S IN T CELL ACTIVATION

Antigen stimulation of T lymphocytes induces the T cell antigen receptor (TCR) signaling cascade that induces tyrosine and serine phosphorylation of various proteins, leading to the ordered transcription of genes required for T cell proliferation and effector function (Fig. 1). This signaling cascade leads to increases in DNA synthesis and cell proliferation. When rat spleen lymphocytes, murine L1210 lymphoma, and HeLa cells were cultured in GSH-depleted media using diethyl maleate and allowed to utilize ^{35}S methionine, ^{35}S-cystine, or ^{35}S-cysteine for GSH synthesis. The rat spleen lymphocytes exhibited a 30-fold increase in the uptake of ^{35}S cysteine as compared to ^{35}S cystine for GSH synthesis. However, methionine was not incorporated into GSH by the same lymphocytes (Brodie et al., 1982). Human B lymphoblastoid cell lines grew in cystine-deficient medium, but T lymphoblastoid cell lines failed to utilize homocysteine thiolactone or cystathionine for their growth. This suggested that B and T lymphoblastoid cell lines have differences in H_2S biosynthesis and metabolism (Kamatani & Carson, 1982).

Activation of macrophages with lipopolysaccharide or TNFα enhanced the release of thiols, predominantly cysteine, into the culture medium. Dendritic cells also release GSH that is converted to cysteine (Yan, 2009). Thus, antigen-presenting cells (APCs) supply substrates for H_2S biosynthesis in the local microenvironment during antigen presentation to T cells (Fig. 1). Treatment of lymphocytes with exogenous cysteine or syngenic macrophages enhanced their GSH concentration, viability, and DNA synthesis (Gmunder, Eck, Benninghoff, Roth, & Droge, 1990). Conversely, depletion of L-cystine and GSH in the medium leads to growth arrest at late G1 to S phase and regulates adult T cell leukemia-derived factor (ADF) function via redox control in PBMC (Iwata et al., 1994). *N*-acetyl-L-cysteine (NAC), potentially by serving as a precursor for GSH biosynthesis, showed antitumor activity by increasing TNFα and its TNF receptor in stimulated T cells. NAC increased IL-2 production and expression of CD25 and increased T cell proliferation. Consequently, NAC has been proposed to treat tumors by increasing the proliferation of cytotoxic T cells (Delneste, Jeannin, Potier, Romero, & Bonnefoy, 1997). ADF and thioredoxin may act as autocrine growth factors and are implicated in the regulation of NF-κB, AP-1, and glucocorticoid receptor.

We have investigated the physiological role of exogenous and endogenous H_2S in T cell activation (Miller et al., 2012). Treatment of primary murine and Jurkat T cells with 300 n*M* NaHS in an atmosphere containing 1% O_2 while being activated on immobilized anti-CD3 plus anti-CD28 enhanced the induction of mRNAs encoding CD69, CD25, and IL-2. Regulation by H_2S of antigen-specific T cell activation was established using murine $CD4^+$ OT-II T cells in the presence of OVA2 peptide and anti-CD28. Addition of 300 n*M* H_2S enhanced antigen-specific induction of the T cell activation markers CD25 and IL-2. T cell activation leads to attachment of T cells, and enhancement of this adhesion by H_2S was confirmed by measuring cell–substrate electrical impedance. Conversely, knocking down CBS using siRNA decreased T cell activation and proliferation. This demonstrates that H_2S concentrations consistent with current estimates of physiological H_2S concentrations (Furne et al., 2008; Shen et al., 2012) enhance T cell activation.

6. ROLE OF H_2S IN THE CYTOSKELETON

T cells interact with APCs by forming a highly organized cluster of cell membrane receptors at the cell–cell interface known as the immunological synapse, which is supported by components of the underlying cytoskeleton in each cell (Monks, Freiberg, Kupfer, Sciaky, & Kupfer, 1998; Vicente-Manzanares & Sanchez-Madrid, 2004). During immunological synapse formation, TCR-mediated signaling promotes integrin-mediated adhesive interactions and redistribution of membrane proteins on the T cell side into supramolecular activation clusters (SMACs). Both microfilaments and microtubules in the T cell cytoskeleton are required for immunological synapse formation. Actin polymerization also plays a vital role in organizing SMACs and the immunological synapse (Lafouresse, Vasconcelos, Cotta-de-Almeida, & Dupre, 2013; Sampath, Gallagher, & Pavalko, 1998; Sechi & Wehland, 2004).

H_2S can alter the function of target proteins via sulfhydration of protein thiols, which may overlap with protein thiols that are nitrosated by reactive nitrogen species (Mustafa, Gadalla, & Snyder, 2009). Sulfhydration requires the activity of CSE, utilizes L-cysteine, and is a reversible process for the regulation of cellular dynamics (Mustafa, Gadalla, Sen, et al., 2009; Paul & Snyder, 2012). Sulfhydration of multiple endogenous liver proteins was detected in WT mice and much less so in *Cse*$^{-/-}$ mice. Sulfhydrated proteins included glyceraldehyde phosphate dehydrogenase, ß-tubulin, and actin. Transient expression of WT CSE but not a catalytically inactive mutant of CSE in HEK293 cells enhanced actin sulfhydration and actin polymerization and rearrangement of the actin cytoskeleton. H_2S was therefore proposed to increase actin polymerization via protein S-sulfhydration, although the molecular details of this regulation remain unclear. The H_2S-induced actin polymerization could be reversed by reducing with dithiothreitol, consistent with a mechanism involving protein sulhydration (Mustafa, Gadalla, Sen, et al., 2009). However, recent studies suggest that persulfide and polysulfide modifications should also be examined that could not be distinguished by the methods used by Mustafa. A subsequent study reported that H_2S increased endothelial cell migration and induced actin rearrangement, which could be prevented by deleting Rac1. Vascular endothelial growth factor receptor and phosphatidylinositol 3-kinase inhibitors also blocked Rac1 and cell migration and cytoskeleton rearrangements induced by H_2S (Zhang, Tao, Wang, Jin, & Zhu, 2012). Further evidence that H_2S regulates actin comes from a proteomic study of mice lacking the H_2S catabolic enzyme ETHE1. Actin was the sulfhydrated protein identified in this study that exhibited the largest decrease in abundance in the *Ethe1*-null versus WT mouse livers (Hildebrandt, Di Meo, Zeviani, Viscomi, & Braun, 2013).

Consistent with actin and tubulin being sulfhydration targets, we found that physiological concentrations of exogenous H_2S enhance the T cell cytoskeletal reorganization induced by TCR signaling (Miller et al., 2012). During T cell activation, a cytoskeletal structure known as the microtubule-organizing complex (MTOC) reorients to a region just beneath the immunological synapse at the T cell–APC contact site (Billadeau, Nolz, & Gomez, 2007). MTOC reorientation leads to increased TCR signaling and T cell proliferation. Addition of 300 n*M* H_2S increased activation-dependent reorientation of the MTOC induced in Jurkat T cells by anti-CD3 plus anti-CD28. H_2S at 10 n*M*–1 μ*M* enhanced the adhesion of activated T cells in a dose-dependent manner. The translocation of MTOC was also enhanced with treatment of H_2S at 500 n*M*. Using cells expressing fluorescent EYFP-α-tubulin, H_2S was shown to induce polarization of tubulin adjacent to clustered TCR. In parallel, deposition of actin in a collar surrounding the immunological synapse was enhanced by H_2S.

These studies demonstrate that H_2S modulates actin and microtubules in the T cell cytoskeleton during T cell activation. However, it remains unclear that these changes are directly regulated by sulfhydration of actin and tubulin. The same cysteine thiols that can be modified by H_2S may also be subject to S-nitrosation, controlled by levels of reactive nitrogen species and S-denitrosating enzymes such as thioredoxins (Benhar, Forrester, Hess, & Stamler, 2008). Evidence for cross-talk between H_2S and NO signaling suggests that nitrosation and sulfhydration may compete for some of the same cysteine residues to regulate the homeostasis and dynamics of cytoskeletal proteins during T cell activation (Kolluru, Shen, & Kevil, 2013; Paul & Snyder, 2012, 2014). Further studies are needed to characterize the specific cytoskeletal protein targets and modified cysteine sulfhydryls that are modified by S-sulfhydration and nitrosation under physiological conditions to regulate the cytoskeleton dynamics during antigen presentation and T cell activation.

7. T CELL REGULATION BY TSP1/CD47 SIGNALING

TSP1 is a secreted protein that is released by activated platelets and is expressed by many cell types in response to stress or injury (Roberts, Miller,

Rogers, Yao, & Isenberg, 2012). TSP1 is an inhibitor of T cell activation induced by engaging the TCR complex (Li, He, Wilson, & Roberts, 2001). Although TSP1 engages and signals through several receptors on T cells, this inhibitory activity is mediated via the TSP1 receptor CD47 (Kaur et al., 2011; Li et al., 2002). Consistent with our results, both *Thbs1*-null and *Cd47*-null mice exhibited prolonged inflammatory responses in a T cell-dependent oxazolone-induced inflammation model (Lamy et al., 2007). TSP1 signaling through CD47 in T cells controls several signal transduction pathways including the NO/cGMP pathway (Ramanathan, Mazzalupo, Boitano, & Montfort, 2011), which prompted us to examine whether TSP1 can also regulate H_2S signaling in T cells.

Activated murine T cells lacking TSP1 were more sensitive to H_2S and expressed elevated mRNA levels of IL-2 and CD69 and cell proliferation as compared to WT activated murine T cells (Miller et al., 2013). The increase in mRNA expression of IL-2 and CD69 was reversed in a dose-dependent manner by physiological concentrations (0.2–2 n*M*) of exogenous TSP1. Resting and activated *Thbs1*-null cells have upregulated CBS and CSE mRNA expression. This suggested that TSP1 endogenously regulates H_2S signaling by limiting T cell activation and limits the expression of *CBS* and *CSE* genes. A CD47-binding peptide (7N3) derived from TSP1 recapitulated the inhibitory activity of TSP1, whereas *Cd47*-null murine T cells were resistant to inhibition by TSP1. This indicated that TSP1 signaling through CD47 has homeostatic inhibitory role in T cell activation and, therefore, may play a role in inflammatory diseases that are associated with induction of H_2S. Since H_2S and NO shared many common signaling mechanisms, TSP1 could be a useful therapeutic target. Small molecule agonists and antagonists of its CD47-binding site could be good candidates for modulating physiological H_2S concentrations and T cell signaling responses.

8. H_2S REGULATION OF LEUKOCYTE ADHESION

NaHS administered IP to rats at 100 μmol/kg prevented damage to the gastric mucosa induced by acetylsalicylic acid, whereas administering the relatively nonspecific CSE inhibitor DL-propargylglycine increased gastric injury caused by the same treatment (Fiorucci et al., 2005). The effects of NaHS were associated with changes in mRNA expression for the leukocyte adhesion molecules LFA1 and ICAM1. Correspondingly, NaHS administered intraperitoneally at 10 or 100 μmol/kg completely inhibited leukocyte adhesion to the endothelium of mesenteric venules of the rats. This antiadhesive activity of NaHS was blocked by the K-ATPase inhibitor glibenclamide. Additional studies established that NaHS at 100 μmol/kg abolished leukocyte adherence to the mesenteric microcirculation induced by the bacterial chemotactic peptide formyl-Met-Leu-Phe (fMLP) (Zanardo et al., 2006). Lawesson's reagent, an H_2S donor, also inhibited fMLP-induced leukocyte adhesion when administered to the rats at a dose of 1 μmol/kg. Treatment with the CSE inhibitor β-cyano-alanine (BCA) reduced leukocyte rolling velocity and increased leukocyte adherence in the rat mesenteric venules. These results were replicated in an air-pouch inflammation model induced by injection of the polysaccharide carrageenan. Leukocyte recruitment was reduced by NaHS and Lawesson's reagent and increased by pretreatment with the CSE inhibitor BCA. The reduction of leukocyte adhesion in the air-pouch model caused by treatment with *N*-acetylcysteine was reversed by treatment with the K_{ATP} channel antagonist glibenclamide, again suggesting that this channel is the target of H_2S signaling that regulates leukocyte adhesion. Although these studies clearly demonstrate that H_2S inhibits leukocyte adhesion *in vivo*, it is unclear whether the leukocytes or the endothelium are the primary source and/or target of H_2S action.

A recent study compared the roles of three members of the selectin (Sel) family of adhesion molecules in the antiadhesive activity of NaHS (Shimizu et al., 2013). NaHS reduced adhesion in *Sele*- and *Selp*-null mice but not in *Sell*-null mice. Antibody blockade of L-selectin in WT mice confirmed this result. At the protein level, NaHS inhibited TNFα-stimulated induction of L-selectin protein expression. Because L-selectin is expressed by the leukocytes, these data indicate that the antiadhesive activity of H_2S occurs at least in part in the leukocytes.

9. ROLE OF H_2S IN DISEASES ASSOCIATED WITH ALTERED T CELL IMMUNITY

9.1. Inflammatory bowel disease

The intestinal microbiota includes major producers of H_2S, and mitochondrial oxidative metabolism of H_2S plays an important role in regulating H_2S levels in the intestinal mucosa (Chan & Wallace, 2013; Medani et al., 2011). The sulfate-reducing bacterium *Desulfovibrio piger* was recently identified as a major component of the normal human intestinal microbiome (Rey et al., 2013), and sulfate in the diet increased the abundance of this species and H_2S production in mice colonized with *D. piger*. Several studies have demonstrated beneficial effects of H_2S donors in rodent inflammatory bowel disease (IBD) models, although it is unclear that this benefit results directly from suppression of T cell function. Some studies have considered H_2S as a therapeutic target for treating IBD. An H_2S-releasing derivative of mesalamine (ATB-429) was more effective than underivatized mesalamine in a mouse model of colitis induced with trinitrobenzene sulfonic acid (Fiorucci et al., 2007). Production of mRNAs encoding the proinflammatory mediators interferon-γ and TNFα was decreased.

9.2. Renal injury

A T cell-dependent proinflammatory role of H_2S is also implied by a cisplatin-induced renal damage model, which is associated with increased T cell infiltration, H_2S formation, and CSE expression (Della Coletta Francescato et al., 2011). Treatment of challenged rats with the H_2S synthesis inhibitor DL-propargylglycine reduced renal damage and T cell infiltration. Similar renal protection by DL-propargylglycine was reported in a separate study where rats were challenged using adriamycin (Francescato et al., 2011). Protection was reported, but T cell infiltration was not determined.

9.3. Rheumatoid arthritis

Rheumatoid arthritis is an autoimmune disease that may involve altered regulatory and effector T cell functions (Carvalheiro, da Silva, & Souto-Carneiro, 2013; Esensten, Wofsy, & Bluestone, 2009). Elevated concentrations of H_2S were reported in synovial fluid from rheumatoid arthritis patients (Whiteman et al., 2010; Winyard et al., 2011). The zinc-trap method used to quantify H_2S detects all forms of RSH. Plasma H_2S levels did not differ in rheumatoid arthritis patients. The synovial fluid levels measured in patients correlated with inflammatory and clinical scores of their disease activity. These data are consistent with H_2S increasing T cell activation, but causality cannot be determined from this study. In other studies, H_2S at 10–100 μM decreased IL-6 expression by fibroblast-like synoviocytes (Kloesch, Liszt, & Broell, 2010). H_2S at 1 mM transiently decreased phosphorylation of ERK. However, more prolonged exposure to 125 μM H_2S for 6 h increased both IL-6 expression and ERK phosphorylation.

9.4. Asthma and allergy

CSE is a major enzyme for H_2S synthesis in the lung, and its role in allergen hypersensitivity was studied using WT and *Cse*-null mice. *Cse*-null mice challenged with ovalbumin produced elevated levels of Th2 cytokines such as IL-5, IL-13, and eotaxin-1 (Zhang, Wang, Yang, Cao, & Wang, 2013). Intraperitoneal administration of 14 μmol/kg NaHS prevented IL-5 and eotaxin-1 induction without altering IL-4 induction. *In vitro*, cytokine production was higher in lymphocytes from lung-draining lymph nodes of challenged *Cse*-null mice, and treatment with 50–500 μ*M* H_2S inhibited cytokine production by WT and *Cse*-null lymphocytes.

9.5. Psoriasis

Psoriasis is a chronic inflammatory skin disease that is characterized by hyperproliferative keratinocytes and infiltration of T cells, dendritic cells, macrophages, and neutrophils (Cai, Fleming, & Yan, 2012). A study comparing *in vitro* activated T cells of psoriasis patients with activated T cells from healthy controls implicated H_2S metabolism in the pathogenesis of this disease (Palau et al., 2013). Gene ontology pathways for H_2S biosynthesis and H_2S metabolic processes showed significant activation in the psoriasis patient T cells ($p = 0.008$). Elevated expression of CTH (CSE) and CBS was found in T cells from psoriasis patients, and both genes were linked to the KLF6 coexpression network, which was upregulated in psoriasis. In other studies, KLF6 was found to induce iNOS (*NOS2*) promoter activity in Jurkat T cells and primary T lymphocytes under conditions of hypoxia, heat shock, and serum starvation (Warke et al., 2003).

9.6. Systemic lupus erythematosus

SLE is a chronic autoimmune disease with symptoms that include the appearance of autoantigen-reactive T cells in peripheral blood. Abnormalities in T cell signaling may contribute to the pathogenesis of this disease (Moulton & Tsokos, 2011). Cell viability and cell proliferation were enhanced in healthy and SLE patients at low doses of H_2S (Han et al., 2013). However, higher doses of H_2S decreased cell proliferation induced by PHA. H_2S also showed different effects on healthy and SLE patients. H_2S further altered the cell cycle of stimulated lupus lymphocytes by increasing expression of CDK2, p27Kip1, and p21WAF1/CIP1.

10. FUTURE PROSPECTIVE

In T cells, H_2S acts as a double-edge sword, yielding biphasic responses that are presumably mediated by different biochemical targets of H_2S. Both pro- and anti-inflammatory roles of H_2S have been reported in T cells. Similarly, H_2S has a dual role for increasing or inhibiting T cell activation and proliferation. The development of methodology for improved detection and quantitation of endogenous H_2S and the posttranslational modification of proteins that mediate its signaling are critical for defining the protein targets that mediate each of these roles of H_2S in T cells.

Although it is clear that submicromolar concentrations of H_2S positively regulate T cell activation, and this parallels the stimulatory role of CBS and CSE induced in response to TCR signaling, we do not know the identities of all of the signaling targets in T cells that are regulated by H_2S. Furthermore, several recent studies indicate that H_2S is not directly the biologically active signaling molecule. Protein sulfhydration is presumed to mediate many signaling effects of H_2S, but it is unclear that H_2S can react readily with cysteine sulfhydryls on the relevant proteins. Sulfane sulfur in the form of persulfides and polysulfides was reported to mediate the ability of H_2S and known H_2S donors to cause sulfhydration of cysteine residues in the lipid phosphatase PTEN (Greiner et al., 2013). Another recent study demonstrated that cysteine persulfide (CysSSH) is the major product produced by both CBS and CSE from the substrate cystine (CysSSCys) (Ida et al., 2014). In the intracellular environment, the majority of CysSSH reacts with reduced GSH to generate GSSH and GSSSH. Based on *in vitro* studies, polysulfides may also result from the direct reaction of H_2S with superoxide (Wedmann et al., 2014). These persulfides and polysulfides were responsible for the increase in sulfhydration of specific target proteins that was demonstrated when CSE was overexpressed in A549 cells (Ida et al., 2014). Based on these studies, we predict that GSH persulfides, rather than H_2S, may be the direct mediator of protein sulfhydration induced when CBS and CSE are activated in T cells. The specific protein sulfhydration targets that promote T cell activation remain to be identified.

Our studies also demonstrate that H_2S is among a growing list of signal transduction pathways that should not be studied under classic "normoxic" tissue culture conditions (Ivanovic, 2009; Kang, Brown, & Chung, 2007; Quinn & Payne, 1985). T cell responses to H_2S in tissue culture are enhanced under 1% O_2 hypoxic conditions, and this can be explained by the rapid oxygen-dependent catabolism of H_2S by mitochondria. Tissue culture in room air with a $PO_2 = 149$ mmHg is generally considered normoxic, but capillaries in well-perfused tissues have $PO_2 = 10–70$ mmHg (Tsai, Johnson, & Intaglietta, 2003), and the mean PO_2 in tissues relevant to T cells such as spleen and thymus ranges from 5 to 20 mmHg (Hale, Braun,

Gwinn, Greer, & Dewhirst, 2002), which corresponds to 0.7–2.7% O_2. Therefore, we propose that 1–2% O_2 should be considered normoxic for studying cellular H_2S signaling under physiological conditions, whereas room air is hyperoxic and suppresses long-term responses to H_2S.

ACKNOWLEDGMENTS

This work was supported by the Intramural Research Program of the National Institutes of Health, National Cancer Institute, Center for Cancer Research.

REFERENCES

Ang, A. D., Rivers-Auty, J., Hegde, A., Ishii, I., & Bhatia, M. (2013). The effect of CSE gene deletion in caerulein-induced acute pancreatitis in the mouse. *American Journal of Physiology. Gastrointestinal and Liver Physiology*, *305*, G712–G721.

Barathi, S., Vadhana, P., Angayarkanni, N., & Ramakrishnan, S. (2007). Estimation of hydrogen sulphide in the human lymphocytes. *Indian Journal of Biochemistry & Biophysics*, *44*, 179–182.

Baskar, R., & Bian, J. S. (2011). Hydrogen sulfide gas has cell growth regulatory role. *European Journal of Pharmacology*, *656*, 5–9.

Benhar, M., Forrester, M. T., Hess, D. T., & Stamler, J. S. (2008). Regulated protein denitrosylation by cytosolic and mitochondrial thioredoxins. *Science*, *320*, 1050–1054.

Bhatia, M., Wong, F. L., Fu, D., Lau, H. Y., Moochhala, S. M., & Moore, P. K. (2005). Role of hydrogen sulfide in acute pancreatitis and associated lung injury. *FASEB Journal*, *19*, 623–625.

Billadeau, D. D., Nolz, J. C., & Gomez, T. S. (2007). Regulation of T-cell activation by the cytoskeleton. *Nature Reviews. Immunology*, *7*, 131–143.

Bjorgo, E., Moltu, K., & Tasken, K. (2011). Phosphodiesterases as targets for modulating T-cell responses. *Handbook of Experimental Pharmacology*, *204*, 345–363.

Brodie, A. E., Potter, J., & Reed, D. J. (1982). Unique characteristics of rat spleen lymphocyte, L1210 lymphoma and HeLa cells in glutathione biosynthesis from sulfur-containing amino acids. *European Journal of Biochemistry*, *123*, 159–164.

Broome, J. D., & Jeng, M. W. (1973). Promotion of replication in lymphoid cells by specific thiols and disulfides in vitro. Effects on mouse lymphoma cells in comparison with splenic lymphocytes. *Journal of Experimental Medicine*, *138*, 574–592.

Brune, V., Tiacci, E., Pfeil, I., Doring, C., Eckerle, S., van Noesel, C. J., et al. (2008). Origin and pathogenesis of nodular lymphocyte-predominant Hodgkin lymphoma as revealed by global gene expression analysis. *Journal of Experimental Medicine*, *205*, 2251–2268.

Bruzzese, L., Fromonot, J., By, Y., Durand-Gorde, J. M., Condo, J., Kipson, N., et al. (2014). NF-kappaB enhances hypoxia-driven T-cell immunosuppression via upregulation of adenosine A(2A) receptors. *Cellular Signalling*, *26*, 1060–1067.

Cai, Y., Fleming, C., & Yan, J. (2012). New insights of T cells in the pathogenesis of psoriasis. *Cellular & Molecular Immunology*, *9*, 302–309.

Carvalheiro, H., da Silva, J. A., & Souto-Carneiro, M. M. (2013). Potential roles for CD8(+) T cells in rheumatoid arthritis. *Autoimmunity Reviews*, *12*, 401–409.

Chan, M. V., & Wallace, J. L. (2013). Hydrogen sulfide-based therapeutics and gastrointestinal diseases: Translating physiology to treatments. *American Journal of Physiology. Gastrointestinal and Liver Physiology*, *305*, G467–G473.

Chattopadhyay, M., Nath, N., Kodela, R., Sobocki, T., Metkar, S., Gan, Z. Y., et al. (2013). Hydrogen sulfide-releasing aspirin inhibits the growth of leukemic Jurkat cells and modulates beta-catenin expression. *Leukemia Research, 37*, 1302–1308.
Chen, X., Jhee, K. H., & Kruger, W. D. (2004). Production of the neuromodulator H2S by cystathionine beta-synthase via the condensation of cysteine and homocysteine. *Journal of Biological Chemistry, 279*, 52082–52086.
Della Coletta Francescato, H., Cunha, F. Q., Costa, R. S., Barbosa Junior, F., Boim, M. A., Arnoni, C. P., et al. (2011). Inhibition of hydrogen sulphide formation reduces cisplatin-induced renal damage. *Nephrology, Dialysis, Transplantation, 26*, 479–488.
Delneste, Y., Jeannin, P., Potier, L., Romero, P., & Bonnefoy, J. Y. (1997). N-acetyl-L-cysteine exhibits antitumoral activity by increasing tumor necrosis factor alpha-dependent T-cell cytotoxicity. *Blood, 90*, 1124–1132.
Dickhout, J. G., Carlisle, R. E., Jerome, D. E., Mohammed-Ali, Z., Jiang, H., Yang, G., et al. (2012). Integrated stress response modulates cellular redox state via induction of cystathionine gamma-lyase: Cross-talk between integrated stress response and thiol metabolism. *Journal of Biological Chemistry, 287*, 7603–7614.
Du, S. X., Xiao, J., Guan, F., Sun, L. M., Wu, W. S., Tang, H., et al. (2011). Predictive role of cerebrospinal fluid hydrogen sulfide in central nervous system leukemia. *Chinese Medical Journal, 124*, 3450–3454.
Esensten, J. H., Wofsy, D., & Bluestone, J. A. (2009). Regulatory T cells as therapeutic targets in rheumatoid arthritis. *Nature Reviews. Rheumatology, 5*, 560–565.
Eto, K., & Kimura, H. (2002). A novel enhancing mechanism for hydrogen sulfide-producing activity of cystathionine beta-synthase. *Journal of Biological Chemistry, 277*, 42680–42685.
Fanger, M. W., Hart, D. A., Wells, J. V., & Nisonoff, A. (1970). Enhancement by reducing agents of the transformation of human and rabbit peripheral lymphocytes. *Journal of Immunology, 105*, 1043–1045.
Fiorucci, S., Antonelli, E., Distrutti, E., Rizzo, G., Mencarelli, A., Orlandi, S., et al. (2005). Inhibition of hydrogen sulfide generation contributes to gastric injury caused by anti-inflammatory nonsteroidal drugs. *Gastroenterology, 129*, 1210–1224.
Fiorucci, S., Orlandi, S., Mencarelli, A., Caliendo, G., Santagada, V., Distrutti, E., et al. (2007). Enhanced activity of a hydrogen sulphide-releasing derivative of mesalamine (ATB-429) in a mouse model of colitis. *British Journal of Pharmacology, 150*, 996–1002.
Francescato, H. D., Marin, E. C., Cunha Fde, Q., Costa, R. S., Silva, C. G., & Coimbra, T. M. (2011). Role of endogenous hydrogen sulfide on renal damage induced by adriamycin injection. *Archives of Toxicology, 85*, 1597–1606.
Furne, J., Saeed, A., & Levitt, M. D. (2008). Whole tissue hydrogen sulfide concentrations are orders of magnitude lower than presently accepted values. *American Journal of Physiology. Regulatory, Integrative and Comparative Physiology, 295*, R1479–R1485.
Glode, L. M., Epstein, A., & Smith, C. G. (1981). Reduced gamma-cystathionase protein content in human malignant leukemia cell lines as measured by immunoassay with monoclonal antibody. *Cancer Research, 41*, 2249–2254.
Gmunder, H., Eck, H. P., Benninghoff, B., Roth, S., & Droge, W. (1990). Macrophages regulate intracellular glutathione levels of lymphocytes. Evidence for an immunoregulatory role of cysteine. *Cellular Immunology, 129*, 32–46.
Greiner, R., Palinkas, Z., Basell, K., Becher, D., Antelmann, H., Nagy, P., et al. (2013). Polysulfides link H2S to protein thiol oxidation. *Antioxidants & Redox Signaling, 19*, 1749–1765.
Hale, L. P., Braun, R. D., Gwinn, W. M., Greer, P. K., & Dewhirst, M. W. (2002). Hypoxia in the thymus: Role of oxygen tension in thymocyte survival. *American Journal of Physiology. Heart and Circulatory Physiology, 282*, H1467–H1477.

Han, Y., Zeng, F., Tan, G., Yang, C., Tang, H., Luo, Y., et al. (2013). Hydrogen sulfide inhibits abnormal proliferation of lymphocytes via AKT/GSK3beta signal pathway in systemic lupus erythematosus patients. *Cellular Physiology and Biochemistry*, *31*, 795–804.
Harding, H. P., Zhang, Y., Zeng, H., Novoa, I., Lu, P. D., Calfon, M., et al. (2003). An integrated stress response regulates amino acid metabolism and resistance to oxidative stress. *Molecular Cell*, *11*, 619–633.
Hellmich, M. R., Coletta, C., Chao, C., & Szabo, C. (2014). The therapeutic potential of cystathionine β-synthetase/hydrogen sulfide inhibition in cancer. *Antioxidants & Redox Signaling*. http://dx.doi.org/10.1089/ars.2014.5933.
Hildebrandt, T. M., Di Meo, I., Zeviani, M., Viscomi, C., & Braun, H. P. (2013). Proteome adaptations in Ethe1-deficient mice indicate a role in lipid catabolism and cytoskeleton organization via post-translational protein modifications. *Bioscience Reports*, *33*.
Hongfang, J., Cong, B., Zhao, B., Zhang, C., Liu, X., Zhou, W., et al. (2006). Effects of hydrogen sulfide on hypoxic pulmonary vascular structural remodeling. *Life Sciences*, *78*, 1299–1309.
Hughes, M. N., Centelles, M. N., & Moore, K. P. (2009). Making and working with hydrogen sulfide The chemistry and generation of hydrogen sulfide in vitro and its measurement in vivo: A review. *Free Radical Biology & Medicine*, *47*, 1346–1353.
Hyspler, R., Ticha, A., Indrova, M., Zadak, Z., Hysplerova, L., Gasparic, J., et al. (2002). A simple, optimized method for the determination of sulphide in whole blood by GC-mS as a marker of bowel fermentation processes. *Journal of Chromatography. B, Analytical Technologies in the Biomedical and Life Sciences*, *770*, 255–259.
Ida, T., Sawa, T., Ihara, H., Tsuchiya, Y., Watanabe, Y., Kumagai, Y., et al. (2014). Reactive cysteine persulfides and S-polythiolation regulate oxidative stress and redox signaling. *Proceedings of the National Academy of Sciences of the United States of America*, *111*, 7606–7611.
Isenberg, J. S., Jia, Y., Field, L., Ridnour, L. A., Sparatore, A., Del Soldato, P., et al. (2007). Modulation of angiogenesis by dithiolethione-modified NSAIDs and valproic acid. *British Journal of Pharmacology*, *151*, 63–72.
Ivanovic, Z. (2009). Hypoxia or in situ normoxia: The stem cell paradigm. *Journal of Cellular Physiology*, *219*, 271–275.
Iwata, S., Hori, T., Sato, N., Ueda-Taniguchi, Y., Yamabe, T., Nakamura, H., et al. (1994). Thiol-mediated redox regulation of lymphocyte proliferation. Possible involvement of adult T cell leukemia-derived factor and glutathione in transferrin receptor expression. *Journal of Immunology*, *152*, 5633–5642.
Julian, D., Statile, J. L., Wohlgemuth, S. E., & Arp, A. J. (2002). Enzymatic hydrogen sulfide production in marine invertebrate tissues. *Comparative Biochemistry and Physiology. Part A, Molecular & Integrative Physiology*, *133*, 105–115.
Kabil, O., & Banerjee, R. (2014). Enzymology of H2S biogenesis, decay and signaling. *Antioxidants & Redox Signaling*, *20*, 770–782.
Kamatani, N., & Carson, D. A. (1982). Differential cyst(e)ine requirements in human T and B lymphoblastoid cell lines. *International Archives of Allergy and Applied Immunology*, *68*, 84–89.
Kang, S. G., Brown, A. L., & Chung, J. H. (2007). Oxygen tension regulates the stability of insulin receptor substrate-1 (IRS-1) through caspase-mediated cleavage. *Journal of Biological Chemistry*, *282*, 6090–6097.
Kanno, S., Hirano, S., Sagi, M., Chiba, S., Takeshita, H., Ikawa, T., et al. (2013). Sulfide induces apoptosis and Rho kinase-dependent cell blebbing in Jurkat cells. *Archives of Toxicology*, *87*, 1245–1256.
Kaur, S., Kuznetsova, S. A., Pendrak, M. L., Sipes, J. M., Romeo, M. J., Li, Z., et al. (2011). Heparan sulfate modification of the transmembrane receptor CD47 is necessary for

inhibition of T cell receptor signaling by thrombospondin-1. *Journal of Biological Chemistry, 286,* 14991–15002.
Kloesch, B., Liszt, M., & Broell, J. (2010). H2S transiently blocks IL-6 expression in rheumatoid arthritic fibroblast-like synoviocytes and deactivates p44/42 mitogen-activated protein kinase. *Cell Biology International, 34,* 477–484.
Kolluru, G. K., Shen, X., & Kevil, C. G. (2013). A tale of two gases: NO and H2S, foes or friends for life? *Redox Biology, 1,* 313–318.
Krishnan, N., Fu, C., Pappin, D. J., & Tonks, N. K. (2011). H2S-induced sulfhydration of the phosphatase PTP1B and its role in the endoplasmic reticulum stress response. *Science Signaling, 4,* ra86.
Lafouresse, F., Vasconcelos, Z., Cotta-de-Almeida, V., & Dupre, L. (2013). Actin cytoskeleton control of the comings and goings of T lymphocytes. *Tissue Antigens, 82,* 301–311.
Lamy, L., Foussat, A., Brown, E. J., Bornstein, P., Ticchioni, M., & Bernard, A. (2007). Interactions between CD47 and thrombospondin reduce inflammation. *Journal of Immunology, 178,* 5930–5939.
Levonen, A. L., Lapatto, R., Saksela, M., & Raivio, K. O. (2000). Human cystathionine gamma-lyase: Developmental and in vitro expression of two isoforms. *Biochemical Journal, 347*(Pt 1), 291–295.
Li, Z., Calzada, M. J., Sipes, J. M., Cashel, J. A., Krutzsch, H. C., Annis, D. S., et al. (2002). Interactions of thrombospondins with alpha4beta1 integrin and CD47 differentially modulate T cell behavior. *Journal of Cell Biology, 157,* 509–519.
Li, Z. Q., He, L. S., Wilson, K. E., & Roberts, D. D. (2001). Thrombospondin-1 inhibits TCR-mediated T lymphocyte early activation. *Journal of Immunology, 166,* 2427–2436.
Li, Q., Lan, Q., Zhang, Y., Bassig, B. A., Holford, T. R., Leaderer, B., et al. (2013). Role of one-carbon metabolizing pathway genes and gene-nutrient interaction in the risk of non-Hodgkin lymphoma. *Cancer Causes & Control, 24,* 1875–1884.
Li, L., Rose, P., & Moore, P. K. (2011). Hydrogen sulfide and cell signaling. *Annual Review of Pharmacology and Toxicology, 51,* 169–187.
Li, L., Salto-Tellez, M., Tan, C. H., Whiteman, M., & Moore, P. K. (2009). GYY4137, a novel hydrogen sulfide-releasing molecule, protects against endotoxic shock in the rat. *Free Radical Biology & Medicine, 47,* 103–113.
Lund, R., Aittokallio, T., Nevalainen, O., & Lahesmaa, R. (2003). Identification of novel genes regulated by IL-12, IL-4, or TGF-beta during the early polarization of CD4+ lymphocytes. *Journal of Immunology, 171,* 5328–5336.
Medani, M., Collins, D., Docherty, N. G., Baird, A. W., O'Connell, P. R., & Winter, D. C. (2011). Emerging role of hydrogen sulfide in colonic physiology and pathophysiology. *Inflammatory Bowel Diseases, 17,* 1620–1625.
Metayer, C., Scelo, G., Chokkalingam, A. P., Barcellos, L. F., Aldrich, M. C., Chang, J. S., et al. (2011). Genetic variants in the folate pathway and risk of childhood acute lymphoblastic leukemia. *Cancer Causes & Control, 22,* 1243–1258.
Miller, T. W., Kaur, S., Ivins-O'Keefe, K., & Roberts, D. D. (2013). Thrombospondin-1 is a CD47-dependent endogenous inhibitor of hydrogen sulfide signaling in T cell activation. *Matrix Biology, 32,* 316–324.
Miller, T. W., Wang, E. A., Gould, S., Stein, E. V., Kaur, S., Lim, L., et al. (2012). Hydrogen sulfide is an endogenous potentiator of T cell activation. *Journal of Biological Chemistry, 287,* 4211–4221.
Mirandola, P., Gobbi, G., Sponzilli, I., Pambianco, M., Malinverno, C., Cacchioli, A., et al. (2007). Exogenous hydrogen sulfide induces functional inhibition and cell death of cytotoxic lymphocytes subsets. *Journal of Cellular Physiology, 213,* 826–833.
Monks, C. R., Freiberg, B. A., Kupfer, H., Sciaky, N., & Kupfer, A. (1998). Three-dimensional segregation of supramolecular activation clusters in T cells. *Nature, 395,* 82–86.

Moulton, V. R., & Tsokos, G. C. (2011). Abnormalities of T cell signaling in systemic lupus erythematosus. *Arthritis Research & Therapy*, *13*, 207.

Mustafa, A. K., Gadalla, M. M., Sen, N., Kim, S., Mu, W., Gazi, S. K., et al. (2009). H2S signals through protein S-sulfhydration. *Science Signaling*, *2*, ra72.

Mustafa, A. K., Gadalla, M. M., & Snyder, S. H. (2009). Signaling by gasotransmitters. *Science Signaling*, *2*, re2.

Nagahara, N., Nagano, M., Ito, T., Shimamura, K., Akimoto, T., & Suzuki, H. (2013). Antioxidant enzyme, 3-mercaptopyruvate sulfurtransferase-knockout mice exhibit increased anxiety-like behaviors: A model for human mercaptolactate-cysteine disulfiduria. *Scientific Reports*, *3*, 1986.

Noelle, R. J. (1980). Modulation of T-cell functions. I. Effect of 2-mercaptoethanol and macrophages on T-cell proliferation. *Cellular Immunology*, *50*, 416–431.

Olson, K. R. (2009). Is hydrogen sulfide a circulating "gasotransmitter" in vertebrate blood? *Biochimica et Biophysica Acta*, *1787*, 856–863.

Ottaviano, G., Marioni, G., Staffieri, C., Giacomelli, L., Marchese-Ragona, R., Bertolin, A., et al. (2011). Effects of sulfurous, salty, bromic, iodic thermal water nasal irrigations in nonallergic chronic rhinosinusitis: A prospective, randomized, double-blind, clinical, and cytological study. *American Journal of Otolaryngology*, *32*, 235–239.

Palau, N., Julia, A., Ferrandiz, C., Puig, L., Fonseca, E., Fernandez, E., et al. (2013). Genome-wide transcriptional analysis of T cell activation reveals differential gene expression associated with psoriasis. *BMC Genomics*, *14*, 825.

Paul, B. D., & Snyder, S. H. (2012). H(2)S signalling through protein sulfhydration and beyond. *Nature Reviews. Molecular Cell Biology*, *13*, 499–507.

Paul, B. D., & Snyder, S. H. (2014). Modes of physiologic H_2S signaling in the brain and peripheral tissues. *Antioxidants & Redox Signaling*. http://dx.doi.org/10.1089/ars.2014.5917.

Peter, E. A., Shen, X., Shah, S. H., Pardue, S., Glawe, J. D., Zhang, W. W., et al. (2013). Plasma free H2S levels are elevated in patients with cardiovascular disease. *Journal of the American Heart Association*, *2*, e000387.

Quinn, P. G., & Payne, A. H. (1985). Steroid product-induced, oxygen-mediated damage of microsomal cytochrome P-450 enzymes in Leydig cell cultures. Relationship to desensitization. *Journal of Biological Chemistry*, *260*, 2092–2099.

Ramanathan, S., Mazzalupo, S., Boitano, S., & Montfort, W. R. (2011). Thrombospondin-1 and angiotensin II inhibit soluble guanylyl cyclase through an increase in intracellular calcium concentration. *Biochemistry*, *50*, 7787–7799.

Rey, F. E., Gonzalez, M. D., Cheng, J., Wu, M., Ahern, P. P., & Gordon, J. I. (2013). Metabolic niche of a prominent sulfate-reducing human gut bacterium. *Proceedings of the National Academy of Sciences of the United States of America*, *110*, 13582–13587.

Richardson, C. J., Magee, E. A., & Cummings, J. H. (2000). A new method for the determination of sulphide in gastrointestinal contents and whole blood by microdistillation and ion chromatography. *Clinica Chimica Acta*, *293*, 115–125.

Ridnour, L. A., Isenberg, J. S., Espey, M. G., Thomas, D. D., Roberts, D. D., & Wink, D. A. (2005). Nitric oxide regulates angiogenesis through a functional switch involving thrombospondin-1. *Proceedings of the National Academy of Sciences of the United States of America*, *102*, 13147–13152.

Rinaldi, L., Gobbi, G., Pambianco, M., Micheloni, C., Mirandola, P., & Vitale, M. (2006). Hydrogen sulfide prevents apoptosis of human PMN via inhibition of p38 and caspase 3. *Laboratory Investigation*, *86*, 391–397.

Roberts, D. D., Miller, T. W., Rogers, N. M., Yao, M., & Isenberg, J. S. (2012). The matricellular protein thrombospondin-1 globally regulates cardiovascular function and responses to stress via CD47. *Matrix Biology*, *31*, 162–169.

Saadat, M., & Bahaoddini, A. (2004). Hematological changes due to chronic exposure to natural gas leakage in polluted areas of Masjid-i-Sulaiman (Khozestan province, Iran). *Ecotoxicology and Environmental Safety, 58*, 273–276.
Sampath, R., Gallagher, P. J., & Pavalko, F. M. (1998). Cytoskeletal interactions with the leukocyte integrin beta2 cytoplasmic tail. Activation-dependent regulation of associations with talin and alpha-actinin. *Journal of Biological Chemistry, 273*, 33588–33594.
Sechi, A. S., & Wehland, J. (2004). Interplay between TCR signalling and actin cytoskeleton dynamics. *Trends in Immunology, 25*, 257–265.
Shen, X., Pattillo, C. B., Pardue, S., Bir, S. C., Wang, R., & Kevil, C. G. (2011). Measurement of plasma hydrogen sulfide in vivo and in vitro. *Free Radical Biology & Medicine, 50*, 1021–1031.
Shen, X. G., Peter, E. A., Bir, S., Wang, R., & Kevil, C. G. (2012). Analytical measurement of discrete hydrogen sulfide pools in biological specimens. *Free Radical Biology & Medicine, 52*, 2276–2283.
Shimizu, K., Ogawa, F., Hara, T., Yoshizaki, A., Muroi, E., Yanaba, K., et al. (2013). Exogenous application of hydrogen sulfide donor attenuates inflammatory reactions through the L-selectin-involved pathway in the cutaneous reverse passive Arthus reaction. *Journal of Leukocyte Biology, 93*, 573–584.
Sidhapuriwala, J. N., Ng, S. W., & Bhatia, M. (2009). Effects of hydrogen sulfide on inflammation in caerulein-induced acute pancreatitis. *Journal of Inflammation (London), 6*, 35.
Sparatore, A., Perrino, E., Tazzari, V., Giustarini, D., Rossi, R., Rossoni, G., et al. (2009). Pharmacological profile of a novel H(2)S-releasing aspirin. *Free Radical Biology & Medicine, 46*, 586–592.
Switzer, C. H., Cheng, R. Y., Ridnour, L. A., Murray, M. C., Tazzari, V., Sparatore, A., et al. (2012). Dithiolethiones inhibit NF-kappaB activity via covalent modification in human estrogen receptor-negative breast cancer. *Cancer Research, 72*, 2394–2404.
Switzer, C. H., Ridnour, L. A., Cheng, R. Y., Sparatore, A., Del Soldato, P., Moody, T. W., et al. (2009). Dithiolethione compounds inhibit Akt signaling in human breast and lung cancer cells by increasing PP2A activity. *Oncogene, 28*, 3837–3846.
Tiranti, V., Viscomi, C., Hildebrandt, T., Di Meo, I., Mineri, R., Tiveron, C., et al. (2009). Loss of ETHE1, a mitochondrial dioxygenase, causes fatal sulfide toxicity in ethylmalonic encephalopathy. *Nature Medicine, 15*, 200–205.
Tokuda, K., Kida, K., Marutani, E., Crimi, E., Bougaki, M., Khatri, A., et al. (2012). Inhaled hydrogen sulfide prevents endotoxin-induced systemic inflammation and improves survival by altering sulfide metabolism in mice. *Antioxidants & Redox Signaling, 17*, 11–21.
Tsai, A. G., Johnson, P. C., & Intaglietta, M. (2003). Oxygen gradients in the microcirculation. *Physiological Reviews, 83*, 933–963.
Twiner, M. J., Ryan, J. C., Morey, J. S., Smith, K. J., Hammad, S. M., Van Dolah, F. M., et al. (2008). Transcriptional profiling and inhibition of cholesterol biosynthesis in human T lymphocyte cells by the marine toxin azaspiracid. *Genomics, 91*, 289–300.
Valitutti, S., Castellino, F., & Musiani, P. (1990). Effect of sulfurous (thermal) water on T lymphocyte proliferative response. *Annals of Allergy, 65*, 463–468.
Vicente-Manzanares, M., & Sanchez-Madrid, F. (2004). Role of the cytoskeleton during leukocyte responses. *Nature Reviews. Immunology, 4*, 110–122.
Wallace, J. L., Vong, L., McKnight, W., Dicay, M., & Martin, G. R. (2009). Endogenous and exogenous hydrogen sulfide promotes resolution of colitis in rats. *Gastroenterology, 137*, 569–578, 578 e561.
Warke, V. G., Nambiar, M. P., Krishnan, S., Tenbrock, K., Geller, D. A., Koritschoner, N. P., et al. (2003). Transcriptional activation of the human inducible nitric-oxide synthase promoter by Kruppel-like factor 6. *Journal of Biological Chemistry, 278*, 14812–14819.

Wedmann, R., Bertlein, S., Macinkovic, I., Boltz, S., Miljkovic, J. L., Munoz, L. E., et al. (2014). Working with "H_2S": Facts and apparent artifacts. *Nitric Oxide, 41*, 85–96.

Weiner, A. S., Beresina, O. V., Voronina, E. N., Voropaeva, E. N., Boyarskih, U. A., Pospelova, T. I., et al. (2011). Polymorphisms in folate-metabolizing genes and risk of non-Hodgkin's lymphoma. *Leukemia Research, 35*, 508–515.

Whiteman, M., Haigh, R., Tarr, J. M., Gooding, K. M., Shore, A. C., & Winyard, P. G. (2010). Detection of hydrogen sulfide in plasma and knee-joint synovial fluid from rheumatoid arthritis patients: Relation to clinical and laboratory measures of inflammation. *Annals of the New York Academy of Sciences, 1203*, 146–150.

Winyard, P. G., Ryan, B., Eggleton, P., Nissim, A., Taylor, E., Lo Faro, M. L., et al. (2011). Measurement and meaning of markers of reactive species of oxygen, nitrogen and sulfur in healthy human subjects and patients with inflammatory joint disease. *Biochemical Society Transactions, 39*, 1226–1232.

Yan, Z. (2009). Extracellular redox modulation by regulatory T cells. *Nature Chemical Biology, 5*, 721–723.

Yuan, Y., Hilliard, G., Ferguson, T., & Millhorn, D. E. (2003). Cobalt inhibits the interaction between hypoxia-inducible factor-alpha and von Hippel-Lindau protein by direct binding to hypoxia-inducible factor-alpha. *Journal of Biological Chemistry, 278*, 15911–15916.

Yusuf, M., Kwong Huat, B. T., Hsu, A., Whiteman, M., Bhatia, M., & Moore, P. K. (2005). Streptozotocin-induced diabetes in the rat is associated with enhanced tissue hydrogen sulfide biosynthesis. *Biochemical and Biophysical Research Communications, 333*, 1146–1152.

Zanardo, R. C., Brancaleone, V., Distrutti, E., Fiorucci, S., Cirino, G., & Wallace, J. L. (2006). Hydrogen sulfide is an endogenous modulator of leukocyte-mediated inflammation. *FASEB Journal, 20*, 2118–2120.

Zhang, J., Sio, S. W., Moochhala, S., & Bhatia, M. (2010). Role of hydrogen sulfide in severe burn injury-induced inflammation in mice. *Molecular Medicine, 16*, 417–424.

Zhang, L. J., Tao, B. B., Wang, M. J., Jin, H. M., & Zhu, Y. C. (2012). PI3K p110alpha isoform-dependent Rho GTPase Rac1 activation mediates H2S-promoted endothelial cell migration via actin cytoskeleton reorganization. *PLoS One, 7*, e44590.

Zhang, G., Wang, P., Yang, G., Cao, Q., & Wang, R. (2013). The inhibitory role of hydrogen sulfide in airway hyperresponsiveness and inflammation in a mouse model of asthma. *American Journal of Pathology, 182*, 1188–1195.

CHAPTER NINE

Anti-inflammatory and Cytoprotective Properties of Hydrogen Sulfide

Burcu Gemici*, John L. Wallace[†,‡,1]
*Near East University, Nicosia, Northern Cyprus, Turkey
[†]Department of Physiology & Pharmacology, University of Calgary, Calgary, Alberta, Canada
[‡]Department of Pharmacology & Toxicology, University of Toronto, Toronto, Ontario, Canada
[1]Corresponding author: e-mail address: altapharm@hotmail.com

Contents

Abstract

Hydrogen sulfide is an endogenous gaseous mediator that plays important roles in many physiological processes in microbes, plants, and animals. This chapter focuses on the important roles of hydrogen sulfide in protecting tissues against injury, promoting the repair of damage, and downregulating the inflammatory responses. The chapter focuses largely, but not exclusively, on these roles of hydrogen sulfide in the gastrointestinal tract. Hydrogen sulfide is produced throughout the gastrointestinal tract, and it contributes to maintenance of mucosal integrity. Suppression of hydrogen sulfide synthesis renders the tissue more susceptible to injury and it impairs repair. In contrast, administration of hydrogen sulfide donors can increase resistance to injury and accelerate repair. Hydrogen sulfide synthesis is rapidly and dramatically enhanced in the gastrointestinal tract after injury is induced. These increases occur specifically at the site of tissue injury. Hydrogen sulfide also plays an important role in promoting resolution of

Methods in Enzymology, Volume 555
ISSN 0076-6879
http://dx.doi.org/10.1016/bs.mie.2014.11.034

inflammation, and restoration of normal tissue function. In recent years, these beneficial actions of hydrogen sulfide have provided the basis for development of novel hydrogen sulfide-releasing drugs. Nonsteroidal anti-inflammatory drugs that release small amounts of hydrogen sulfide are among the most advanced of the hydrogen sulfide-based drugs. Unlike the parent drugs, these modified drugs do not cause injury in the gastrointestinal tract, and do not interfere with healing of preexisting damage. Because of the increased safety profile of these drugs, they can be used in circumstances in which the toxicity of the parent drug would normally limit their use, such as in chemoprevention of cancer.

1. INTRODUCTION

In the past two decades, there has been an exponential growth of publications related to hydrogen sulfide (H_2S). Studies in the 1990s on identifying roles for H_2S in neuromodulation (Abe & Kimura, 1996) were a catalyst for others to examine this novel gasotransmitter, and studies in the early 2000s on cardiovascular effects of H_2S (Wang, 2002; Zhao, Zhang, Lu, & Wang, 2001) triggered an even greater degree of research on this molecule. H_2S plays a broad range of roles in the function of bacteria, plants, and animals, and its role as an electron donor for mitochondrial respiration dates back to the prephotosynthesis period (Goubern, Andriamihaja, Nubel, Blachier, & Bouillaud, 2007; Olson, Donald, Dombkowski, & Perry, 2012). In mammals, H_2S appears to play a major role as a "rescue molecule" (Wallace, Ferraz, & Muscara, 2012). Thus, the focus of this chapter is on the actions of H_2S in prevention and repair of tissue injury (cytoprotection) and in reducing and promoting the resolution of inflammation. We also review some of the ongoing attempts to develop novel therapeutics for cytoprotection and anti-inflammatories that employ hydrogen sulfide release as a key action of the drugs.

2. ENZYMATIC SYNTHESIS OF H_2S

As described in more detail in other chapters, there are three main pathways for enzymatic synthesis of H_2S (Fig. 1). L-Cysteine is the major substrate for H_2S synthesis, although there is evidence that D-cysteine can also be metabolized to H_2S, particularly in the kidney (Kimura, Shibuya, & Kimura, 2012). The vitamin B_6-dependent enzymes, cystathionine γ-lyase (CSE) and cystathionine β-synthase (CBS), play major

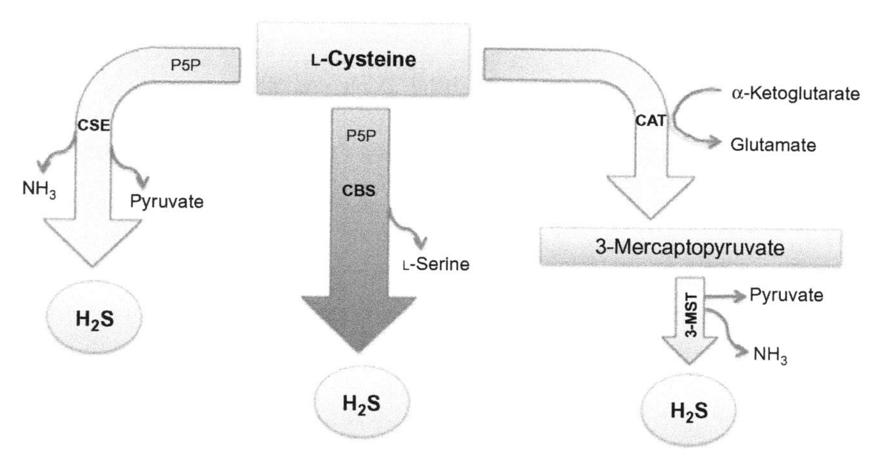

Figure 1 The primary pathways for enzymatic synthesis of hydrogen sulfide (H_2S). L-Cysteine can be converted to H_2S via cystathionine γ-lyase (CSE) or cystathionine β-synthase (CBS), both of which require pyroxidal-5-phosphate (P5P) for their activity. L-Cysteine can also be converted to 3-mercaptopyruvate via cysteine aminotransferase (CAT), which can in turn be metabolized by 3-mercaptopyruvate sulfurtransferase (3MST) to generate H_2S. CAT activity requires α-ketoglutarate as a cofactor. *Reproduced from Chan and Wallace (2013), with permission of the publisher.*

roles in H_2S synthesis in various tissues. In mice deficient of these enzymes, there are distinctive phenotypes that include systemic inflammation and, in the case of the CSE-deficient mice, elevated blood pressure (Cheng et al., 2011; Yang et al., 2008). These phenotypes can be recapitulated with inhibitors of these enzymes. CBS also metabolizes homocysteine to cysteine. Lack of such metabolism, as can occur in certain genetic disorders or when a patient is deficient of B vitamins, results in elevated levels of homocysteine in the blood (hyperhomocysteinemia). This condition is characterized by systemic inflammation, increased susceptibility to thrombosis, and aggravation of preexisting inflammatory conditions. Indeed, mice genetically deficient of CBS display a hyperhomocysteinemic phenotype (Cheng et al., 2011). At least some of the features of these phenotypes may be a consequence of reduced capacity of H_2S synthesis. This has recently been demonstrated in rat and mouse models of hyperhomocysteinemia (Flannigan et al., 2014). In rodent models of inflammatory bowel disease (IBD), prior induction of hyperhomocysteinemia resulted in a marked exacerbation of the severity of colitis, which was due to impaired colonic synthesis of H_2S. These research findings are consistent with the clinical scenario of exacerbation of IBD in patients with hyperhomocysteinemia

(Oldenburg, Fijnheer, van der Griend, vanBerge-Henegouwen, & Koningsberger, 2000; Vagianos & Bernstein, 2012).

3. HEALING AND RESOLUTION OF INFLAMMATION

H_2S is synthesized throughout the gastrointestinal (GI) tract of rodents and humans, with contributions from all three enzymatic pathways (Martin et al., 2010; Wallace, Vong, McKnight, Dicay, & Martin, 2009). We have studied the role of H_2S in several models of GI tissue injury. A common feature in these models is the rapid upregulation of H_2S synthesis, specifically at sites of injury. Thus, shortly after induction of an ulcer in the stomach, there was a marked upregulation of CSE and CBS at the ulcer margin, which is the region where angiogenesis and increased epithelial turnover can be observed (Wallace, Dicay, McKnight, & Martin, 2007). Inhibiting the activity of these enzymes resulted in a significant impairment of ulcer healing. Administration of H_2S donors or of L-cysteine resulted in a significant acceleration of ulcer healing (Wallace, Dicay, et al., 2007). Similar results were obtained in rodent models of colitis, where expression of CSE and 3-mercaptopyruvate sulfurtransferase (3MST) were markedly upregulated very soon after the induction of injury and remained upregulated over the course of the period of repair (Flannigan, Ferraz, Wang, & Wallace, 2013). The upregulation of these enzymes occurred specifically at sites of ulceration. There was no change in expression of these enzymes in the tissue immediately adjacent to ulcers, despite those tissues being inflamed (Flannigan et al., 2013). We also observed significant downregulation of expression of sulfide quinone reductase (SQR) at sites of ulceration (Flannigan et al., 2013). SQR is a key enzyme for oxidation of H_2S (Mimoun et al., 2012). Elevated synthesis of H_2S and decreased oxidation of H_2S at sites of ulceration would result in increased tissue levels of H_2S at those sites, thereby facilitating repair.

The elevation of H_2S synthesis that occurs after induction of colitis begins to subside gradually, in parallel with resolution of colonic inflammation and healing of ulcers (Wallace et al., 2009). Intracolonic administration of H_2S donors can significantly accelerate this process, as has been shown in several animal models of colitis (Fiorucci et al., 2007; Wallace et al., 2009), and can reduce expression of tumor necrosis factor (TNF)α in human inflamed colonic tissue (Bai, Ouyang, & Hu, 2005). The beneficial effects of H_2S donors include reduction of tissue granulocyte numbers and marked suppression of expression of a number of proinflammatory cytokines (e.g., TNFα, interleukin (IL)-1β, IL-8, and interferon (IFN)γ) but sparing of expression of the proinflammatory cytokine, IL-10 (Fiorucci et al., 2007; Li et al., 2007). In sharp contrast, administration of inhibitors of CSE or 3MST results in a marked exacerbation of colitis, leading to bowel perforation and death if treatment is continued over several days (Wallace et al., 2009).

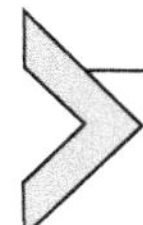

4. MECHANISMS OF ANTI-INFLAMMATORY EFFECTS OF H_2S

The ability of H_2S to reduce pain and inflammation has been recognized for centuries, though not, until recently, at a molecular level. Natural sulfur springs have long been used as a means to alleviate the symptoms of an array of inflammatory diseases, such as rheumatoid arthritis. In the past decade, considerable progress has been made in identifying the molecular mechanisms underlying the anti-inflammatory properties of H_2S (Fig. 2). One of the earliest events in an inflammatory reaction is the recruitment of leukocytes to the site of injury. This involves upregulation of adhesion molecules on the vascular endothelium and on the leukocytes, adhesion of the leukocytes to the endothelium, and then migration of the leukocytes into the interstitial space, from where they move up a chemotaxin gradient to the site of injury/infection (Fig. 3). H_2S plays a key physiological role in regulating these processes. Indeed, H_2S is a tonic inhibitor of leukocyte adherence to the vascular endothelium, as is evident from the observation that administration of inhibitors of H_2S synthesis results in a very rapid increase in leukocyte–endothelial adhesion (Zanardo et al., 2006). This process is mediated via upregulation of intercellular adhesion molecule-1 (ICAM-1) and P-selectin on the endothelium, and lymphocyte function-associated antigen on circulating leukocytes (Fiorucci et al., 2005) (Fig. 3). Administration of H_2S donors has the opposite effect, decreasing leukocyte adherence via activation of ATP-sensitive K^+ channels on leukocytes and endothelial cells. H_2S donors have been shown to cause a marked suppression of inflammatory responses in several animal models (discussed below). H_2S has also been shown to suppress endothelial ICAM-1 expression in response to high blood-glucose concentrations (Guan et al., 2013).

Studies using transgenic mice that produce lower-than-normal levels of H_2S have provided further evidence for a key role of H_2S in modulating leukocyte–endothelial adhesion. Mice that are heterozygous for the CBS gene exhibit increased vascular permeability, slower leukocyte rolling velocity, and increased levels of leukocyte adherence to the vascular endothelium (Kamath et al., 2006). Rats with a diet-induced deficiency of B vitamins

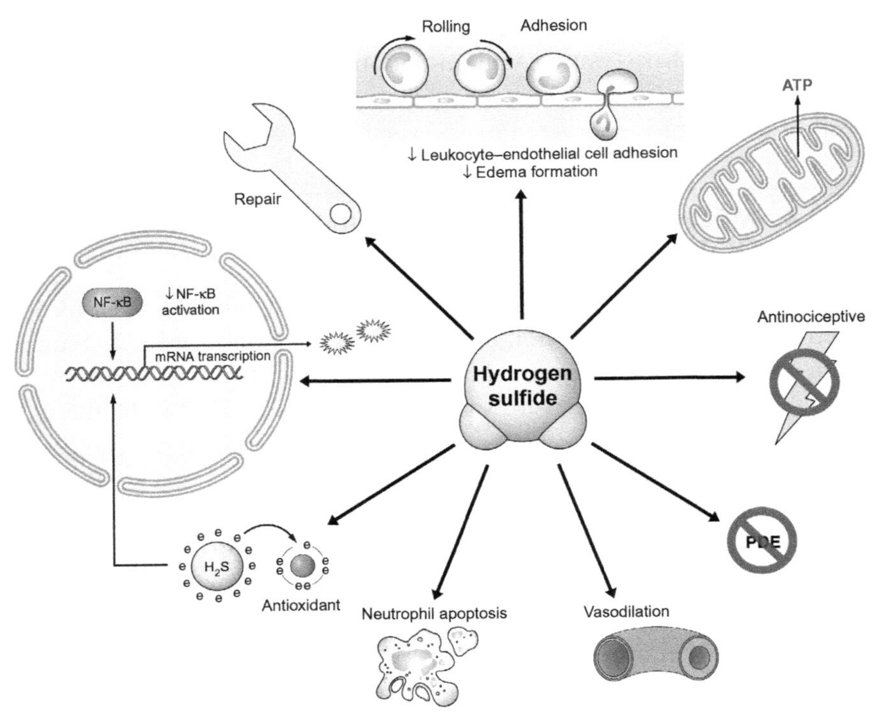

Figure 2 Anti-inflammatory actions of hydrogen sulfide (H_2S). H_2S can affect many aspects of an inflammatory response, through many mechanisms. H_2S is a tonic inhibitor of leukocyte adherence to the vascular endothelium, limiting leukocyte extravasation and edema formation. Mitochondria can utilize H_2S as an electron donor in adenosine triphosphate (ATP) production, particularly during anoxia/hypoxia, and in doing so reduces generation of tissue-damaging oxygen-derived free radicals. Antinociceptive effects of H_2S have been demonstrated in models of visceral pain. By inhibiting phosphodiesterases (PDE), H_2S can elevate tissue cyclic guanylate monophosphate (GMP) levels, which can contribute to vasodilation. H_2S promotes resolution of inflammation through various mechanisms, including promotion of neutrophil apoptosis. The antioxidant actions of H_2S further reduce tissue injury. Several anti-inflammatory and antioxidant systems are activated by H_2S through its effects on transcription factors (including NfκB). Through multiple mechanisms, including induction of cyclooxygenase-2 expression and stimulation of angiogenesis, H_2S can promote repair of damaged tissue. *Reproduced from Chan and Wallace (2013), with permission of the publisher.* (See the color plate.)

have reduced capacity to synthesize H_2S (via CSE and CBS, which require vitamin B_6 for their activity) also exhibit significantly enhanced inflammatory responses, including accumulation of leukocytes in the affected tissues (Flannigan et al., 2014).

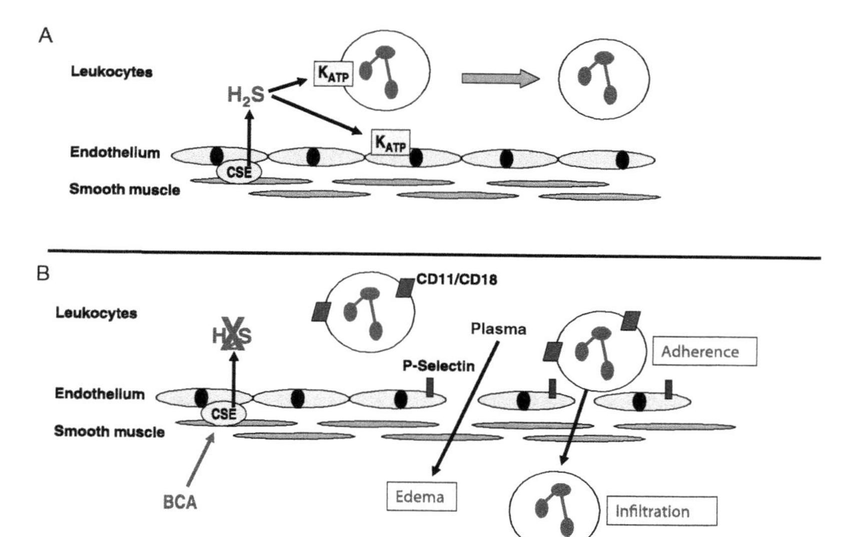

Figure 3 Hydrogen sulfide (H_2S) regulates leukocyte adhesion to the vascular endothelium. (A) H_2S produced via cystathione γ-lyase (CSE) tonically inhibits adherence of leukocytes to the endothelium, via activation of ATP-dependent K^+ channels on the leukocytes and on the endothelium. This activity downregulates expression of CD11/CD18 on leukocytes and both P-selectin and ICAM-1 on the endothelium. (B) When H_2S synthesis is inhibited, such as by β-cyano-alanine (BCA), CD11/CD18, and P-selectin expression increases, leading to leukocyte rolling, adhesion, and extravasation. Inhibition of H_2S synthesis also leads to enhanced edema formation. *Reproduced from Zanardo et al. (2006), with permission of the publisher.*

The mechanisms underlying the ability of H_2S to inhibit leukocyte adherence to the vascular endothelium also involve the anti-inflammatory protein, annexin-1. Annexin-1 is contained within neutrophils and is released during inflammatory reactions as part of the process of induction appropriate resolution of inflammation (Perretti & D'Acquisto, 2009; Serhan et al., 2007; Vong et al., 2007). Micromolar concentrations of NaHS trigger a striking translocation of annexin-1 from the cytosol to the membrane of human neutrophils (Brancaleone, Sampaio, Cirino, Flower, & Perretti, 2011). In a mesenteric venule preparation, the H_2S donor was able to suppress IL-1-induced leukocyte adhesion and emigration. The dependency of the H_2S response on annexin-1 was further demonstrated by lack of an effect of NaHS when similar experiments were performed in annexin-1-deficient mice (Brancaleone et al., 2011). Annexin-1-deficient mice exhibit a marked upregulation of CBS and CSE in a variety of tissues, further supporting a role for this peptide in regulating H_2S synthesis. A role of annexin-1 in mediating anti-inflammatory effects of H_2S in macrophages has also been demonstrated. H_2S can downregulate endotoxin-induced expression of iNOS and cyclooxygenase (COX)-2, but these effects were absent in annexin-1-deficient mice (Brancaleone et al., 2011).

5. EFFECTS OF H_2S ON VISCERAL PAIN

The role of H_2S in pain has been controversial, because of conflicting reports of its antinociceptive versus pronociceptive actions (Distrutti, Sediari, Mencarelli, Renga, Orlandi, Antonelli, et al., 2006; Donatti et al., 2014; Ekundi-Valentim et al., 2010; Fiorucci et al., 2005; Matsunami, Kirishi, Okui, & Kawabata, 2012; Wallace et al., 2014). To some extent, these differences may be related to the models used and the types and doses of H_2S donors that were used (Wallace et al., 2014). There is evidence that H_2S can contribute to pain through activation of T-type calcium channels (Sekiguchi & Kawabata, 2013). On the other hand, several groups have demonstrated that H_2S can reduce visceral pain (Distrutti, Sediari, Mencarelli, Renga, Orlandi, Antonelli, et al., 2006, Distrutti, Sediari, Mencarelli, Renga, Orlandi, & Russo, 2006; Matsunami et al., 2009; Wallace et al., 2014) and may have utility for treatment of conditions such as irritable bowel syndrome and IBD. Colonic distention-induced pain in rats was significantly reduced by several H_2S donors, and these actions were mediated largely through activation of K_{ATP} channels (Distrutti, Sediari, Mencarelli, Renga, Orlandi, Antonelli, et al., 2006). Figure 4 demonstrates the ability of two structurally unrelated H_2S donors to reduce visceral pain induced by gastric distention in rats. The effect was not reversed by perivagal application of capsaicin, which could be mimicked by administration of L-cysteine (precursor for H_2S synthesis), and the latter effect was blocked by administration of an inhibitor of CSE activity (Wang & Wallace, 2011). Of note, an H_2S-releasing salt of trimebutine is being tested in phase 2 clinical trials as a visceral analgesic (Cukier-Meisner, 2013).

While largely studied in models of visceral pain, analgesic effects of H_2S-releasing agents have also been shown to be effective in models of peripheral pain (Cunha et al., 2008; Ekundi-Valentim et al., 2010, 2013).

6. CYTOPROTECTIVE ACTIONS OF H_2S

Damage to the lining of the stomach occurs on a regular basis, but there are mechanisms in place for rapid repair through the process of restitution (Wallace, 2008; Wallace & McKnight, 1990). Such damage can be

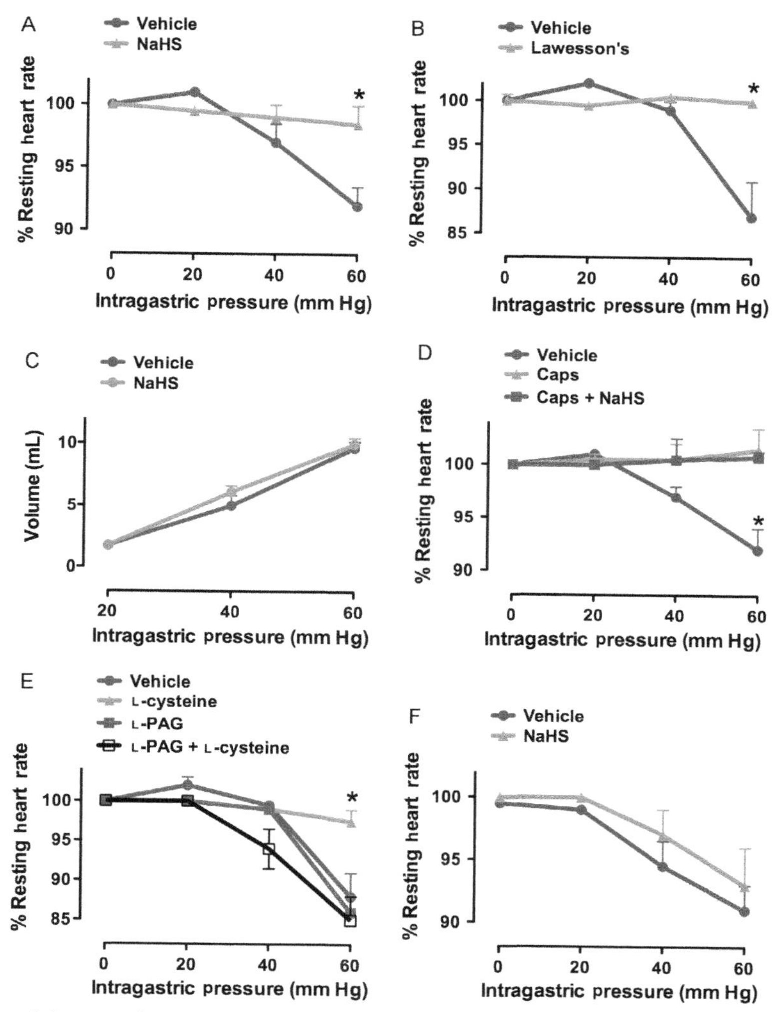

Figure 4 Increased sensitivity to visceral stimuli is one of several mechanisms for symptom generation such as visceral pain in patients with gastrointestinal disorders. Cardioautonomic responses to gastric distention have been recognized as a model to study visceral nociception. In this model, distention of a balloon within the stomach results in pain, and a corresponding decrease in heart rate. (A) Pretreatment with a H_2S donor (NaHS) inhibited the pain response. (B) Similarly, pretreatment with another H_2S donor, Lawesson's reagent, blocked the pain response to gastric distention. (C) Treatment with NaHS did not change gastric compliance. A significant reduction in compliance would result in a reduction of gastric distention, and in turn a lack of effect on heart rate, which could be misinterpreted as an antinociceptive effect. (D) The antinociceptive effect of NaHS was not affected by capsaicin-ablation of sensory afferent nerves in the stomach. (E) An antinociceptive effect was observed by administration of the precursor for H_2S synthesis and was abolished by pretreatment with an inhibitor of H_2S synthesis (L-PAG; propargylglycine). (F) Local topical application of NaHS on the subdiaphragmatic vagus had no effect on cardioautonomic responses to gastric distention. $^*p < 0.05$ versus the vehicle-treated group.

related to stress, ischemia–reperfusion, consumption of alcohol or use of nonsteroidal anti-inflammatory drugs (NSAIDs) (Wallace, 2008). Several endogenous substances contribute to the ability of the GI tract to resist damage in such circumstances, including prostaglandins and nitric oxide. H_2S is another very important mediator of GI mucosal defense (Fiorucci et al., 2005; Wallace, 2010). Inhibition of H_2S synthesis increases the susceptibility of the stomach to injury, whereas H_2S donors can protect the stomach from injury (Fiorucci et al., 2005; Mard et al., 2012; Wallace, Caliendo, Santagada, & Cirino, 2010). The underlying mechanisms of the cytoprotective action of H_2S probably are multifactorial (Table 1). The ability of H_2S to inhibit leukocyte adherence to the vascular endothelium (Mard et al., 2012) is very important in prevention of injury induced by NSAIDs or ischemia–reperfusion (Wallace, Keenan, & Granger, 1990). H_2S can also trigger gastric and duodenal secretion of bicarbonate, which neutralizes gastric acid, thereby limiting its damaging effects as well as those of pepsin (Blackler, Gemici, et al., 2014; Takeuchi et al., 2012). Via its vasodilator effects, H_2S can increase gastric mucosal blood flow, which increases mucosal resistance to injury (Fiorucci et al., 2005).

Damage induced by NSAIDs in the small intestine is more complicated, in terms of its pathogenesis, than the damage these drugs cause in the stomach (Wallace, 2012). Inhibition of COX activity does not appear to be the primary driver of injury. Rather, it is the secretion of the NSAID into bile (enterohepatic circulation of the NSAID), the topical irritant properties of that NSAID-containing bile, and the microbiota of the intestine that play key roles in producing damage (Wallace, 2012). Administration of an H_2S donor prevents NSAID enteropathy in rats in a dose-dependent manner (Blackler, Motta, et al., 2014). While the mechanism of action of the H_2S is not fully understood, this treatment did result in a significant decrease in the cytotoxicity of bile and produced significant changes in the intestinal microbiota (Blackler, Motta, et al., 2014). These effects were not attributable to reduced biliary excretion of the NSAID (Blackler, Motta, et al., 2014). Treatment with the H_2S donor did trigger a significant increase in COX-2 expression and prostaglandin E_2 synthesis by intestinal tissue, which may have contributed to the observed protective effects against NSAID enteropathy (Blackler, Motta, et al., 2014).

When ulcers do form, H_2S plays an important role in promoting healing. Healing of ulcers can be accelerated by drugs that suppress gastric acid secretion. This healing is partially dependent upon H_2S, the synthesis of which is increased at the margins of the ulcer, where there is an increased expression

Table 1 Mechanisms underlying cytoprotective actions of hydrogen sulfide

Primary mechanisms	Protective action	References
Stimulates bicarbonate secretion	Reduces gastroduodenal acidity: reduced tissue damage and enhanced repair	Ise et al. (2011) and Blackler, Gemici, Manko, and Wallace (2014)
Stimulates mucus secretion	Enhances resistance to damage and enhances repair	Motta et al. (2014)
Reduces cytotoxicity of bile	Reduces epithelial injury induced by luminal agents such as nonsteroidal anti-inflammatory drugs	Blackler, Motta, et al. (2014)
Reduces mitochondrial damage/death	Acts as an electron donor in generation of ATP, in circumstances of low oxygen levels; reduces oxygen radical production	Elrod et al. (2007), Kimura, Goto, and Kimura (2010), and Mimoun et al. (2012)
Activates antioxidant response elements	Inactivates Keap1, thereby activating Nrf2 activity	Guo, Liang, Shah Masood, and Yan (2014)
Antioxidant activity	Scavenger of oxygen-derived free radicals	Whiteman et al. (2004)
Reduces neutrophil-mediated tissue injury	Inhibits myeloperoxidase activity	Palinkas et al. (2014)
Inhibits proinflammatory cytokine production	Reduces expression of TNFα, IL-1β, IL-8, IFNγ, etc.	Fiorucci et al. (2007), Li et al. (2007), Wallace et al. (2009), and Flannigan et al. (2014)
Increases or maintains anti-inflammatory cytokine production	Maintenance or increase of expression of IL-10	Fiorucci et al. (2007), Li et al. (2007), Zayachkivska et al. (2014), and Flannigan et al. (2014)
Inhibits leukocyte adherence to the vascular endothelium	Reduces leukocyte-mediated tissue injury; reduces edema formation	Zanardo et al. (2006) and Brancaleone et al. (2011)
Enhances antimicrobial defense	Increases antimicrobial peptide expression; promotes macrophage phagocytosis of bacteria; stabilizes biofilms	Dufton, Natividad, Verdu, and Wallace (2012) and Motta et al. (2014)

of CSE and CBS (Wallace, Dicay, et al., 2007). There is also an increased expression of COX-2 at the ulcer margins (Jones et al., 1999; Ma, del Soldato, & Wallace, 2002; Mizuno et al., 1997). Through the production of prostaglandins that promote angiogenesis and epithelial proliferation, COX-2 contributes significantly to ulcer healing and resolution of inflammation in the GI tract (Jones et al., 1999; Ma et al., 2002; Mizuno et al., 1997; Wallace & Devchand, 2005). H_2S promotes expression of COX-2, while inhibition of H_2S synthesis leads to a reduction of COX-2 expression (Wallace et al., 2014, 2009). Administration of L-cysteine or H_2S donors to rats with gastric ulcers resulted in a significant acceleration of ulcer healing (Wallace, Dicay, et al., 2007). While NSAIDs are known to retard the healing of gastric ulcers in humans and animals, H_2S-releasing NSAIDs have been shown to accelerate ulcer healing in mice (Wallace et al., 2010).

7. THERAPEUTIC APPLICATIONS OF H_2S-RELEASING DRUGS

In recent years, there has been considerable activity aimed at developing novel therapies that exploit the anti-inflammatory and/or cytoprotective properties of H_2S (Chan & Wallace, 2013; Szabo, 2007; Wallace, 2007).

7.1. Inflammation and pain

NSAIDs are among the most commonly used drugs. They are used on a chronic, daily basis by hundreds of millions of patients suffering from disorders, such as osteoarthritis, rheumatoid arthritis, and ankylosing spondylitis, and on an acute basis for a range of disorders characterized by pain (gout, dental pain, dysmenorrhea, injuries, postsurgery, etc.). Several companies and academics have focused on the development of H_2S-releasing NSAIDs, based on early reports that these compounds produced anti-inflammatory effects similar to or superior to conventional NSAIDs (Li et al., 2007; Wallace, Caliendo, Santagada, Cirino, & Fiorucci, 2007; Wallace, Dicay, et al., 2007). NSAIDs are most commonly used for treatment of osteoarthritis, rheumatoid arthritis, ankylosing spondylitis, gout, dysmenorrhea, postsurgical pain, injuries, dental pain, and headaches. NSAIDs are also widely used for veterinary indications, particularly for treating arthritis and post-injury inflammation in companion animals and horses. Extensive preclinical results have been performed to demonstrate the effectiveness of H_2S-NSAIDs in models of inflammation/pain and to demonstrate improved

GI tolerability. H_2S-NSAIDs have been shown to inhibit COX-1 and COX-2 *in vivo*, as effectively as the parent NSAID (Blackler, Syer, Bolla, Ongini, & Wallace, 2012; Ekundi-Valentim et al., 2010, 2013; Wallace, 2013a; Wallace, Caliendo, et al., 2007; Wallace et al., 2010). In models of inflammation, including carrageenan-induced paw edema, zymosan-induced leukocyte infiltration into a subdermal airpouch, and adjuvant-induced arthritis, H_2S-NSAIDs have exhibited anti-inflammatory effects that were comparable or superior to those of the parent NSAID (Wallace, Caliendo, et al., 2007). Figure 5 shows the effects of ATB-346, an H_2S-releasing derivative of naproxen, in a model of adjuvant-induced arthritis in rats. Equimolar doses of naproxen and ATB-346 produced almost identical reductions in joint swelling over a 2-week treatment period. However, at the highest doses tested, the naproxen-treated rats died of intestinal perforation/bleeding after 1 week of treatment, whereas the ATB-346-treated rats survived the full 2 weeks of treatment with a significant reduction of joint inflammation and negligible GI damage.

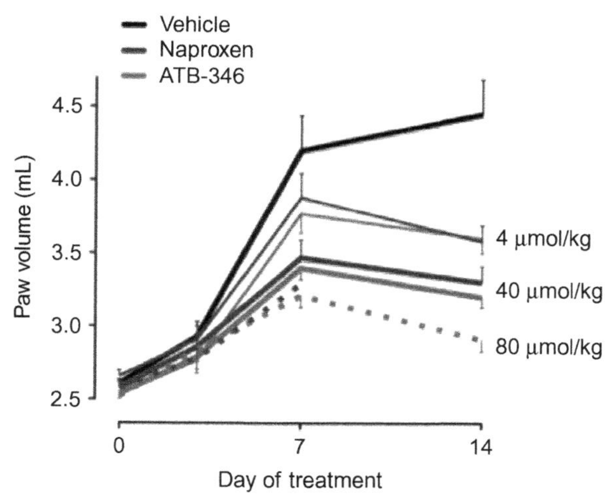

Figure 5 Effectiveness and increased safety of the H_2S-releasing drug, ATB-346, versus naproxen in a rat model of adjuvant-induced arthritis. Groups of 12 rats were treated twice daily with vehicle, naproxen, or ATB-346 for 14 days after induction of arthritis. ATB-346 was administered at doses equimolar to the doses of naproxen shown. There were no significant differences in the anti-inflammatory effects of naproxen versus ATB-346. However, at the highest dose tested, all 12 rats treated with naproxen died of small intestinal perforation by the end of the first week of treatment, while there were no deaths among the rats treated with ATB-346.

ATB-346 has also been shown to exert significantly enhanced anti-nociceptive and anti-inflammatory effects in carrageenan-induced model of knee inflammation in the rats (Ekundi-Valentim et al., 2010, 2013). As a proof-of-concept, this group demonstrated that an H_2S donor (Lawesson's reagent) significantly reduced pain responses in this model, as well as reducing the associated inflammatory changes (Ekundi-Valentim et al., 2010). They then compared a range of doses of naproxen and ATB-346 in this model. Both drugs significantly reduced tactile allodynia and leukocyte infiltration of the knee joints in a dose-dependent manner. However, gastric damage and leukocyte infiltration of the gastric mucosa was significant in rats treated with naproxen, but not in rats treated with ATB-346 (Ekundi-Valentim et al., 2013).

The main beneficial effect of H_2S-NSAIDs over NSAIDs is the greatly reduced GI toxicity. Extensive studies have been carried out on a number of H_2S-NSAIDs and in a variety of models. In healthy rats, for example, ATB-346 was found to produce a low level of gastric hemorrhagic erosion formation, but at a dose 90-times the dose of naproxen that produced a similar level of damage (Wallace et al., 2010). In models of compromised gastric mucosal defense, gastric damage induced by naproxen was consistently greater than that produced in healthy animals, but ATB-346 did not produce significant injury (Wallace et al., 2010). The patients who are at greatest risk for NSAID gastroenteropathy are the elderly patients taking drugs such as glucocorticoids, aspirin, or other anticoagulants, and patients with comorbidities such as obesity and diabetes (Wallace, 2013b). When tested in obese rats, aged rats, and diabetic rats, ATB-346 consistently produced negligible GI damage, while naproxen produced more extensive GI damage than was seen in healthy animals (Blackler et al., 2012).

After significant cardiovascular events were identified as a major risk associated with use of NSAIDs, it became common clinical practice to coprescribe low-dose aspirin together with the NSAID for patients being treated on a chronic basis. Moreover, proton pump inhibitors are often coprescribed to protect the stomach and duodenum from the ulcerogenic effects of NSAIDs. In a rat model, we demonstrated that intestinal damage induced by naproxen or celecoxib was significantly increased when those drugs were coadministered with low-dose aspirin, or when coadministered with a proton pump inhibitor (Blackler et al., 2012; Wallace et al., 2011). The greatest intestinal damage was observed when the NSAIDs were coadministered with both low-dose aspirin and a proton pump inhibitor. However, ATB-346 alone did not cause intestinal damage, and when coadministered with low-dose aspirin, a proton pump inhibitor, or both, the small intestine was spared of hemorrhagic damage (Blackler et al., 2012).

What evidence is there that H_2S accounts for the protective effects of H_2S-NSAIDs? There are several lines of evidence to support this. First, one can observe dose-dependent protection of the stomach and small intestine with a number of different H_2S donors (Blackler, Gemici, et al., 2014; Fiorucci et al., 2005; Wallace et al., 2012), but a worsening of GI damage when inhibitors of H_2S synthesis are administered (Blackler, Gemici, et al., 2014; Mard et al., 2012; Wallace et al., 2010, 2009; Zayachkivska et al., 2014). As shown in Fig. 6, administration of ATB-346 does not produce significant gastric damage in rats, but when a structurally similar compound (naproxen 4-hydroxybenzamide) that lacks the sulfur group of ATB-346 is administered, significant gastric damage is produced. These two compounds exhibited comparable effects on gastric prostaglandin synthesis. Interestingly, when naproxen and the H_2S-releasing moiety of ATB-346 (4-hydroxy-thiobenzamide) were administered to rats as separate entities, the gastric damage produced is similar to what would be seen with naproxen alone (Wallace et al., 2010). Indeed, even if four times the dose of the

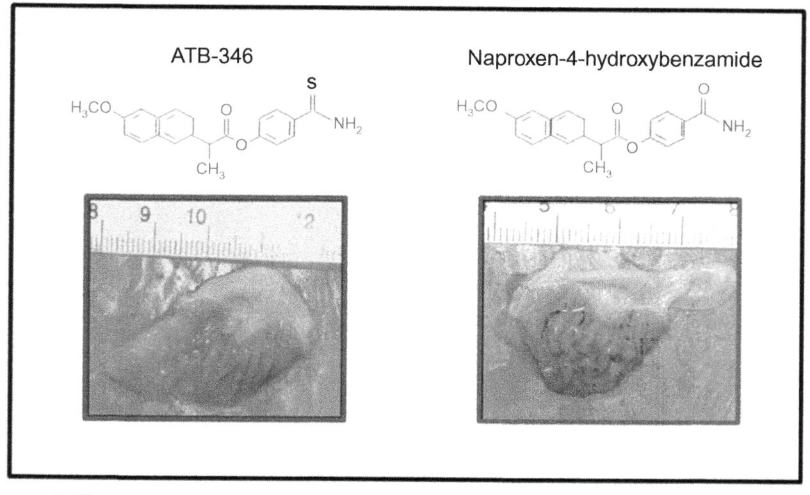

Figure 6 The gastric-sparing property of ATB-346 is due to the ability of this drug to release H_2S. Administration of ATB-346 at a dose of 30 mg/kg did not elicit significant gastric damage, despite markedly suppressing gastric prostaglandin synthesis. However, administration of naproxen-4-hydroxybenzamide, which lacks the sulfur group and therefore cannot generate H_2S, elicits significant hemorrhagic damage in the stomach, with similar suppression of gastric prostaglandin synthesis as observed with the equimolar dose of ATB-346. (See the color plate.)

4-hydroxy-thiobenzamide was administered together with naproxen, no reduction of gastric damage was observed (Wallace et al., 2010). The explanation for this observation may lie in the observation that a much lower amount of H_2S is released from 4-hydroxy-thiobenzamide than is released from ATB-346 (Wallace, Cirino, Santagada, & Caliendo, 2008). The same observation has been made for other H_2S-releasing NSAIDs, including those with a different H_2S-releasing moiety than 4-hydroxy-thiobenzamide (Wallace et al., 2014, 2008). We also observed that while ATB-346 does not produce significant GI damage, coadministration of ATB-346 with an ulcerogenic dose of naproxen did not reduce the damaging effects of naproxen, nor did it exacerbate those effects (Wallace et al., 2014).

H_2S-releasing NSAID derivatives are also being pursued for use in chemoprevention of various types of cancer. The limiting factor for use of currently marketed NSAIDs for this purpose is their GI toxicity. In addition to increased GI safety, H_2S-releasing NSAIDs have been shown to be more potent than the parent NSAIDs in several models (Chattopadhyay, Kodela, Olson, & Kashfi, 2012; Elsheikh, Blackler, Flannigan, & Wallace, 2014; Kashfi, 2014). Figure 7 shows an example of enhanced

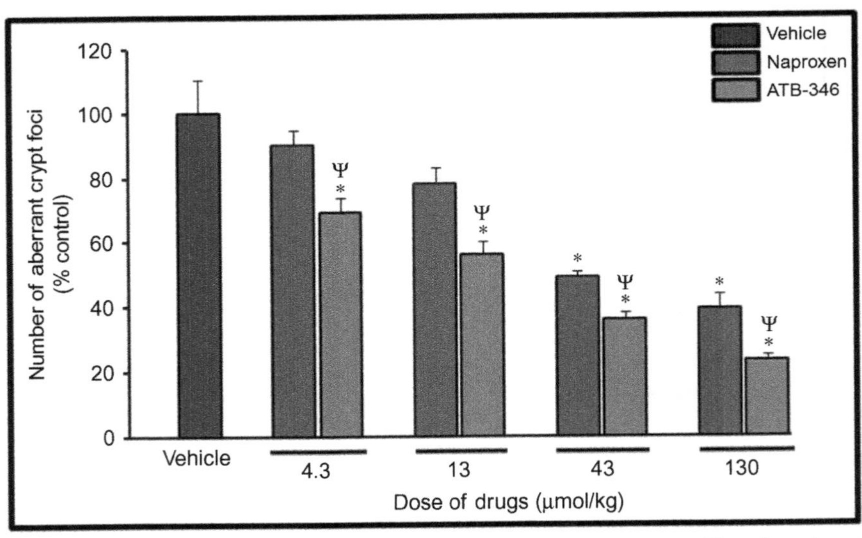

Figure 7 At the highest two doses, daily treatment with naproxen significantly reduced the incidence of aberrant crypt foci formation in the colons of rats that had received the carcinogen azoxymethane (*$p < 0.05$). In contrast, treatment with ATB-346, an H_2S-releasing derivative of naproxen, produced significant reductions of aberrant crypt formation at all doses tested, and significantly greater than the effects of naproxen ($^{\Psi}p < 0.05$ vs. naproxen). *This figure was constructed from data reported by Elsheikh et al. (2014).*

chemopreventative actions of ATB-346 (H_2S-releasing naproxen) as compared to naproxen in a rat model of precancerous lesions. The rats received azoxymethane to induce the formation of aberrant crypt foci in the colon. Daily treatment with ATB-346 for 2 weeks produced a dose-dependent reduction in aberrant crypt foci formation that was significantly enhanced over that of the corresponding doses of naproxen.

7.2. Cardiovascular disease

Several groups are attempting to exploit the cytoprotective effects of H_2S in disease conditions characterized by oxidative stress and the associated tissue injury, such as myocardial dysfunction (Elrod et al., 2007). For example, Sulfagenix is attempting to commercialize zerovalent sulfur as a medicinal food, with the initial clinical target being heart failure. Preclinical studies in heart failure models demonstrated that SG-1002 decreased infarct size, improved cardiac function, increased angiogenesis, reduced inflammation, and downregulated oxidative stress (www.sulfagenix.com/#!sg1002/csny).

In a Phase 1 trial of SG-1002 performed in healthy volunteers, doses of 200–800 mg per day were evaluated initially to determine if the drug was safe and could elevate serum H_2S levels. Having met those two objectives, a second study is underway in patients with heart failure, aimed at confirming the ability of SG-1002 to increase plasma H_2S levels and to produce a reduction of several biomarkers of heart failure.

Reducing oxidative stress, as occurs in the heart during cardiac arrest, is the target of series of compounds being developed by researchers at the University of Exeter. These compounds are H_2S donors, but they are specifically releasing the H_2S inside of mitochondria. Of course, mitochondria play a crucial role in determining if cells live or die (Trionnaire et al., 2014). H_2S can act as an electron donor in mitochondrial respiration and can downregulate the antioxidant response pathway. Thus, they may be useful for treatment of disorders like hypertension, myocardial infarction, and hemorrhagic shock. One of these compounds, AP39, has been shown to increase H_2S levels within endothelial mitochondria and to protect cells against oxidant-induced damage (Szczesny et al. 2014).

7.3. Spinal cord injury and neurodegenerative diseases

Spinal cord injuries often leave the individual with permanent loss of function. Some of the neuronal damage is caused by the trauma itself (Serhan et al., 2007). However, in some cases, considerable damage is caused by the inflammatory reaction to the tissue injury. We examined the potential use of an H_2S-releasing NSAID (ATB-346) in a mouse model of spinal cord injury (Campolo et al., 2013), with the hypothesis that the combination of the anti-inflammatory effect of the NSAID moiety and the anti-inflammatory/antioxidant effects of H_2S would reduce the inflammatory component of the injury, leading to accelerated recovery of motor function. Following induction of spinal cord trauma, the mice were treated daily with naproxen, an equimolar dose of ATB-346, or vehicle. In addition to monitoring recovery of motor function, several indices of spinal cord inflammation were measured over a 10-day recovery period. Mice that were treated with the vehicle developed extensive spinal cord inflammation, with only a very modest recovery of motor function (Fig. 8). As well as clear histological evidence of damage to and inflammation of the spinal cord tissue, there were markedly greater numbers of activated microglia, dense infiltration of granulocytes, and increased expression of TNFα, IL-1β, COX-2, and iNOS (Campolo et al., 2013). Treatment of the mice with naproxen resulted in a significant improvement of motor function recovery and reductions in several indices of spinal cord injury and inflammation. However, a striking and significant improvement of motor function recovery was observed in mice treated with ATB-346, accompanied by dramatic reductions in the various markers of spinal cord inflammation and injury (Campolo et al., 2013). Treatment with the H_2S-releasing moiety alone (4-hydroxythiobenzamide)

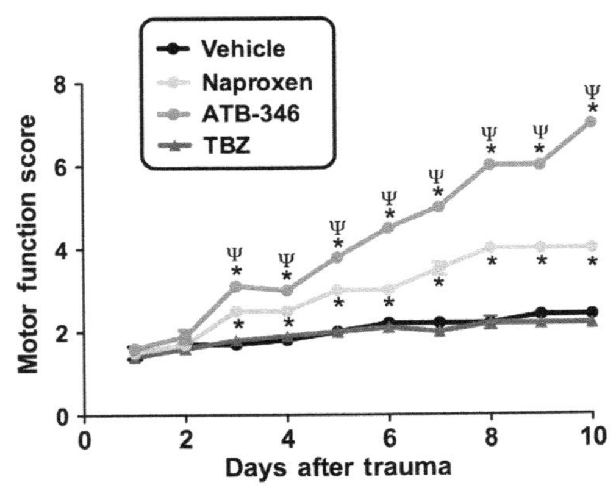

Figure 8 The recovery of motor function in mice after spinal cord trauma was markedly accelerated by daily treatment with ATB-346, a hydrogen sulfide-releasing derivative of naproxen. While naproxen itself accelerated recovery of motor function ($^*p < 0.05$ vs. vehicle-treated), a significantly greater effect ($^\Psi p < 0.05$ vs. naproxen) was seen with ATB-346 at an equimolar dose. Both the NSAID and H_2S-releasing activities were required for the enhanced activity of ATB-346 (note the lack of effect of 4-hydroxythiobenzamide, TBZ, when given alone at a dose equimolar to that of ATB-346). *This figure was constructed using data reported by Campolo et al. (2013).*

at an equimolar dose did not produce significant beneficial effects in this model (Campolo et al., 2013), suggesting that the combination of H_2S release and the NSAID was required to achieve the desired effects.

There has been considerable interest in the use of H_2S-releasing drugs for treatment or prevention of several central nervous system disorders that are characterized by inflammation. For example, H_2S donors have been shown to reduce amyloid peptide-induced neuronal injury in rats by reducing the associated inflammatory response (Fan et al. 2013). H_2S-releasing derivatives of antagonists of *N*-methyl-D-aspartate (NMDA) receptors have recently been suggested to exert cytoprotective effects in an *in vitro* model of Parkinson's disease (Marutani et al., 2014). Cell death in several neurodegenerative diseases may be mediated via activation of NMDA receptors. Marutani et al. (2014) demonstrated that their H_2S-releasing derivatives significantly reduced cell death induced by NMDA activation, and the effects of the derivatives correlated with their ability to increase cellular sulfane sulfur, but not H_2S levels.

7.4. Inflammatory bowel disease

Mesalamine, also known as 5-aminosalicylic acid, is an anti-inflammatory drug, but as it does not have potent inhibitory effects on COX activity, is not considered an NSAID. For decades, mesalamine has been a first-line therapy for ulcerative colitis and Crohn's disease (collectively known as inflammatory bowel disease, or IBD). Mesalamine is a relatively weak anti-inflammatory drug, but it also has minimal adverse effects in IBD patients. The mechanism of action of mesalamine is not completely understood, but its antioxidant actions may contribute most to its beneficial effects (Miles & Grisham, 1995). An H_2S-releasing derivative of mesalamine, ATB-429, has been shown to have significantly enhanced effects in rodent models of colitis (Fiorucci et al., 2007; Wallace et al., 2009). This drug was effective when given orally or intrarectally, eliciting a significant acceleration of healing of colitis in rats and mice. In animals with chemically induced colitis, treatment with ATB-429 markedly reduced tissue levels of granulocytes, much more effectively than mesalamine itself (Fiorucci et al., 2007). Moreover, ATB-429, but not mesalamine, significantly reduced tissue expression of a number of proinflammatory cytokines (IL-1, IL-8, IL-12, and TNFα), while sparing expression of IL-10, an anti-inflammatory cytokine (Fiorucci et al., 2007). ATB-429 also exhibits significantly increased visceral antinociceptive effects over those of mesalamine in a rat model (Distrutti, Sediari, Mencarelli, Renga, Orlandi, & Russo, 2006). A further, unintended benefit of

ATB-429 over mesalamine is that the former is very poorly absorbed (demonstrated in rodents and dogs). If administered orally in its native form, mesalamine is very rapidly absorbed in the upper GI tract. This is undesirable, because IBD primarily affects the lower GI tract, and high luminal concentrations of mesalamine in the affected area are necessary for beneficial effects to be produced. For this reason, mesalamine is sold in various formulations that prevent its absorption in the upper GI tract, or it is administered by enema. The poor absorption of ATB-429 is a benefit for two reasons. First, no formulation is required to prevent its absorption. Second, ATB-429 given orally could treat inflammation of the mucosa throughout the GI tract.

7.5. Visceral analgesia

As mentioned above, an H_2S-releasing compound (GIC-1001) that produces antinociceptive effects in visceral pain models is now in Phase 2 clinical trials as a visceral analgesic. This compound is being developed by GIcare Pharma, and it is a salt of thiobenzamide and trimebutine. Trimebutine has been in use as a treatment for various GI conditions, including irritable bowel syndrome, for over four decades. It is an opioid antispasmodic. Preclinical studies confirmed the enhanced visceral analgesic effects of GIC-1001 versus trimebutine (Cukier-Meisner, 2013). It is being assessed in Phase 2 trials as an analgesic for use prior to colonoscopy. GIC-1001 is also proposed to be a treatment for irritable bowel syndrome, although clinical trials for that indication have not yet been initiated.

ACKNOWLEDGMENTS

The work described in this chapter was supported by grants to Dr. Wallace from the Crohn's and Colitis Foundation of Canada and the Canadian Institutes of Health Research.

REFERENCES

Abe, K., & Kimura, H. (1996). The possible role of hydrogen sulfide as an endogenous neuromodulator. *The Journal of Neuroscience*, *16*, 1066–1071.

Bai, A. P., Ouyang, Q., & Hu, R. W. (2005). Diallyl trisulfide inhibits tumor necrosis factor-alpha expression in inflammed mucosa of ulcerative colitis. *Digestive Diseases and Sciences*, *50*, 1426–1431.

Blackler, R. W., Gemici, B., Manko, A., & Wallace, J. L. (2014). NSAID-gastroenteropathy: New aspects of pathogenesis and prevention. *Current Opinion in Pharmacology*, *19C*, 11–16.

Blackler, R. W., Motta, J. P., Manko, A., Workentine, M., Bercik, P., Surette, M. G., et al. (2014). Hydrogen sulfide protects against NSAID-enteropathy through modulation of bile and the microbiota. *British Journal of Pharmacology*, in press.

Blackler, R., Syer, S., Bolla, M., Ongini, E., & Wallace, J. L. (2012). Gastrointestinal-sparing effects of novel NSAIDs in rats with compromised mucosal defence. *PLoS One*, *7*, e35196.

Brancaleone, V., Sampaio, A. L., Cirino, C., Flower, R. J., & Perretti, M. (2011). A novel cross-talk in resolution: H_2S activates the annexin A1 pathway. *Inflammation Research*, *60*, S291–S293.

Campolo, M., Esposito, E., Ahmad, A., Di Paola, R., Wallace, J. L., & Cuzzocrea, S. (2013). A hydrogen sulfide-releasing cyclooxygenase inhibitor markedly accelerates recovery from experimental spinal cord injury. *The FASEB Journal*, *27*, 4489–4499.

Chan, M. V., & Wallace, J. L. (2013). Hydrogen sulfide-based therapeutics and gastrointestinal diseases: Translating physiology to treatments. *American Journal of Physiology. Gastrointestinal and Liver Physiology*, *305*, G467–G473.

Chattopadhyay, M., Kodela, R., Olson, K. R., & Kashfi, K. (2012). NOSH-aspirin (NBS-1120), a novel nitric oxide- and hydrogen sulfide-releasing hybrid is a potent inhibitor of colon cancer cell growth in vitro and in a xenograft mouse model. *Biochemical and Biophysical Research Communications*, *419*, 523–528.

Cheng, Z., Jiang, X., Kruger, W. D., Pratico, D., Gupta, S., Mallilankaraman, K., et al. (2011). Hyperhomocysteinemia impairs endothelium-derived hyperpolarizing factor-mediated vasorelaxation in transgenic cystathionine beta synthase-deficient mice. *Blood*, *118*, 1998–2006.

Cukier-Meisner, E. (2013). GIcare: Colonoscopy comfort. *BioCentury*, A13, March 18.

Cunha, T. M., Dal-Secco, D., Verri, W. A., Jr., Guerrero, A. T., Souza, G. R., Vieira, S. M., et al. (2008). Dual role of hydrogen sulfide in mechanical inflammatory hypernociception. *European Journal of Pharmacology*, *590*, 127–135.

Distrutti, E., Sediari, L., Mencarelli, A., Renga, B., Orlandi, S., Antonelli, E., et al. (2006a). Evidence that hydrogen sulfide exerts antinociceptive effects in the gastrointestinal tract by activating KATP channels. *The Journal of Pharmacology and Experimental Therapeutics*, *316*, 325–335.

Distrutti, E., Sediari, L., Mencarelli, A., Renga, B., Orlandi, S., Russo, G., et al. (2006b). 5-Amino-2-hydroxybenzoic acid 4-(5-thioxo-5H-[1,2]dithiol-3yl)-phenyl ester (ATB-429), a hydrogen sulfide-releasing derivative of mesalamine, exerts antinociceptive effects in a model of postinflammatory hypersensitivity. *The Journal of Pharmacology and Experimental Therapeutics*, *319*, 447–458.

Donatti, A. F., Araujo, R. M., Soriano, R. N., Azevedo, L. U., Leite-Panissi, C. A., & Branco, L. G. (2014). Role of hydrogen sulfide in the formalin-induced orofacial pain in rats. *European Journal of Pharmacology*, *738C*, 49–56.

Dufton, N., Natividad, J., Verdu, E. F., & Wallace, J. L. (2012). Hydrogen sulfide and resolution of acute inflammation: A comparative study utilizing a novel fluorescent probe. *Scientific Reports*, *2*, 499.

Ekundi-Valentim, E., Mesquita, F. P., Santos, K. T., de Paula, M. A., Florenzano, J., Zanoni, C. I., et al. (2013). A comparative study on the anti-inflammatory effects of single oral doses of naproxen and its hydrogen sulfide (H_2S)-releasing derivative ATB-346 in rats with carrageenan-induced synovitis. *Medical Gas Research*, *3*, 24.

Ekundi-Valentim, E., Santos, K. T., Camargo, E. A., Denadai-Souza, A., Teixeira, S. A., Zanoni, C. I., et al. (2010). Differing effects of exogenous and endogenous hydrogen sulphide in carrageenan-induced knee joint synovitis in the rat. *British Journal of Pharmacology*, *159*, 1463–1474.

Elrod, J. W., Calvert, J. W., Morrison, J., Doeller, J. E., Kraus, D. W., Tao, L., et al. (2007). Hydrogen sulfide attenuates myocardial ischemia-reperfusion injury by preservation of mitochondrial function. *Proceedings of the National Academy of Sciences of the United States of America*, *104*, 15560–15565.

Elsheikh, W., Blackler, R. W., Flannigan, K. L., & Wallace, J. L. (2014). Enhanced chemopreventative effects of a hydrogen sulfide-releasing NSAID (ATB-346) in experimental colorectal cancer. *Nitric Oxide*, *41*, 131–137.

Fan, H., Guo, Y., Liang, X., Yuan, Y., Qi, X., Wang, M., et al. (2013). Hydrogen sulfide protects against amyloid beta-peptide induced neuronal injury via attenuating inflammatory responses in a rat model. *Journal of Biomedical Research*, *27*, 296–304.

Fiorucci, S., Antonelli, E., Distrutti, E., Rizzo, G., Mencarelli, A., Orlandi, S., et al. (2005). Inhibition of hydrogen sulfide generation contributes to gastric injury caused by anti-inflammatory nonsteroidal drugs. *Gastroenterology*, *129*, 1210–1224.

Fiorucci, S., Orlandi, S., Mencarelli, A., Caliendo, G., Santagada, V., Distrutti, E., et al. (2007). Enhanced activity of a hydrogen sulphide-releasing derivative of mesalamine (ATB-429) in a mouse model of colitis. *British Journal of Pharmacology*, *150*, 996–1002.

Flannigan, K. L., Agbor, T. A., Blackler, R. W., Kim, J. J., Khan, W. I., Verdu, E., et al. (2014). Impaired hydrogen sulfide synthesis and IL-10 signaling underlie hyperhomocysteinemia-associated exacerbation of colitis. *Proceedings of the National Academy of Sciences of the United States of America*, *111*, 13559–13564.

Flannigan, K. L., Ferraz, J. G., Wang, R., & Wallace, J. L. (2013). Enhanced synthesis and diminished degradation of hydrogen sulfide in experimental colitis: A site-specific, pro-resolution mechanism. *PLoS One*, *8*, e71962.

Goubern, M., Andriamihaja, M., Nubel, T., Blachier, F., & Bouillaud, F. (2007). Sulfide, the first inorganic substrate for human cells. *The FASEB Journal*, *21*, 1699–1706.

Guan, Q., Wang, X., Gao, L., Chen, J., Liu, Y., Yu, C., et al. (2013). Hydrogen sulfide suppresses high glucose-induced expression of intercellular adhesion molecule-1 in endothelial cells. *Journal of Cardiovascular Pharmacology*, *62*, 278–284.

Guo, C., Liang, F., Shah Masood, W., & Yan, X. (2014). Hydrogen sulfide protected gastric epithelial cell from ischemia/reperfusion injury by Keap1 s-sulfhydration, MAPK dependent anti-apoptosis and NF-kappaB dependent anti-inflammation pathway. *European Journal of Pharmacology*, *725*, 70–78.

Ise, F., Takasura, S., Hayashi, K., Takahashi, K., Koyama, M., Aihara, E., et al. (2011). Stimulation of duodental HCO_3^- secretion by hydrogen sulphide in rats: Relation to prostaglandins, nitric oxide and sensory neurones. *Acta Physiologica*, *201*, 117–126.

Jones, M. K., Wang, H., Peskar, B. M., Levin, E., Itani, R. M., Sarfeh, I. J., et al. (1999). Inhibition of angiogenesis by nonsteroidal anti-inflammatory drugs: Insight into mechanisms and implications for cancer growth and ulcer healing. *Nature Medicine*, *5*, 1418–1423.

Kamath, A. F., Chauhan, A. K., Kisucka, J., Dole, V. S., Loscalzo, J., Handy, D. E., et al. (2006). Elevated levels of homocysteine compromise blood–brain barrier integrity in mice. *Blood*, *107*, 591–593.

Kashfi, K. (2014). Anti-cancer activity of new designer hydrogen sulfide-donating hybrids. *Antioxidants & Redox Signaling*, *20*, 831–846.

Kimura, Y., Goto, Y., & Kimura, H. (2010). Hydrogen sulfide increases glutathione production and suppresses oxidative stress in mitochrondria. *Antioxidants & Redox Signaling*, *12*, 1–13.

Kimura, H., Shibuya, N., & Kimura, Y. (2012). Hydrogen sulfide is a signaling molecule and a cytoprotectant. *Antioxidants & Redox Signaling*, *17*, 45–57.

Li, L., Rossoni, G., Sparatore, A., Lee, L. C., Del Soldato, P., & Moore, P. K. (2007). Anti-inflammatory and gastrointestinal effects of a novel diclofenac derivative. *Free Radical Biology & Medicine*, *42*, 706–719.

Ma, L., del Soldato, P., & Wallace, J. L. (2002). Divergent effects of new cyclooxygenase inhibitors on gastric ulcer healing: Shifting the angiogenic balance. *Proceedings of the National Academy of Sciences of the United States of America*, *99*, 13243–13247.

Mard, S. A., Neisi, N., Solgi, G., Hassanpour, M., Darbor, M., & Maleki, M. (2012). Gastroprotective effect of NaHS against mucosal lesions induced by ischemia-reperfusion injury in rat. *Digestive Diseases and Sciences*, *57*, 1496–1503.

Martin, G. R., McKnight, G. W., Dicay, M. S., Coffin, C. S., Ferraz, J. G., & Wallace, J. L. (2010). Hydrogen sulphide synthesis in the rat and mouse gastrointestinal tract. *Digestive and Liver Disease*, *42*, 103–109.

Marutani, E., Sakaguchi, M., Chen, W., Sasakura, K., Liu, J., Xian, M., et al. (2014). Cytoprotective effects of hydrogen sulfide-releasing N-methyl-D-aspartate receptor antagonists mediated by intracellular sulfane sulphur. *Medicinal Chemistry Communications*, *5*, 1577–1583.

Matsunami, M., Tarui, T., Mitani, K., Nagasawa, K., Fukushima, O., Okubo, K., et al. (2009). Luminal hydrogen sulfide plays a pronociceptive role in mouse colon. *Gut*, *58*, 751–761.

Matsunami, M., Kirishi, S., Okui, T., & Kawabata, A. (2012). Hydrogen sulfide-induced colonic mucosal cytoprotection involves T-type calcium channel-dependent neuronal excitation in rats. *Journal of Physiology and Pharmacology*, *63*, 61–68.

Miles, A. M., & Grisham, M. B. (1995). (1995) Antioxidant properties of 5-aminosalicylic acid: Potential mechanism for its protective effect in ulcerative colitis. *Advances in Experimental Medicine and Biology*, *371B*, 1317–1321.

Mimoun, S., Andriamihaja, M., Chaumontet, C., Atanasiu, C., Benamouzig, R., Blouin, J. M., et al. (2012). Detoxification of H_2S by differentiated colonic epithelial cells: Implication of the sulfide oxidizing unit and of the cell respiratory capacity. *Antioxidants & Redox Signaling*, *17*, 1–10.

Mizuno, H., Sakamoto, C., Matsuda, K., Wada, K., Uchida, T., Noguchi, H., et al. (1997). Induction of cyclooxygenase 2 in gastric mucosal lesions and its inhibition by the specific antagonist delays healing in mice. *Gastroenterology*, *112*, 387–397.

Motta, J. P., Flannigan, K. L., Agbor, T. A., Beatty, J. K., Blackler, R. W., Workentine, M. L., et al. (2014). Hydrogen sulfide protects from colitis and restores intestinal microbiota biofilm and mucus production. *Inflammatory Bowel Diseases*, in press.

Oldenburg, B., Fijnheer, R., van der Griend, R., vanBerge-Henegouwen, G. P., & Koningsberger, J. C. (2000). Homocysteine in inflammatory bowel disease: A risk factor for thromboembolic complications? *The American Journal of Gastroenterology*, *95*, 2825–2830.

Olson, K. R., Donald, J. A., Dombkowski, R. A., & Perry, S. F. (2012). Evolutionary and comparative aspects of nitric oxide, carbon monoxide and hydrogen sulfide. *Respiratory Physiology & Neurobiology*, *184*, 117–129.

Palinkas, Z., Furtmuller, P. G., Nagy, A., Jakopitsch, C., Pirker, K. F., Magierowski, M., et al. (2014). Interactions of hydrogen sulfide with myeloperoxidase. *British Journal of Pharmacology*.

Perretti, M., & D'Acquisto, F. (2009). Annexin A1 and glucocorticoids as effectors of the resolution of inflammation. *Nature Reviews. Immunology*, *9*, 62–70.

Sekiguchi, F., & Kawabata, A. (2013). T-type calcium channels: Functional regulation and implication in pain signalling. *Journal of Pharmacological Sciences*, *122*, 244–250.

Serhan, C. N., Brain, S. D., Buckley, C. D., Gilroy, D. W., Haslett, C., O'Neill, L. A., et al. (2007). Resolution of inflammation: State of the art, definitions and terms. *The FASEB Journal*, *21*, 325–332.

Szabo, C. (2007). Hydrogen sulphide and its therapeutic potential. *Nature Reviews. Drug Discovery*, *6*, 917–935.

Szczesny, B., Modis, K., Yanagi, K., Coletta, C., Le Trionnaire, S., Perry, A., et al. (2014). AP39, a novel mitochondria-targeted hydrogen sulfide donor, stimulates cellular bioenergetics, exerts cytoprotective effects and protects against the loss of mitochondrial DNA integrity in oxidatively stressed endothelial cells in vitro. *Nitric Oxide*, *41*, 120–130.

Takeuchi, K., Aihara, E., Kimura, M., Dogishi, K., Hara, T., & Hayashi, S. (2012). Gas mediators involved in modulating duodenal HCO3(−) secretion. *Current Medicinal Chemistry*, *19*, 43–54.

Trionnaire, S., Perry, A., Szczesny, B., Szabo, C., Winyard, P. G., Whatmore, J. L., et al. (2014). The synthesis and functional evaluation of a mitochondria-targeted hydrogen sulfide donor, (10-oxo-10-(4-(3-thioxo-3H-1,2-dithiol-5-yl)phenoxy)decyl) triphenylphosphonium bromide (AP39). *Medicinal Chemistry Communications*, *5*, 728–736.

Vagianos, K., & Bernstein, C. N. (2012). Homocysteinemia and B vitamin status among adult patients with inflammatory bowel disease: A one-year prospective follow-up study. *Inflammatory Bowel Diseases*, *18*, 718–724.

Vong, L., D'Acquisto, F., Pederzoli-Ribeil, M., Lavagno, L., Flower, R. J., Witko-Sarsat, V., et al. (2007). Annexin-2 cleavage in activated neutrophils: A pivotal role for proteinase 3. *The Journal of Biological Chemistry*, *282*, 29998–30004.

Wallace, J. L. (2007). Hydrogen sulfide-releasing anti-inflammatory drugs. *Trends in Pharmacological Sciences*, *28*, 501–505.

Wallace, J. L. (2008). Prostaglandins, NSAIDs, and gastric mucosal protection: Why doesn't the stomach digest itself? *Physiological Reviews*, *88*, 1547–1565.

Wallace, J. L. (2010). Physiological and pathophysiological roles of hydrogen sulfide in the gastrointestinal tract. *Antioxidants & Redox Signaling*, *12*, 1125–1133.

Wallace, J. L. (2012). NSAID gastropathy and enteropathy: Distinct pathogenesis likely necessitates distinct prevention strategies. *British Journal of Pharmacology*, *165*, 67–74.

Wallace, J. L. (2013a). Mechanisms, prevention and clinical implications of nonsteroidal anti-inflammatory drug-enteropathy. *World Journal of Gastroenterology*, *19*, 1861–1876.

Wallace, J. L. (2013b). Polypharmacy of osteoarthritis: The perfect intestinal storm. *Digestive Diseases and Sciences*, *58*, 3088–3093.

Wallace, J. L., Blackler, R. W., Chan, M. V., Da Silva, G. J., Elsheikh, W., Flannigan, K. L., et al. (2014). Anti-inflammatory and cytoprotective actions of hydrogen sulfide: Translation to therapeutics. *Antioxidants & Redox Signaling*, in press.

Wallace, J. L., Caliendo, G., Santagada, V., & Cirino, G. (2010). Markedly reduced toxicity of a hydrogen sulphide-releasing derivative of naproxen (ATB-346). *British Journal of Pharmacology*, *159*, 1236–1246.

Wallace, J. L., Caliendo, G., Santagada, V., Cirino, G., & Fiorucci, S. (2007). Gastrointestinal safety and anti-inflammatory effects of a hydrogen sulfide-releasing diclofenac derivative in the rat. *Gastroenterology*, *132*, 261–271.

Wallace, J. L., Cirino, G., Santagada, V., & Caliendo, G. (2008). Hydrogen sulfide derivatives of nonsteroidal anti-inflammatory drugs. United States Patent Application No. WO/2008/009127.

Wallace, J. L., & Devchand, P. R. (2005). Emerging roles for cyclooxygenase-2 in gastrointestinal mucosal defense. *British Journal of Pharmacology*, *145*, 275–282.

Wallace, J. L., Dicay, M., McKnight, W., & Martin, G. R. (2007). Hydrogen sulfide enhances ulcer healing in rats. *The FASEB Journal*, *21*, 4070–4076.

Wallace, J. L., Ferraz, J. G., & Muscara, M. N. (2012). Hydrogen sulfide: An endogenous mediator of resolution of inflammation and injury. *Antioxidants & Redox Signaling*, *17*, 58–67.

Wallace, J. L., Keenan, C. M., & Granger, D. N. (1990). Gastric ulceration induced by nonsteroidal anti-inflammatory drugs is a neutrophil-dependent process. *The American Journal of Physiology*, *259*, G462–G467.

Wallace, J. L., & McKnight, G. W. (1990). The mucoid cap over superficial gastric damage in the rat. A high-pH microenvironment dissipated by nonsteroidal antiinflammatory drugs and endothelin. *Gastroenterology*, *99*, 295–304.

Wallace, J. L., Syer, S., Denou, E., de Palma, G., Vong, L., McKnight, W., et al. (2011). Proton pump inhibitors exacerbate NSAID-induced small intestinal injury by inducing dysbiosis. *Gastroenterology*, *141*, 1314–1322.
Wallace, J. L., Vong, L., McKnight, W., Dicay, M., & Martin, G. R. (2009). Endogenous and exogenous hydrogen sulfide promotes resolution of colitis in rats. *Gastroenterology*, *137*, 569–578.
Wang, R. (2002). Two's company, three's a crowd: Can H_2S be the third endogenous gaseous transmitter? *The FASEB Journal*, *16*, 1792–1798.
Wang, L., & Wallace, J. L. (2011). Hydrogen sulfide reduces cardio-autonomic responses to gastric distention (GD) in an ATP-sensitive potassium channel K_{ATP}-dependent manner. *Neurogastroenterology and Motility*, *23*(Suppl. 1), 59.
Whiteman, M., Armstrong, J. S., Chu, S. H., Jia-Ling, S., Wong, B. S., Cheung, N. S., et al. (2004). The novel neuromodulator hydrogen sulfide: An endogenous peroxynitrite 'scavenger'? *Journal of Neurochemistry*, *90*, 765–768.
Yang, G., Wu, L., Jiang, B., Yang, W., Qi, J., Cao, K., et al. (2008). H_2S as a physiologic vasorelaxant: Hypertension in mice with deletion of cystathionine gamma-lyase. *Science*, *322*, 587–590.
Zanardo, R. C., Brancaleone, V., Distrutti, E., Fiorucci, S., Cirino, G., & Wallace, J. L. (2006). Hydrogen sulfide is an endogenous modulator of leukocyte-mediated inflammation. *The FASEB Journal*, *20*, 2118–2120.
Zayachkivska, O., Havryluk, O., Hrycevych, N., Bula, N., Grushka, O., & Wallace, J. L. (2014). Cytoprotective effects of hydrogen sulfide in novel rat models of non-erosive esophagitis. *PLoS One*, *9*, e110688.
Zhao, W., Zhang, J., Lu, Y., & Wang, R. (2001). The vasorelaxant effect of H_2S as a novel endogenous gaseous K_{ATP} channel opener. *The EMBO Journal*, *20*, 6008–6016.

CHAPTER TEN

H_2S and Substance P in Inflammation

Madhav Bhatia[1]
Department of Pathology, University of Otago, Christchurch, New Zealand
[1]Corresponding author: e-mail address: madhav.bhatia@otago.ac.nz

Contents

Abstract

Hydrogen sulfide (H_2S) and substance P play a key role in inflammation. Using animal models of inflammation of different etiologies such as acute pancreatitis, sepsis, burns, and joint inflammation, studies have recently shown an important role of the proinflammatory action of H_2S and substance P. Also, H_2S contributes to inflammation in different conditions via substance P. This chapter reviews methods and key data that have led to our current understanding of the role of H_2S and substance P in inflammation.

1. INTRODUCTION

Inflammation is a normal response to injury and a useful physiological event. However, uncontrolled inflammation can lead to disease. Recent research has shown a key role of hydrogen sulfide (H_2S) and substance P in inflammation. Endogenously synthesized H_2S acts as a vasodilator, as does substance P. Substance P is an important mediator of neurogenic

Methods in Enzymology, Volume 555
ISSN 0076-6879
http://dx.doi.org/10.1016/bs.mie.2014.11.024

inflammation. This chapter reviews methods and key data that have led to our current understanding of the role of H_2S and substance P in inflammation.

2. DISEASE MODELS USED TO STUDY THE ROLE OF H_2S AND SUBSTANCE P

In vivo experimental models of disease have been key to our understanding of the role played by mediators, such as H_2S and substance P, in inflammation. Diseases in which H_2S and substance P have been shown to act as inflammatory mediators and their experimental models are described below.

2.1. Caerulein-induced acute pancreatitis

Acute pancreatitis is a common clinical condition, the incidence of which has been increasing worldwide over recent years (Bhatia, 2012). For example, in the United States alone, acute pancreatitis is the most common reason for hospitalization among all gastrointestinal disease, causing significant economic burden as well (Peery et al., 2012).

Administration of high doses of an exogenous secretagogue leads to abnormally high digestive enzyme secretion, resulting in acute pancreatitis. The cholecystokinin (CCK) analogue caerulein has been used to induce acute pancreatitis in rodents (Bhatia et al., 1998; Lampel & Kern, 1977). Rapid induction, noninvasiveness, high reproducibility, high applicability, and acute pancreatitis-like pancreatic histological changes have made caerulein the best characterized and the most favored model for acute pancreatitis. Caerulein, acting through the CCK receptors, yields exaggerated stimulation of acinar cells, which leads to activation of trypsinogen. Caerulein can be administrated to mice via intraperitoneal injection (50 μg/kg body weight) (Fig. 1). Variation in the amount of caerulein, by adjusting the number of injections, results in differences in degree of severity. Pancreatic injury in this model of acute pancreatitis is characterized by hyperamylasemia, an increase in pancreatic water content (an indicator of pancreatic edema), an increase in pancreatic myeloperoxidase (MPO) activity (an indicator of neutrophil infiltration into the pancreas), and histological evidence of pancreatic injury (Bhatia et al., 1998). When multiple injections of caerulein are given, quite severe acute pancreatitis can be observed in mice as evidenced by acute pancreatitis-associated lung injury, which is a common complication in severe acute pancreatitis (Bhatia et al., 1998). Lung injury can be determined

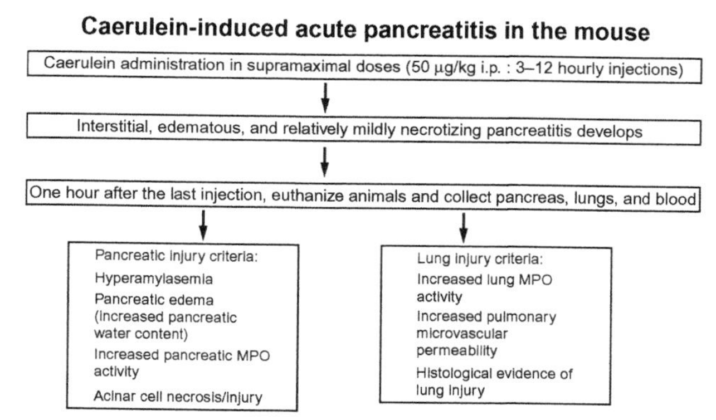

Figure 1 Caerulein-induced acute pancreatitis.

by an increase in pulmonary microvascular permeability, an increase in lung MPO activity, and histological evidence of lung injury.

Cystathionine-β-synthetase and cystathionine-γ-lyase (CSE), the two major H_2S synthesizing enzymes, are highly expressed in the pancreas. Endogenously produced H_2S acts as a mediator of inflammation in acute pancreatitis induced by caerulein in mice (Bhatia et al., 2005). Treatment of mice with propargylglycine (PAG), a CSE inhibitor, reduces the severity of acute pancreatitis and associated lung injury. The proinflammatory role of H_2S synthesized by CSE has been confirmed recently using knockout mice deficient in CSE, which are protected against acute pancreatitis and associated lung injury, when compared to wild-type controls (Ang, Rivers-Auty, Hegde, Ishii, & Bhatia, 2013). H_2S contributes to inflammation in acute pancreatitis via activation of cytokines, chemokines, and adhesion molecules (Tamizhselvi, Koh, Sun, Zhang, & Bhatia, 2010; Tamizhselvi, Moore, & Bhatia, 2008).

Substance P has been shown to play a key role in inflammation in acute pancreatitis and associated lung injury. This has been shown using different experimental approaches—gene knockout for substance P (Bhatia, Slavin, Cao, Basbaum, & Neoptolemos, 2003) and its receptor (Bhatia et al., 1998), pharmacological inhibition of its action by using specific receptor antagonist (Lau, Wong, & Bhatia, 2005), and inhibitors for neutral endopeptidase, the enzyme responsible for its inactivation (Koh, Moochhala, & Bhatia, 2011). Substance P contributes to inflammation in acute pancreatitis via activation of cytokines, chemokines, and adhesion molecules that results in leukocyte recruitment to the site of inflammation (Lau & Bhatia, 2006, 2007; Sun & Bhatia, 2007).

2.2. Lipopolysaccharide-induced endotoxemia and cecal ligation and puncture-induced sepsis

Sepsis is a major health problem worldwide. The incidence of sepsis in North America, for example, is 3.0 per 1000 population, or 750,000 cases per year, 210,000 of which are fatal (Bhatia, 2012; Levy et al., 2010; Martin, Mannino, Eaton, & Moss, 2003).

Lipopolysaccharide (LPS) is a purified macromolecular glycolipid extracted from the cell walls of Gram-negative bacteria (Buras, Holzmann, & Sitkovsky, 2005; Howard, Rowley, & Wardclaw, 1957). Most of the toxicity of LPS resides in the lipid A moiety. Although it has been used in different species as it is an easy model to perform (Fig. 2), and highly reproducible, it cannot reproduce the features of Gram-negative sepsis in the human.

The cecal ligation and puncture (CLP)-induced sepsis (Fig. 2) has been suggested to be the golden standard for sepsis research. It successfully simulates clinical problems like perforated appendicitis and diverticulitis (Buras et al., 2005; Wichterman, Baue, & Chaudry, 1980). The procedures include midline laparotomy, exteriorization of the cecum, ligation of the cecum distal to the ileocecal valve, and puncture of the ligated cecum. This surgical process produces a bowel perforation with leakage of feces into peritoneum and thus causes peritoneal contamination with mixed flora in devitalized tissue. The severity of sepsis, as evaluated by mortality rate, is related to the needle puncture size or the number of punctures and reproduces the inflammatory response observed in clinical sepsis.

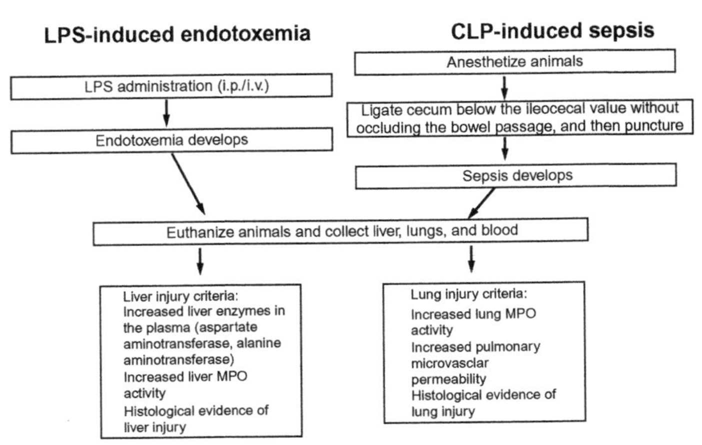

Figure 2 LPS-induced endotoxemia and CLP-induced sepsis.

In both these models, liver injury is characterized by an increase in liver enzymes aspartate amino transferase and alanine amino transferase, an increase in liver MPO activity, and histological evidence of liver injury. Lung injury is characterized by an increase in pulmonary microvascular permeability, an increase in lung MPO activity, and histological evidence of lung injury.

H_2S has been shown to act as a mediator of inflammation in both LPS-induced endotoxemia (Li et al., 2005) and CLP-induced sepsis (Zhang, Zhi, Moore, & Bhatia, 2006). H_2S has also been shown to modulate sinusoidal constriction in the liver and contribute to hepatic microcirculatory dysfunction during endotoxemia (Norris et al., 2013). H_2S contributes to inflammation in sepsis via activation of cytokines, chemokines, and adhesion molecules, resulting in increased leukocyte recruitment to the site of inflammation (Zhang, Moochhala, & Bhatia, 2008; Zhang, Zhi, Moochhala, Moore, & Bhatia, 2007a, 2007b).

Substance P has also been shown to act as a mediator of inflammation in both LPS-induced endotoxemia and CLP-induced sepsis, using complementary strategies including targeted gene deletion of substance P gene, and specific receptor antagonists (Hegde, Zhang, Moochhala, & Bhatia, 2007; Ng, Zhang, Hegde, & Bhatia, 2008; Puneet et al., 2006). Substance P contributes to inflammation in acute pancreatitis via activation of cytokines, chemokines, and adhesion molecules (Hegde, Koh, Moochhala, & Bhatia, 2010; Hegde et al., 2010; Hegde, Uttamchandani, Moochhala, & Bhatia, 2010).

2.3. Burn injury

Severe burn injuries are among the leading causes of morbidity and mortality worldwide (Bhatia, 2012; Endorf & Ahrenholz, 2011). Mouse is an excellent model for studying the postburn immune dysfunction as its immune system is well characterized. Mouse burn injury models vary with regard to the source as well as size of the injury, and studies have shown a strong correlation between burn size and mortality, with increasing burn size being associated with higher mortality rates. All these models are performed with appropriate anesthesia in order to minimize suffering to animals. Scalding is the easiest mechanism of provoking a dermal experimental burn. The possibility of varying water temperature, time of exposure, and the burned area makes this method ideal for reproducing almost every kind of burn. Constant temperature water scald burn models have been created in several

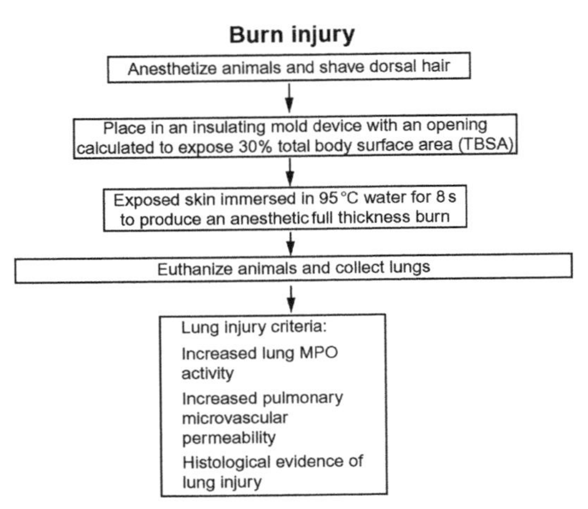

Figure 3 Burn injury.

strains of mice (Cribbs, Luquette, & Besner, 1998). Severe burn injury is associated with systemic organ damage, such as lung and liver injury.

Using a model of full-thickness burn injury (Fig. 3), H_2S has been shown to play an important role in inflammation (Zhang, Sio, Moochhala, & Bhatia, 2010). Burn injury in mice results in an increase in plasma H_2S levels, liver and lung CSE mRNA expression, and liver H_2S-synthesizing activity. There is a protection against systemic inflammation and multiple organ damage by prophylactic or therapeutic treatment of PAG. Administration of NaHS at the same time of burn injury results in a more severe tissue injury in the lung and liver (Zhang et al., 2010).

Substance P acts as an important mediator of burn-induced acute inflammation (Sio, Ang, Lu, Moochhala, & Bhatia, 2010; Sio, Lu, Moochhala, & Bhatia, 2010; Sio, Puthia, Lu, Moochhala, & Bhatia, 2008). Burn injury in mice subjected to full-thickness burn augments production of substance P, and its gene expression, which correlates with exacerbated lung damage after burn. Substance P gene deletion or pharmacological inhibition by using specific receptor antagonist protects mice against inflammation following burn injury.

2.4. Carrageenan-induced hindpaw edema

Joint inflammation/arthritis is a major health problem worldwide. In the United States, for example, back pain and arthritis are among the most common and costly conditions, affecting more than 100 million individuals and costing more than $200 billion per year (Ma, Chan, & Carruthers, 2014).

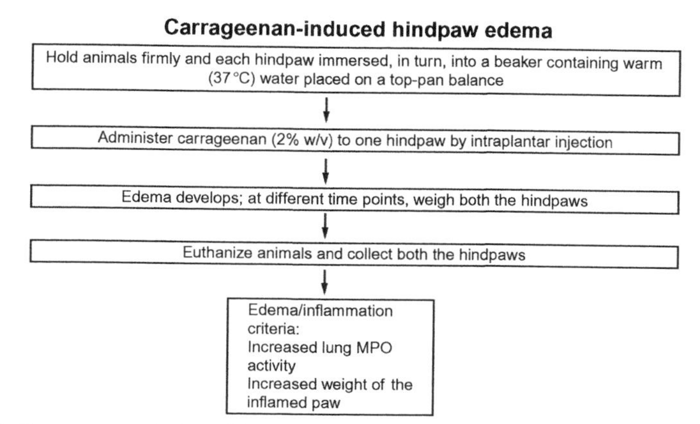

Figure 4 Carrageenan-induced hindpaw edema.

Carrageenan-induced hindpaw edema in the rat (Fig. 4) is a classical model of joint inflammation (Winter, Risley, & Nuss, 1962). It is easy to perform and highly reproducible. Due to this, carrageenan-induced hindpaw edema is an ideal model with which to study joint inflammation.

In carrageenan-induced hindpaw edema, an increase in H_2S synthesis has been seen, suggesting a localized overproduction of H_2S during inflammation. Pretreatment with PAG results in a dose-dependent inhibition of hindpaw edema as well as of hindpaw MPO activity (Bhatia, Sidhapuriwala, Moochhala, & Moore, 2005). These results suggest that H_2S acts as a mediator of inflammation in joint inflammatory disease.

Substance P has also been shown to act as a mediator of inflammation in carrageenan-induced hindpaw edema (Gilligan, Lovato, Erion, & Jeng, 1994). Results from this study suggested that substance P may play an important role in carrageenan-induced paw edema and that a reduction in the biosynthesis of substance P may lessen the severity of this inflammatory disease.

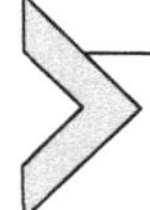

3. H_2S AND SUBSTANCE P—WHAT ARE THEY DOING TOGETHER?

Intraperitoneal administration of the H_2S donor NaHS to mice results in a significant increase in circulating levels of substance P (Bhatia, Zhi, Zhang, Ng, & Moore, 2006). H_2S, by itself, causes lung inflammation, as evidenced by a significant increase in lung MPO activity and histological evidence of lung injury. In substance P knockout mice, H_2S does not cause any lung inflammation. Also, pretreatment of mice with a substance

P receptor antagonist protects mice against lung inflammation caused by H_2S (Bhatia et al., 2006). These results show that H_2S contributes to inflammation through substance P. Studies with caerulein-induced acute pancreatitis (Bhatia, Sidhapuriwala, Ng, Tamizhselvi, & Moochhala, 2008; Tamizhselvi, Moore, & Bhatia, 2007; Tamizhselvi, Shrivastava, Koh, Zhang, & Bhatia, 2011) and CLP-induced sepsis (Ang, Moochhala, & Bhatia, 2010; Ang, Moochhala, MacAry, & Bhatia, 2011; Ang, Sio, Moochhala, MacAry, & Bhatia, 2011; Zhang et al., 2007) further show that H_2S acts upstream of substance P and contributes to neurogenic inflammation in this manner.

4. SUMMARY

Experimental animal models have been invaluable in the research on the role of H_2S and substance P in inflammation. With the help of these models, we now have a good understanding of the contribution of H_2S and substance P to inflammation, and the mechanism by which they act in the pathogenesis of inflammatory diseases. More research into the mechanisms by which H_2S and substance P contribute to inflammation would enable us to take these studies to the clinic.

ACKNOWLEDGMENTS

The author's laboratory is supported by research grants from the Lottery Health, Arthritis New Zealand, Maurice and Phyllis Paykel Trust, and University of Otago (Establishment Grant and University of Otago Research Grant).

REFERENCES

Ang, S. F., Moochhala, S., & Bhatia, M. (2010). Hydrogen sulfide promotes transient receptor potential vanilloid 1-mediated neurogenic inflammation in polymicrobial sepsis. *Critical Care Medicine*, *38*, 619–628.

Ang, S. F., Moochhala, S. M., MacAry, P. A., & Bhatia, M. (2011). Hydrogen sulfide and neurogenic inflammation in polymicrobial sepsis: Involvement of substance P and ERK-NF-κB signaling. *PLoS One*, *6*, e24535.

Ang, A. D., Rivers-Auty, J., Hegde, A., Ishii, I., & Bhatia, M. (2013). The effect of CSE gene deletion in caerulein-induced acute pancreatitis in the mouse. *American Journal of Physiology. Gastrointestinal and Liver Physiology*, *305*, G712–G721.

Ang, S. F., Sio, S. W., Moochhala, S. M., MacAry, P. A., & Bhatia, M. (2011). Hydrogen sulfide upregulates cyclooxygenase-2 and prostaglandin E metabolite in sepsis-evoked acute lung injury *via* transient receptor potential vanilloid type 1 channel activation. *Journal of Immunology*, *187*, 4778–4787.

Bhatia, M. (2012). Role of hydrogen sulfide in the pathology of inflammation. *Scientifica*, *2012*, Article ID 159680.

Bhatia, M., Saluja, A. K., Hofbauer, B., Lee, H.-S., Frossard, J.-L., Castagliuolo, I., et al. (1998). Role of NK1 receptor in the development of acute pancreatitis and pancreatitis-associated lung injury. *Proceedings of the National Academy of Sciences of the United States of America, 95*, 4760–4765.

Bhatia, M., Sidhapuriwala, J., Moochhala, S., & Moore, P. K. (2005). Hydrogen sulphide is a mediator of carrageenan-induced hindpaw oedema in the rat. *British Journal of Pharmacology, 145*, 141–144.

Bhatia, M., Sidhapuriwala, J., Ng, S. W., Tamizhselvi, R., & Moochhala, S. (2008). Proinflammatory effects of hydrogen sulfide on substance P in caerulein-induced acute pancreatitis. *Journal of Cellular and Molecular Medicine, 12*, 580–590.

Bhatia, M., Slavin, J., Cao, Y., Basbaum, A. I., & Neoptolemos, J. (2003). Preprotachykinin-A gene deletion protects mice against acute pancreatitis and associated lung injury. *American Journal of Physiology Gastrointestinal and Liver Physiology, 284*, G830–G836.

Bhatia, M., Wong, F. L., Fu, D., Lau, H. Y., Moochhala, S., & Moore, P. K. (2005). Role of hydrogen sulfide in acute pancreatitis and associated lung injury. *FASEB Journal, 19*, 623–625.

Bhatia, M., Zhi, L., Zhang, H., Ng, S. W., & Moore, P. K. (2006). Role of substance P in hydrogen sulfide-induced pulmonary inflammation in mice. *American Journal of Physiology. Lung Cellular and Molecular Physiology, 291*, L896–L904.

Buras, J. A., Holzmann, B., & Sitkovsky, M. (2005). Animal models of sepsis: Setting the stage. *Nature Reviews. Drug Discovery, 4*, 854–865.

Cribbs, R. K., Luquette, M. H., & Besner, G. E. (1998). A standardized model of partial thickness scald burns in mice. *Journal of Surgical Research, 80*, 69–74.

Endorf, F. W., & Ahrenholz, D. (2011). Burn management. *Current Opinion in Critical Care, 17*, 601–605.

Gilligan, J. P., Lovato, S. J., Erion, M. D., & Jeng, A. Y. (1994). Modulation of carrageenan-induced hind paw edema by substance P. *Inflammation, 18*, 285–292.

Hegde, A., Koh, Y. H., Moochhala, S. M., & Bhatia, M. (2010). Neurokinin-1 receptor antagonist treatment in polymicrobial sepsis: Molecular insights. *International Journal of Inflammation, 2010*, 601098.

Hegde, A., Tamizhselvi, R., Manikandan, J., Melendez, A. J., Moochhala, S. M., & Bhatia, M. (2010). Substance P in polymicrobial sepsis: Molecular fingerprint of lung injury in preprotachykinin-$A^{-/-}$ mice. *Molecular Medicine, 16*, 188–198.

Hegde, A., Uttamchandani, M., Moochhala, S. M., & Bhatia, M. (2010). Plasma cytokine profiles in preprotachykinin-A knockout mice subjected to polymicrobial sepsis. *Molecular Medicine, 16*, 45–52.

Hegde, A., Zhang, H., Moochhala, S., & Bhatia, M. (2007). Neurokinin-1 receptor antagonist treatment protects mice against lung injury in polymicrobial sepsis. *Journal of Leukocyte Biology, 82*, 678–685.

Howard, J. G., Rowley, D., & Wardclaw, A. C. (1957). Stimulation of non-specific immunity by the lipid A component of bacterial lipopolysaccharide. *Nature, 179*, 314–315.

Koh, Y. H., Moochhala, S., & Bhatia, M. (2011). The role of neutral endopeptidase in caerulein-induced acute pancreatitis. *Journal of Immunology, 187*, 5429–5439.

Lampel, M., & Kern, H. F. (1977). Acute interstitial pancreatitis in the rat induced by excessive doses of a pancreatic secretagogue. *Virchows Archiv. A, Pathological Anatomy and Histology, 373*, 97–117.

Lau, H. Y., & Bhatia, M. (2006). Effect of CP96,345 on the expression of tachykinins and neurokinin receptors in acute pancreatitis. *Journal of Pathology, 208*, 364–371.

Lau, H. Y., & Bhatia, M. (2007). Effect of CP 96,345 on the expression of adhesion molecules in acute pancreatitis in mice. *American Journal of Physiology. Gastrointestinal and Liver Physiology, 292*, G1283–G1292.

Lau, H. Y., Wong, F. L., & Bhatia, M. (2005). A key role of neurokinin 1 receptors in acute pancreatitis and associated lung injury. *Biochemical and Biophysical Research Communications, 327*, 509–515.

Levy, M. M., Dellinger, R. P., Townsend, S. R., Linde-Zwirble, W. T., Marshall, J. C., Bion, J., et al. (2010). The surviving sepsis campaign: Results of an international guideline-based performance improvement program targeting severe sepsis. *Critical Care Medicine, 38*, 367–374.

Li, L., Bhatia, M., Zhu, Y. Z., Ramnath, R. D., Wang, Z. J., Anuar, F., et al. (2005). Hydrogen sulfide is a novel mediator of lipopolysaccharide-induced inflammation in the mouse. *FASEB Journal, 19*, 1196–1198.

Ma, V. Y., Chan, L., & Carruthers, K. J. (2014). Incidence, prevalence, costs, and impact on disability of common conditions requiring rehabilitation in the United States: Stroke, spinal cord injury, traumatic brain injury, multiple sclerosis, osteoarthritis, rheumatoid arthritis, limb loss, and back pain. *Archives of Physical and Medical Rehabilitation, 95*, 986–995.

Martin, G. S., Mannino, D. M., Eaton, S., & Moss, M. (2003). The epidemiology of sepsis in the United States from 1979 through 2000. *New England Journal of Medicine, 348*, 1546–1554.

Ng, S. W., Zhang, H., Hegde, A., & Bhatia, M. (2008). Role of preprotachykinin-A gene products on multiple organ injury in LPS-induced endotoxemia. *Journal of Leukocyte Biology, 83*, 288–295.

Norris, E. J., Feilen, N., Nguyen, N. H., Culberson, C. R., Shin, M. C., Fish, M., et al. (2013). Hydrogen sulfide modulates sinusoidal constriction and contributes to hepatic microcirculatory dysfunction during endotoxemia. *American Journal of Physiology. Gastrointestinal and Liver Physiology, 304*, G1070–G1078.

Peery, A. F., Dellon, E. S., Lund, J., Crockett, S. D., McGowan, C. E., Bulsiewicz, W. J., et al. (2012). Burden of gastrointestinal disease in the United States: 2012 update. *Gastroenterology, 143*, 1179–1187.

Puneet, P., Hegde, A., Ng, S. W., Lau, H. Y., Lu, J., Moochhala, S., et al. (2006). Preprotachykinin-A gene products are key mediators of lung injury in polymicrobial sepsis. *Journal of Immunology, 176*, 3813–3820.

Sio, S. W., Ang, S. F., Lu, J., Moochhala, S., & Bhatia, M. (2010). Substance P up-regulates cyclooxygenase-2 and prostaglandin E metabolite by activating ERK1/2 and NF-κB in a mouse model of burn-induced remote acute lung injury. *Journal of Immunology, 185*, 6265–6276.

Sio, S., Lu, J., Moochhala, S., & Bhatia, M. (2010). Early protection from burn-induced acute lung injury by deletion of preprotachykinin-A gene. *American Journal of Respiratory and Critical Care Medicine, 181*, 36–46.

Sio, S., Puthia, M. K., Lu, J., Moochhala, S., & Bhatia, M. (2008). The neuropeptide substance P is a critical mediator of burn-induced acute lung injury. *Journal of Immunology, 180*, 8333–8341.

Sun, J., & Bhatia, M. (2007). Blockade of neurokinin 1 receptor attenuates CC and CXC chemokine production in experimental acute pancreatitis and associated lung injury. *American Journal of Physiology. Gastrointestinal and Liver Physiology, 292*, G143–G153.

Tamizhselvi, R., Koh, Y. H., Sun, J., Zhang, H., & Bhatia, M. (2010). Hydrogen sulfide-induces ICAM-1 expression and neutrophil adhesion to caerulein treated pancreatic acinar cells through NF-κB and Src-family kinases pathway. *Experimental Cell Research, 316*, 1625–1636.

Tamizhselvi, R., Moore, P. K., & Bhatia, M. (2007). Hydrogen sulfide acts as a mediator of inflammation in acute pancreatitis: In vitro studies using isolated mouse pancreatic acinar cells. *Journal of Cellular and Molecular Medicine, 11*, 315–326.

Tamizhselvi, R., Moore, P. K., & Bhatia, M. (2008). Inhibition of hydrogen sulfide synthesis attenuates chemokine production and protects mice against acute pancreatitis and associated lung injury. *Pancreas*, *36*, e24–e31.

Tamizhselvi, R., Shrivastava, P., Koh, Y. H., Zhang, H., & Bhatia, M. (2011). Preprotachykinin-A gene deletion regulates hydrogen sulfide induced toll-like receptor 4 signaling pathway in cerulein-treated pancreatic acinar cells. *Pancreas*, *40*, 444–452.

Wichterman, K. A., Baue, A. E., & Chaudry, I. H. (1980). Sepsis and septic shock—A review of laboratory models and a proposal. *Journal of Surgical Research*, *29*, 189–201.

Winter, C. A., Risley, E. A., & Nuss, G. W. (1962). Carrageenin-induced edema in hind paw of the rat as an assay for antiiflammatory drugs. *Proceedings of the Society for Experimental Biology and Medicine*, *111*, 544–547.

Zhang, H., Hegde, A., Ng, S. W., Adhikari, S., Moochhala, S. M., & Bhatia, M. (2007). Hydrogen sulfide up-regulates substance P in polymicrobial sepsis associated lung injury. *Journal of Immunology*, *179*, 4153–4160.

Zhang, H., Moochhala, S., & Bhatia, M. (2008). Endogenous hydrogen sulfide regulates inflammatory response by activating the extracellular signal-regulated kinase (ERK) pathway in polymicrobial sepsis. *Journal of Immunology*, *181*, 4320–4331.

Zhang, J., Sio, S. W., Moochhala, S., & Bhatia, M. (2010). Role of hydrogen sulfide in severe burn injury-induced inflammation in the mouse. *Molecular Medicine*, *16*, 417–424.

Zhang, H., Zhi, L., Moochhala, S., Moore, P. K., & Bhatia, M. (2007a). Hydrogen sulfide acts as an inflammatory mediator in cecal ligation and puncture induced sepsis in mice by up-regulating the production of cytokines and chemokines via NF-κB. *American Journal of Physiology. Lung Cellular and Molecular Physiology*, *292*, L960–L971.

Zhang, H., Zhi, L., Moochhala, S., Moore, P. K., & Bhatia, M. (2007b). Endogenous hydrogen sulfide regulates leukocyte trafficking in cecal ligation and puncture induced sepsis. *Journal of Leukocyte Biology*, *82*, 894–905.

Zhang, H., Zhi, L., Moore, P. K., & Bhatia, M. (2006). The role of hydrogen sulfide in cecal ligation and puncture induced sepsis in the mouse. *American Journal of Physiology. Lung Cellular and Molecular Physiology*, *290*, L1193–L1201.

CHAPTER ELEVEN

Role of Hydrogen Sulfide in Brain Synaptic Remodeling

Pradip Kumar Kamat, Anuradha Kalani, Neetu Tyagi[1]
Department of Physiology and Biophysics, School of Medicine, University of Louisville, Louisville, Kentucky, USA
[1]Corresponding author: e-mail address: n0tyag01@louisville.edu

Contents

1. Introduction 208
2. Pharmacological and Physiological Effect of H_2S 210
3. Effect of H_2S on the CNS 212
4. Effect of H_2S on Brain Cells (Astrocyte, Microglia, and Oligodendrocyte) 214
5. Synapse 216
6. Glia and Neurons Interactions 217
7. Effect of H_2S on Neuronal Redox Stress 218
8. Effect of H_2S on Glutamate Neurotransmission 219
9. Effect of H_2S on NMDA Receptor Regulation 220
10. Effect of H_2S on GABA-Mediated Neurotransmission 221
11. Effect of H_2S on Calmodulin Kinase 222
12. Conclusion 222
Conflict of Interest 223
Acknowledgment 223
References 223

Abstract

Synapses are the functional connection between neurons which are necessary for the transfer of electric activity or chemical activity from one cell to another. Synapses are formed by the pre- and postsynaptic membrane which communicates between pre- and postneurons while a neurochemical modulator is operated in this process. H_2S has been known as a toxic gas with rotten eggs smell. However, increasing number of researches show that it regulate a variety of physiological and pathological processes in mammals. Hence, H_2S is a physiologically important molecule and has been referred to as the third gaseous molecule alongside carbon monoxide and nitric oxide. The previous era has made an exponential development in the physiological and pathological significance of H_2S. Specifically, in the central nervous system, H_2S facilitates long-term potentiation and regulates intracellular calcium concentration in brain cells. We as well as others have also shown that H_2S has antioxidant, antiapoptotic, and anti-inflammatory properties against various neurodegenerative disorders such as stroke, Alzheimer's disease, and vascular dementia. In this chapter, we highlight the current knowledge of H_2S and its neuroprotective effects with a special emphasis on synaptic remodeling.

Methods in Enzymology, Volume 555
ISSN 0076-6879
http://dx.doi.org/10.1016/bs.mie.2014.11.025

ABBREVIATIONS

3MST 3-mercaptopyruvate sulfurtransferase
AMPA α-amino-3-hydroxy-5-methyl-4-isoxazolepropionic acid receptor
Ca^{2+}/calmodulin calcium ion-dependent calmodulin kinase
CBS cystathionine beta-synthase
CNS central nervous system
CSE cystathionine gamma-lyase
GSH reduced form of glutathione
GSSG oxidized form of glutathione
H_2S hydrogen sulfide
LTD long-term depression
LTP long-term potentiation
NMDAR *N*-methyl-D-aspartate receptor
PAG DL-propargylglycine
SAM *S*-adenosyl-L-methionine

1. INTRODUCTION

Hydrogen sulfide (H_2S) was found to be produced endogenously in various parts of the body such as the heart (Geng et al., 2004), blood (Zhao, Chen, Shen, Kahn, & Lipke, 2001), and central nervous system (CNS) (Warenycia et al., 1989). H_2S is synthesized endogenously by a variety of mammalian tissues by two pyridoxal-5′-phosphate-dependent enzymes responsible for metabolism of L-cysteine which is a by-product of L-methionine, homocysteine, and cystathione. Cystathionine beta-synthase (CBS), cystathionine gamma-lyase (CSE), and a newly identified enzyme, 3-mercaptopyruvate sulfurtransferase (3MST) (Sen et al., 2012) are involved in generation of H_2S. The substrate of CBS and CSE can be derived from alimentary sources or can be liberated from endogenous proteins (Rezessy-Szabo et al., 2007; Zhu, Song, Li, & Dao, 2008). In the CNS, CBS was found highly expressed in the hippocampus and cerebellum (Abe & Kimura, 1996). CBS is mainly confined to astrocytes (Enokido et al., 2005; Ichinohe et al., 2005) and microglial cells. CSE is mainly expressed in the cardiovascular system, but was also found in microglial cells (Oh et al., 2006), spinal cord (Distrutti et al., 2006), and cerebellar granule neurons (Garcia-Bereguiain, Samhan-Arias, Martin-Romero, & Gutierrez-Merino, 2008). However, 3MST is also an important enzyme for the synthesis of H_2S in the brain which is localized within neurons and astrocytes (Shibuya et al., 2009; Zhao, Chan, Ng, & Wong, 2013).

3-Mercaptopyruvate is converted from cysteine by the action of cysteine aminotransferase (Tanizawa, 2011) (Fig. 1). By comparing the production of H_2S in different brain cells, Lee, Kim, Kim, and Ahn (2009) found that H_2S production in astrocytes was 7.9-fold higher than in cultured microglial cells, 9.7-fold higher than in neuron-committed teratocarcinoma NT2 cell line (NT-2 cells), and 11.5-fold higher than in neuroblastoma cell line (SH-SY5Y cells) (Lee et al., 2009). These data clearly indicate that astrocytes may be the main cells that produce H_2S in the brain. The estimated physiological concentration of H_2S was recently measured to be around 14–30 μ*M*, based on measurements of the brains of mice (Furne, Saeed, & Levitt, 2008) and consistent with values reported by another group (Ishigami et al., 2009). Above information regarding H_2S in the brain indicates the impact of

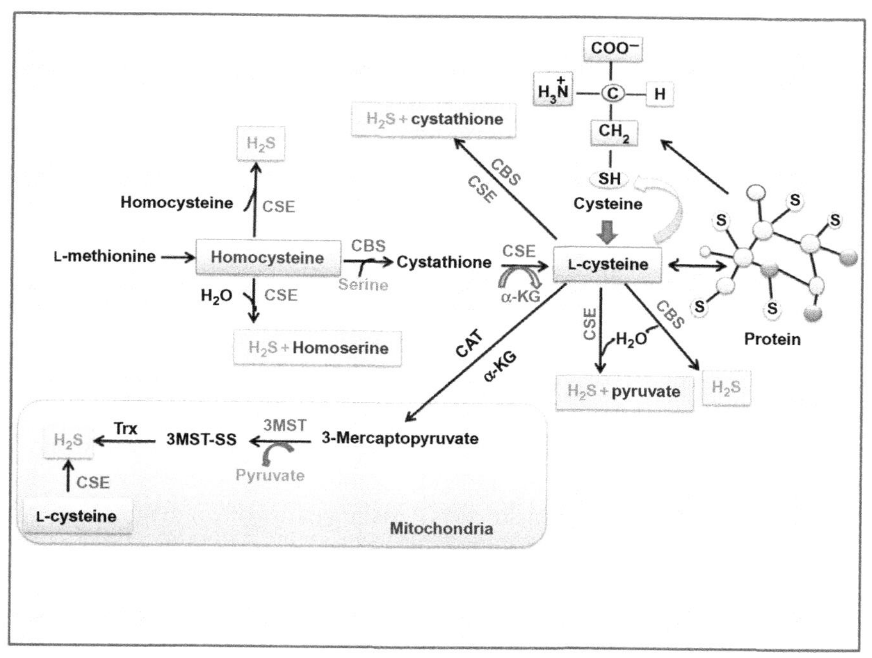

Figure 1 H_2S formation in cytosol as well as in mitochondria. The three enzymes CBS, CSE, and 3MST are responsible for H_2S generation in cells. CBS and CSE are usually present in cytosol while 3MST is present in mitochondria. The sources of cytosolic H_2S generation are L-methionine, homocysteine, cystathione, and L-cysteine. L-Cysteine also originates from a protein and cysteine and moreover also produces pyruvate and cystathione. In mitochondria, the source of H_2S synthesis are mercaptopyruvate and L-cysteine. CBS, cystathionine β-synthase; 3MST, 3-mercaptopyruvate sulfurtransferase; CSE, cystathionine γ-lyase; α-KG, α-ketoglutarate; CAT, cysteine aminotransferase; H_2S, hydrogen sulfide; TRX, thioredoxin. (See the color plate.)

H_2S sulfide on neuronal function. Neural circuits are composed of mainly glutamatergic and GABAergic neurons, which communicate through synaptic connections. GABAergic synapse maturation occurs in many brain regions. In addition, changes in GABAergic output are cell wide and not target-cell specific. Chang et al. (2014) found that glutamatergic neuronal activity also determined the AMPA receptor (a non-NMDA-type ionotropic transmembrane receptor) for glutamate that mediates fast synaptic transmission in the CNS. AMPA receptor also determines the properties of synapses on the partner with GABAergic neuron. The *N*-methyl-D-aspartate (NMDA) receptor is a major type of ionotropic glutamate receptor that plays a pivotal role in the CNS under both physiological and pathological conditions. The functional diversity of NMDA receptors can be mainly attributed to their different subunit compositions that perform multiple functions in various situations. Recent reports have indicated that synaptic NMDA receptors have a distinct role in neuronal cell survival. NMDA receptors modulate the LTP and synaptic plasticity which is responsible for learning and memory function and therefore maintains synapse function. Although H_2S enhances the induction of hippocampal LTP, the mechanism by which H_2S modulates synaptic activity is still in debate. H_2S enhances the responses of neurons to glutamate in hippocampal slices, and H_2S alone induces the increase in intracellular Ca^{2+} in astrocytes (Nagai, Tsugane, Oka, & Kimura, 2004). In the presence of H_2S, the induction of LTP is enhanced at the synapse and Ca^{2+} waves are induced in the surrounding astrocytes. Ca^{2+} waves propagate and reach another synapse and may modulate it. H_2S may therefore modulate synaptic activity by enhancing the responses to glutamate in neurons and inducing Ca^{2+} waves in astrocytes that propagate and modulate the neighboring synapse. Some of the reports showed that H_2S controls the neuronal signaling. As signaling passes through neuronal synapse, it is also modulated by H_2S. Thus, understanding the role of H_2S in the brain will help in gaining a mechanism for controlling synapse function. This review presents an overview of the current evidence that H_2S probably acts as a neuromodulator and/or as an intracellular messenger and plays an important role in synaptic remodeling.

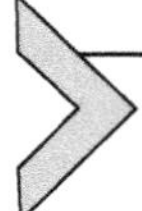

2. PHARMACOLOGICAL AND PHYSIOLOGICAL EFFECT OF H_2S

Dietary garlic has long been known for its beneficial effects on the cardiovascular system. Their components are the main source of H_2S donor naturally. The paste of garlic cloves allows the enzyme alliinase to metabolize the amino acid alliin, producing allicin (diallyl thiosulfinate), which decomposes to polysulfides, including diallyl disulfide and diallyl trisulfide (Banerjee & Maulik, 2002). When these compounds react with reduced thiol groups, including reduced glutathione, in cells, they produce H_2S. In experimental set of condition, H_2S or H_2S donors (most commonly NaHS) provide protection in many physiological systems including the cardiovascular system and central nervous system (Wallace, 2007). Most notably, H_2S mainly attenuate vasoconstriction and reduce damage via dilating blood vessels (e.g., myocardial infarct size) in several animal models of cardiovascular disease. However, in the brain, H_2S acts as anti-inflammatory, antioxidant, and finally as a neuroprotective agent which can also be designated as neuromodulator. Most researchers prefer to use pharmacological inhibitors that inhibit H_2S synthesis such as DL-propargylglycine (PAG); a CSE inhibitor and β-cyano-L-alanine (BCA), L-amino ethoxyvinylglycine, aminooxyacetic acid, trifluoroalanine, and hydroxylamine. Though BCA is a selective inhibitor for CSE, it also inhibits CBS at high concentrations. In contrast, none of the compounds tested exhibited significant selectivity toward CBS. In addition, the above-mentioned compounds did not inhibit 3MST. The effects of H_2S on cells studied using H_2S donors are variable in nature. Lower (micromolar) (lower micromolar concentrations) of H_2S are generally cytoprotective, with protection often ascribed to a reduction of free radicals generation such as neutralization of reactive oxygen and nitrogen species. These effects have been reported in neurons, myoblasts, neutrophils, and macrophages. At higher concentration such as millimolar levels, H_2S is often proapoptotic or cytotoxic, via free radical and oxidant generation, glutathione depletion, and thus promotes apoptosis. In addition to that, when sulfide donors or precursors administered systemically can convert into H_2S gas, this gas can be absorbed via the lung into the circulation and after that all over the body. Rezessy-Szabo et al. (2007) have reported the capacity and potential of intravenously (i.v.) administered H_2S donors to induce exhalation of H_2S. Previously, we and others have used H_2S donors such as IK-1001 (sodium sulfide for injection), a parenteral injectable GMP formulation of H_2S, and H_2S donor (NaHS, GYY-4137) which acts as a neuroprotector and vasoprotector. This compound has recently been proven to be protective against various physiological and pathological conditions. Moreover, the effect of H_2S donors were used in many experimental studies to characterize the pharmacological effects of H_2S *in vitro* and *in vivo* (Esechie et al., 2008; Kamat et al., 2013; Mishra, Tyagi, Sen, Givvimani, & Tyagi, 2010; Qipshidze, Metreveli, Mishra, Lominadze, & Tyagi, 2012;

Sen et al., 2012; Simon et al., 2008; Sodha et al., 2008; Tyagi, Vacek, Givvimani, Sen, & Tyagi, 2010).

3. EFFECT OF H_2S ON THE CNS

Several findings suggested that H_2S exists in the CNS at a nanomolar (n*M*) or very low micromolar (μ*M*) concentrations. In contrasts to previous literature, CNS concentrations of H_2S 50–160 μ*M*/l have been reported (Lopez, Prior, Reiffenstein, & Goodwin, 1989; Warenycia et al., 1989). This has significant impact on most previously published papers which used H_2S concentrations in the range of μ*M*. Previous findings may yet be valid based on recent evidence showing that μ*M* concentrations of H_2S rapidly decays *in vitro* to undetectable levels within 30 min suggesting single μ*M* doses of H_2S might have already exerted their effects before they decay to undetectable levels and imply that the action of H_2S is likely a molecular "switch" that activates downstream pathways that persist long after H_2S decay. H_2S is usually stored as bound sulfane sulfur in neurons and astrocytes (Ishigami et al., 2009). Upon neuron excitation or other stimulation, the bound sulfane sulfur then releases free H_2S. Free H_2S is mainly oxidized to thiosulfate, sulfite, and finally sulfate by thiosulfate:cyanide sulfurtransferase in mitochondria (Lowicka & Beltowski, 2007). H_2S can also be methylated by the enzyme thiol-*S*-methyltransferase to methanethiol and dimethylsulfide or be bound to methemoglobin, an oxidized form of hemoglobin (Lowicka & Beltowski, 2007). The clearance of H_2S via transport from the brain to the clearance organs such as kidneys, lungs, or liver is less likely as concentrations of H_2S in the blood are lower than 14 n*M* (Whitfield, Kreimier, Verdial, Skovgaard, & Olson, 2008) and H_2S has a short half-life *in vitro* (Hu et al., 2009). It was proposed that the neurological and cardiovascular actions of H_2S were continuously modulated primarily by circulating sulfide rather than by endogenous production (Olson et al., 2008). This was disproved based on recent studies reporting undetectable H_2S levels in mouse and rat blood samples using sensors that can detect 14 n*M* of H_2S (Whitfield et al., 2008), implying that H_2S found in the CNS is more likely to be derived directly from the CNS than from the blood. This also supports the hypotheses put forth by recent reviews (Li et al., 2005; Rezessy-Szabo et al., 2007) that H_2S does not circulate in the plasma at measurable conditions (Whitfield et al., 2008) which is consistent with speculations that H_2S has neuromodulatory functions *in vivo* (Abe & Kimura, 1996; Kim, Lee, Jang, Han, & Kim, 2011). In the CNS,

H_2S acts as a messenger in response to specific stimuli (usually noxious) such as febrile seizures (Han et al., 2005), stimuli leading to pain (Kubo, Kajiwara, & Kawabata, 2007), and cerebral ischemia (Qu, Chen, Halliwell, Moore, & Wong, 2006), which do not occur frequently. Therefore, it is clear that H_2S is present in the brain and comes from different sources and is involved in the regulation of intracellular signaling molecules, ion channel function, and the release and function of amino acid neurotransmitters (Fig. 2).

As reports suggesting that H_2S might play a role in synaptic transmission, it also maintains excitatory postsynaptic potentials (EPSPs). EPSPs are necessary for electrical and chemical stimulation in synapse. In the hippocampus

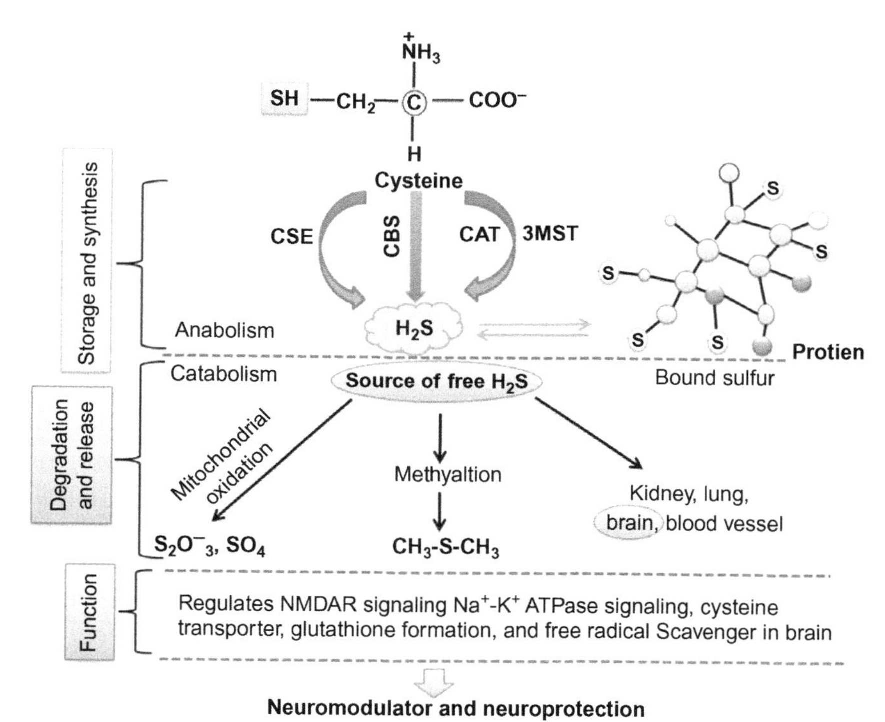

Figure 2 Diagrammatic representation shows the sources of H_2S and moreover anabolism and catabolism of H_2S. The usual sources of H_2S are blood lung, kidney, heart, and brain. Free H_2S gets converted into sulfate and thiosulfate through mitochondrial oxidation and sulfomethane via methylation which becomes a source of H_2S. Diagram also depicts the synthesis, storage, and function of H_2S in neuronal synapse. These free H_2S involved in NMDA receptor signaling, Na^+ K^+ ATPase signaling, and glutathione formation maintains and regulates neuronal function. (See the color plate.)

field and population spikes evoked by the electrical stimulation of the Schafer collaterals in the CA1 region, there is a concentration-dependent sensitivity toward H_2S (Abe & Kimura, 1996). H_2S concentrations greater than 130 μM were found to suppress both field EPSPs and population spikes. The suppression by H_2S was specific to EPSPs and the population spikes as the action potentials generated by direct stimulation of presynaptic fibers were not affected by H_2S. This indicates that the physiological concentration of H_2S is a determining factor for synapse to work properly or improperly.

4. EFFECT OF H_2S ON BRAIN CELLS (ASTROCYTE, MICROGLIA, AND OLIGODENDROCYTE)

Glial cells (astrocytes, oligodendrocytes, and microglia) are significantly abundant in the brain and are pathologically linked to neurodegenerative disorders. Large numbers of available data imply that neurodegeneration in different brain pathology is associated with the alterations in glial cells (Klegeris & McGeer, 2002; McGeer, Yasojima, & McGeer, 2002). Microglia play important roles in responses of the brain to injury and activated microglia congregate around degenerating neurons, and may produce toxins and inflammatory cytokines that contribute to the neurodegenerative process (Li et al., 2005; Rai, Kamat, Nath, & Shukla, 2013, 2014). The severe changes in glial cells and microglial cells in brain disease may promote neuronal degeneration (Jantzen et al., 2002). In addition, although synapses degenerate in vulnerable neuronal circuits, the remaining synapses may increase in size to compensate and astrocytes may play a role in this process (Murai, Nguyen, Irie, Yamaguchi, & Pasquale, 2003). Studies show that glial activation modifies long-term depression (LTD) and potentiation of synaptic transmission in the hippocampus (Albensi & Mattson, 2000). Finally, changes in the mitochondrial membrane permeability in synaptic terminals have been associated with impaired synaptic plasticity in the hippocampus (Albensi, Alasti, & Mueller, 2000). Progressively, accumulating evidence suggested that astrocytes play roles in synaptic transmission through the regulated release of synaptically active molecules including glutamate, purines (ATP and adenosine), GABA, and D-serine (Perea & Araque, 2010; Perea, Navarrete, & Araque, 2009; Shigetomi, Bowser, Sofroniew, & Khakh, 2008). Synaptic stimulation through NMDA receptors is important for learning and memory functions, but excess glutamate can over stimulate these receptors resulting in excitotoxicity and neurodegeneration (Kamat, Rai, Swarnkar, Shukla, & Nath, 2014; Michaels & Rothman, 1990). The release of such gliotransmitters occurs in response to changes in neuronal synaptic activity, which involves astrocyte excitability as reflected by increases in astrocyte Ca^{2+} and can alter neuronal excitability (Halassa, Fellin, & Haydon, 2007; Halassa, Fellin, Takano, Dong, & Haydon, 2007; Nedergaard, Ransom, & Goldman, 2003). Such evidence has given rise to the "tripartite synapse" hypothesis (Perea et al., 2009). Synaptically associated astrocytes are considered as an integral modulatory element of tripartite synapses consisting of the presynapse, the postsynapse, and the glial element (Fig. 3). Astrocytes may secrete glial binding proteins into the synaptic cleft, thus binding free neurotransmitters and thereby reducing the levels of neurotransmitters available for stimulating the postsynapse through receptors. Astrocytes also have membrane-bound receptors for neurotransmitters,

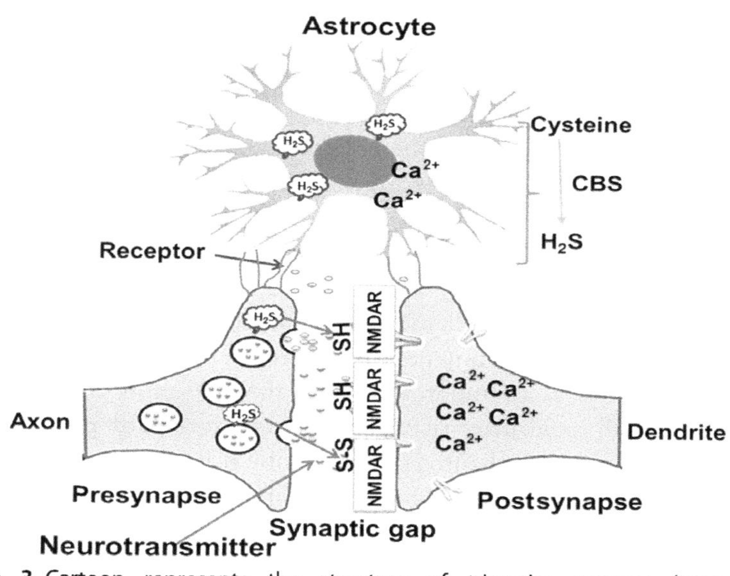

Figure 3 Cartoon represents the structure of tripartite synapse (presynapse, postsynapse, and astrocyte). In astrocyte, H_2S interacts with H_2S and slow down the excessive glutamate and calcium ion release from astrocyte. On the other hand, in pathological condition H_2S moderates the calcium ion influx through NMDA receptor in postsynapse and prevents from excitotoxic cell death. Neuronal synapses showing the presence of H_2S in presynapse and postsynapse. Moreover, it also represents how H_2S interacts with NMDA receptors; actually, H_2S modulates the NMDA receptor and controls synapse function properly. Apart from that H_2S also associated with glutathione production in cells which acts as cellular antioxidant.

and when these bind to neurotransmitters, the astrocytes upregulate the amount of binding protein secreted into the synapse. Thus, astrocytes play an important role in the formation, maintenance, and proper functioning of synapses (Christopherson et al., 2005; Ransom, Behar, & Nedergaard, 2003). Astrocytes exert a powerful influence on the synapse remodeling and pruning of the healthy adult CNS or in response to CNS disorders (Barker & Ullian, 2008). It has been reported that H_2S enhances the induction of hippocampal long-term potentiation (LTP) and induces calcium waves in astrocytes. Based on these observations, it could be strongly suggest that H_2S acts as a synaptic modulator in the brain mediated through astrocytes. Tsugane, Nagai, Kimura, Oka, and Kimura (2007) showed that differentiated astrocytes acquire sensitivity to H_2S that is diminished by their transformation into reactive astrocytes.

5. SYNAPSE

Synapses are the structures where neurons exchange the neurotransmitter which is essential for neuronal functions and mediate signals to individual target cells. At a synapse, the plasma membrane of the signal-passing neuron (the presynaptic neuron) comes into close proximity with the membrane of the target (postsynaptic) cell. Both the presynaptic and postsynaptic sites of neurons contain extensive groups of molecular machinery that link the two membranes together and carry out the signaling process. Astrocytes also exchange information with the synaptic neurons, responding to synaptic activity and, in turn, regulating neurotransmission (Fig. 3). It is well known that the synapse plays an important role in the formation of learning and memory through molecular machinery. Memory formation process occurs by way of synaptic strengthening which is known as LTP. Memory formation is related to altered release of neurotransmitters and plasticity of synapses. The postsynaptic cell can be regulated by altering the function and number of its receptors. Changes in postsynaptic signaling are most commonly associated with *N*-methyl-D-aspartic acid receptor (NMDAR)-dependent LTP and LTD, which are the most analyzed forms of plasticity at excitatory synapses. When astrocytes interact with neuronal synapses, the structure is known as tripartite synapse. Synaptic physiology is usually based on the bidirectional communication between astrocytes and neurons. Since recent evidence has demonstrated that astrocytes integrate and process synaptic information as well as control synaptic transmission and plasticity, astrocytes, being active partners in synaptic function, are cellular elements involved in the processing, transfer, and storage of information by the nervous system. As evidences suggest, H_2S is generated from astrocytes and considered as the main source. Therefore, it may affect neuronal function when it comes in contact with the synapse and may influence the synapse function.

6. GLIA AND NEURONS INTERACTIONS

Glial cells have been considered to be the nonexcitable and supportive elements in the nervous system, but they are now regarded as fundamentals for neuronal activity and modulate synaptic activity (Haydon, 2001). Glial cells, such as microglia and astrocytes, have neurotransmitter and hormone receptors and also integrate neuronal functions. The multiple interactions between neurons and glia strongly suggest that glial cells are integral parts of neurons and are referred to as modulatory elements in synaptic transmission (Araque, Parpura, Sanzgiri, & Haydon, 1999). The observation that H_2S enhances the induction of hippocampal LTP suggests that H_2S may modulate some aspects of synaptic activity. Although H_2S enhances the NMDA receptor-mediated responses to glutamate in neurons, the effects of H_2S in the absence of glutamate on brain cells are not well understood. Interactions between neurons and glia may modulate synaptic transmission, as neuronal activity can evoke glial function; may inhibit the exocytosis of glutamate or some other factor from nerve terminals when neurons are stimulated by NMDA. H_2S released in response to neuronal excitation may increase intracellular Ca^{2+} and induce Ca^{2+} waves in neighboring astrocytes. Physiological concentrations of H_2S specifically potentiate the activity of NMDA receptors and alter the induction of LTP in the hippocampus, a synaptic model for memory (Abe & Kimura, 1996). H_2S can also regulate the release of the corticotrophin-releasing hormone from the hypothalamus (Dello Russo et al., 2000; Walsh et al., 2014). H_2S increases intracellular concentrations of Ca^{2+} in glia and induces Ca^{2+} waves, which mediate glial signal transmission (Nagai et al., 2004). Given the accumulating evidence for reciprocal interactions between glia and neurons, it has been suggested that glia modulate the synaptic transmission. H_2S may regulate the synaptic activity by modulating the activity of both neurons and glia in the brain. Based upon these observations, it has been proposed that H_2S may function as a neuromodulator (Abe & Kimura, 1996).

7. EFFECT OF H_2S ON NEURONAL REDOX STRESS

H_2S has been functioning as an endogenous neuromodulator through the activation of NMDA receptors. NMDA receptors also lead to a sustained rise of neuronal cytosolic calcium ion, and interestingly, NMDA receptors are highly sensitive to oxidative stress (Kamat et al., 2013, Kamat, Tota, Saxena, Shukla, & Nath, 2010; Rai et al., 2013). The reduction potential of H_2S is close to that of the thiol group of reduced glutathione, and H_2S, like reduced glutathione, inhibits the oxidative stress-induced damage (Tyagi, Mishra, & Tyagi, 2009; Zhou & Freed, 2005). The reports from various experiments showed protective mechanisms of H_2S against cellular stress. The protection afforded by H_2S against hydrogen peroxide-induced cellular damage (Wang et al., 2013), against protein nitration induced by peroxynitrite (Whiteman et al., 2004), and against oxidative stress-induced death of neurons (Kalani, Kamat, Chaturvedi, Tyagi, & Tyagi, 2014; Kalani, Kamat, Givvimani, et al., 2014; Kamat et al., 2013). These reports further support a role of H_2S as a significant cellular antioxidant. However, the molecular mechanisms through which H_2S can attenuate the neuronal oxidative stress are still to be settled because H_2S can act both as a sacrificial scavenger of ROS and also as an inhibitor of major ROS production in cells. In addition to the well-known inhibition by H_2S of the mitochondrial respiration (Wang, Guo, & Wang, 2012), it is to be noted that the expression of gp91phox (ROS generation regulatory protein), a plasma membrane-bound NADPH-dependent oxidase which releases superoxide anion, has been recently shown to be downregulated by H_2S level (Dong et al., 2012).

The oxidative stress plays a critical role at the early stages of apoptosis. As we and others have shown in earlier works (Kalani, Kamat, Chaturvedi, et al., 2014; Kalani, Kamat, Givvimani, et al., 2014; Kamat et al., 2010; Tyagi et al., 2009), oxidative stress is largely generated by production of superoxide anions such as reactive oxygen species and nitrogen species. Glutathione (g-glutamylcysteinyl glycine; GSH) is a tripeptide containing cysteine, glutamate, and glycine with the amine group of cysteine forming a peptide bond with the carboxyl group of the side chain found in glutamate. It can exist alone in reduced forms as glutathione or in an oxidized dimer form also known as glutathione disulfide (GSSG) (Monks, Ghersi-Egea, Philbert, Cooper, & Lock, 1999). Glutathione biosynthesis is catalyzed by the enzyme g-glutamylcysteine synthetase and glutathione synthase, while glutathione recovery from GSSG is catalyzed by GSSG reductase

(Kimura & Kimura, 2004). Recent studies have also suggested that H_2S can antagonize apoptosis through inhibiting the production of ROS and thus promotes neuronal survival (Tang et al., 2013, 2011). H_2S can also protect against oxidative stress-induced neuronal damage through increasing the level of intracellular glutathione (Kimura & Kimura, 2004; Yang et al., 2011). Previous evidences also show that the pretreatment of H_2S may attenuate neuronal injury/apoptosis through inhibition of cellular apoptosis (Biermann, Lagreze, Schallner, Schwer, & Goebel, 2011; Elrod et al., 2007). Glutathione evenly distributed throughout the brain occurs low in neuronal cells but high in astrocytes and oligodendrocytes, indicating that these glial cells may be the major source of glutathione generated from H_2S; which is also an indication of the presence of H_2S in astrocyte and microglial cells. Lastly, glutathione is formed by the H_2S in the CNS.

As a reducing agent, glutathione protects neuronal as well as non-neuronal cells from free radical species either by direct action or indirectly by promoting the regeneration of other antioxidant systems (Monks et al., 1999). H_2S was also reported to inhibit peroxynitrite-induced cytotoxicity, intracellular protein nitration, and protein oxidation in human neuroblastoma SH-SY5Y cells. These data suggest that H_2S has the potential to act as an inhibitor of peroxynitrite-mediated processes *in vivo* and the potential antioxidant action of H_2S (Whiteman et al., 2004). Similarly, H_2S inhibits cell toxicity due to oxytosis (a novel form of apoptosis), a form of oxidative glutamate toxicity independent of glutamatergic signaling at ionotropic glutamate receptors, in neuronal cells and primary cultured immature cortical neurons (Umemura & Kimura, 2007). In these cells, increased intracellular cysteine levels were observed to correlate to the glutathione levels (Fig. 3). Therefore, H_2S protects against the activity of peroxynitrite and other damages of the cells from free radicals, presumably through increased glutathione production in the neuronal cells as well as neuronal supportive cells.

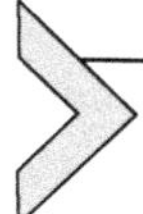

8. EFFECT OF H_2S ON GLUTAMATE NEUROTRANSMISSION

NMDA receptor blockers were reported to inhibit H_2S-induced cell death in neurons (Cheung, Peng, Chen, Moore, & Whiteman, 2007) and infarct volume in an *in vivo* rat stroke model (Qu et al., 2006), suggesting that H_2S may induce cell death through the opening of NMDA receptors. In summary of the properties of H_2S-induced NMDA signaling, H_2S may promote excitation and regulate survival/death decisions of the neurons.

It was reported that H_2S increased glutamate secretion in rat cerebellar granule neurons, which resulted in the neuronal cell death (Garcia-Bereguiain et al., 2008). This was demonstrated by a significant increase in extracellular concentrations of glutamate from physiological concentrations of 2–5 μ*M* (Erecinska, Dagani, Nelson, Deas, & Silver, 1991; Erecinska & Silver, 1990) to supraphysiological (and thus toxic) concentrations of 10–15 μ*M*. This observation was confirmed by blockade of H_2S-induced cell death by NMDA blocker MK-801 and glutamate antagonist DL-2-amino-5-phosphonovaleric acid. A recent paper by Whitfield et al. (2008) suggested the amount of H_2S derived from the plasma is likely to be negligible compared to endogenous production. On the other hand, H_2S is more likely to decay rapidly than persist at micromolar concentrations *in vitro* (Hu et al., 2009), and this suggests that the observed downstream effects of glutamate release by H_2S due to continuous maintenance of high H_2S concentrations (Garcia-Bereguiain et al., 2008) may be reflective of a toxicological situation. If so, a toxic exposure of the cells to glutamate due to a constant high level of H_2S may lead to excitotoxicity and neuronal cell death, thus leading to the compromised neuronal functions, e.g., memory, in addition to neuropathic pain (Hudspith, 1997).

9. EFFECT OF H_2S ON NMDA RECEPTOR REGULATION

The NMDA type of glutamate receptor (NMDAR) plays a key role in neuronal plasticity, learning, and memory in the CNS, most of which is related to its high permeability to Ca^{2+} (Li & Tsien, 2009). Synaptic stimulation through NMDA receptors is important for the learning and memory functions, but excess glutamate can over stimulate these receptors resulting into excitotoxicity and neurodegeneration (Michaels & Rothman, 1990). Glutamate is an important excitatory amino acid that functions as a neurotransmitter in the mammalian brain. Glutamate plays a role in physiological processes including learning and memory, especially with respect to its central role in induction of LTP, perception of pain, and also in pathological processes such as excitatory neuronal injury (Hudspith & Munglani, 1998). NMDA receptors are a class of receptor-operated glutamate receptors mostly expressed in the nervous system (central and peripheral) (Laezza, Doherty, & Dingledine, 1999). Although there is no direct evidence demonstrating agonist activity of H_2S on NMDA receptors, accumulating evidence indicates that H_2S may produce physiological or pathological functions via regulating NMDA receptors. It was found that H_2S stimulates LTP via potentiation of NMDA receptors. This effect was achieved mainly by H_2S-induced activation of cAMP/PKA pathway (Kimura, 2000). Excessive NMDA receptor activation causes calcium overload in the cells leading to cell death (Gagliardi, 2000). Harris, Ganong, and Cotman (1984) reported that hippocampal LTP alone is not facilitated by H_2S alone and it needs a stimulation to activate NMDA receptors. It seems that H_2S alone does not induce any illusive currents, but significantly increases the NMDA current. The enhancing effect of H_2S on the NMDA receptor is also concentration dependent. Therefore, H_2S enhances the induction of LTP by activating NMDA receptors. Some reports also suggest that disulfide bonds play a role in modulating the function of many proteins, including NMDA receptors. It is therefore possible that H_2S interacts with disulfide bonds or free thiol group (S-H) in NMDA receptors (Fig. 3). Therefore, NMDA receptors have important roles in the brain disease (Myers, Dingledine, & Borges, 1999) and H_2S may modulate this disease progression.

10. EFFECT OF H_2S ON GABA-MEDIATED NEUROTRANSMISSION

GABA, an inhibitory neurotransmitter, is present at high concentrations in the mammalian brain, especially in the axons. Within the mammalian CNS, GABA is the major inhibitory neurotransmitter: About 20–30% of all synapses in the CNS employ GABA as their neurotransmitter (Kaila, 1994). GABA-mediated inhibition in the CNS is critical as loss of GABAergic inhibition leads to seizures and neuronal hyperexcitability. There are three types of receptors for GABA in the CNS: GABAA, GABAB, and GABAC receptors, and they produce slow, prolonged inhibitory signals and function to modulate the release of neurotransmitters (Chebib & Johnston, 1999). H_2S was known to promote amelioration of hippocampal damage induced by recurrent seizures via reversing the loss of GABABR1 and GABABR2 which is caused by febrile seizures (Han et al., 2005). This amelioration was outlined to the increased mRNA and protein levels of these GABA receptors, which may be due to acute increases in the Ca^{2+} leading to Ca^{2+}-dependent transcription by H_2S induction (Clapham, 2007; Lipscombe, Helton, & Xu, 2004; Pietrobon, 2002). This may have an effect on the restoration of the excitation/inhibition balance perturbed and affecting slow, prolonged inhibitory signals and neurotransmitter release. It was inferred from the above study that H_2S may have therapeutic use in the treatment of excitatory diseases such as epilepsy (Han et al., 2005).

11. EFFECT OF H_2S ON CALMODULIN KINASE

Calcium/calmodulin-dependent protein kinase II (CaM kinase II or CaMKII) is a serine/threonine-specific protein kinase that is regulated by the Ca^{2+}/calmodulin complex. CaMKII is also necessary for Ca^{2+} homeostasis in the neuronal cells. On the other hand, CBS is the main enzyme for the synthesis of H_2S in the brain. In addition, it is also regulated by *S*-adenosylmethionine which acts as an allosteric activator of CBS. As CBS enzyme is a Ca^{2+} and calmodulin-dependent enzyme, the biosynthesis of H_2S is strongly controlled by the intracellular concentration of the Ca^{2+} ion. In addition to that, CBS is also regulated by *S*-adenosyl-L-methionine (SAM) and pyridoxal-5′-phosphate. It was recently found that Ca^{2+}/calmodulin-mediated pathways are involved in the regulation of CBS activity, which acts as an allosteric activator of CBS. In neurons, H_2S stimulates the production of cAMP probably by direct activation of adenylyl cyclase and thus activates cAMP-dependent processes. Cyclic-AMP-mediated pathways may be involved in the modulation of NMDA receptors by H_2S (Kimura, 2000). Ko and Chu (2005) showed a novel regulatory mechanism for H_2S production by Ca^{2+}/calmodulin. In addition to that, they have also shown that L-glutamate, as well as electrical stimulation, enhances the production of H_2S from brain slices and that LTP is altered in CBS knockout mice. The observations by Ko and Chu (2005) also support that endogenous H_2S is produced when CBS is activated by the Ca^{2+} which occurs with neuronal excitation, and that H_2S may function as a neuromodulator or neurotransmitter (Baranano, Ferris, & Snyder, 2001). Thus, H_2S is produced in response to neuronal excitation and alters hippocampal LTP, a synaptic basis of memory (Fig. 4).

12. CONCLUSION

Recent and previous evidences suggest that H_2S plays an important role in the maintenance of physiological conditions of neurons by its neuroprotective effects. H_2S is originated from both neurons and glial cells in the CNS. It has a potential effect on neurotransmitters such as glutamate, GABA, and AMPA neuronal receptors. This effect of H_2S also has an impact on synapse signaling by interacting with neuronal receptor and thus maintains the neuronal function as well as synapse functions. Under pathological condition, H_2S acts as an anti-inflammatory and anti-oxidative molecule and

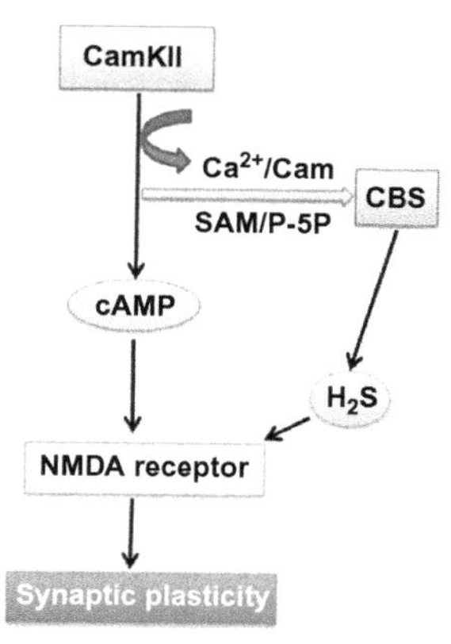

Figure 4 Flow diagram indicates the interaction and regulation of CBS enzyme with the help of CaMKII, calcium ion, *S*-adenosyl methionine, and pyridoxal-5′-phosphate. In other ways, CaMKII also potentiate the NMDA receptor via cAMP pathways. Further, H_2S produced from different sources with the help of CBS enzyme interacts with NMDA receptor and modulates their function and thereby maintains the synaptic plasticity. P-5P, pyridoxal-5′-phosphate; cAMP, cyclic adenosine monophosphate; CamKIIα, calmodulin kinase II alfa; SAM, *S*-adenosyl methionine; NMDA, *N*-methyl-D-aspartate.

hence protects neurons and synapse from abnormal pathology, thereby remodeling the neurons as well as neuronal synapse during or after pathology.

CONFLICT OF INTEREST

The authors declare that they have no conflicts of interest.

ACKNOWLEDGMENT

This work was supported by National Institutes of Health grant HL107640-NT.

REFERENCES

Abe, K., & Kimura, H. (1996). The possible role of hydrogen sulfide as an endogenous neuromodulator. *The Journal of Neuroscience: The Official Journal of the Society for Neuroscience, 16*, 1066–1071.

Albensi, B. C., Alasti, N., & Mueller, A. L. (2000). Long-term potentiation in the presence of NMDA receptor antagonist arylalkylamine spider toxins. *Journal of Neuroscience Research, 62*, 177–185.

Albensi, B. C., & Mattson, M. P. (2000). Evidence for the involvement of TNF and NF-kappaB in hippocampal synaptic plasticity. *Synapse, 35*, 151–159.

Araque, A., Parpura, V., Sanzgiri, R. P., & Haydon, P. G. (1999). Tripartite synapses: Glia, the unacknowledged partner. *Trends in Neurosciences, 22*, 208–215.

Banerjee, S. K., & Maulik, S. K. (2002). Effect of garlic on cardiovascular disorders: A review. *Nutrition Journal, 1*, 4.

Baranano, D. E., Ferris, C. D., & Snyder, S. H. (2001). Atypical neural messengers. *Trends in Neurosciences, 24*, 99–106.

Barker, A. J., & Ullian, E. M. (2008). New roles for astrocytes in developing synaptic circuits. *Communicative & Integrative Biology*, *1*, 207–211.

Biermann, J., Lagreze, W. A., Schallner, N., Schwer, C. I., & Goebel, U. (2011). Inhalative preconditioning with hydrogen sulfide attenuated apoptosis after retinal ischemia/reperfusion injury. *Molecular Vision*, *17*, 1275–1286.

Chang, L. C., Jamain, S., Lin, C. W., Rujescu, D., Tseng, G. C., & Sibille, E. (2014). A conserved BDNF, glutamate- and GABA-enriched gene module related to human depression identified by coexpression meta-analysis and DNA variant genome-wide association studies. *PLoS One*, *9*, e90980.

Chebib, M., & Johnston, G. A. (1999). The 'ABC' of GABA receptors: A brief review. *Clinical and Experimental Pharmacology & Physiology*, *26*, 937–940.

Cheung, N. S., Peng, Z. F., Chen, M. J., Moore, P. K., & Whiteman, M. (2007). Hydrogen sulfide induced neuronal death occurs via glutamate receptor and is associated with calpain activation and lysosomal rupture in mouse primary cortical neurons. *Neuropharmacology*, *53*, 505–514.

Christopherson, K. S., Ullian, E. M., Stokes, C. C., Mullowney, C. E., Hell, J. W., Agah, A., et al. (2005). Thrombospondins are astrocyte-secreted proteins that promote CNS synaptogenesis. *Cell*, *120*, 421–433.

Clapham, D. E. (2007). Calcium signaling. *Cell*, *131*, 1047–1058.

Dello Russo, C., Tringali, G., Ragazzoni, E., Maggiano, N., Menini, E., Vairano, M., et al. (2000). Evidence that hydrogen sulphide can modulate hypothalamo-pituitary-adrenal axis function: In vitro and in vivo studies in the rat. *Journal of Neuroendocrinology*, *12*, 225–233.

Distrutti, E., Sediari, L., Mencarelli, A., Renga, B., Orlandi, S., Antonelli, E., et al. (2006). Evidence that hydrogen sulfide exerts antinociceptive effects in the gastrointestinal tract by activating KATP channels. *The Journal of Pharmacology and Experimental Therapeutics*, *316*, 325–335.

Dong, X. B., Yang, C. T., Zheng, D. D., Mo, L. Q., Wang, X. Y., Lan, A. P., et al. (2012). Inhibition of ROS-activated ERK1/2 pathway contributes to the protection of H2S against chemical hypoxia-induced injury in H9c2 cells. *Molecular and Cellular Biochemistry*, *362*, 149–157.

Elrod, J. W., Calvert, J. W., Morrison, J., Doeller, J. E., Kraus, D. W., Tao, L., et al. (2007). Hydrogen sulfide attenuates myocardial ischemia-reperfusion injury by preservation of mitochondrial function. *Proceedings of the National Academy of Sciences of the United States of America*, *104*, 15560–15565.

Enokido, Y., Suzuki, E., Iwasawa, K., Namekata, K., Okazawa, H., & Kimura, H. (2005). Cystathionine beta-synthase, a key enzyme for homocysteine metabolism, is preferentially expressed in the radial glia/astrocyte lineage of developing mouse CNS. *FASEB Journal: Official Publication of the Federation of American Societies for Experimental Biology*, *19*, 1854–1856.

Erecinska, M., Dagani, F., Nelson, D., Deas, J., & Silver, I. A. (1991). Relations between intracellular ions and energy metabolism: A study with monensin in synaptosomes, neurons, and C6 glioma cells. *The Journal of Neuroscience: The Official Journal of the Society for Neuroscience*, *11*, 2410–2421.

Erecinska, M., & Silver, I. A. (1990). Metabolism and role of glutamate in mammalian brain. *Progress in Neurobiology*, *35*, 245–296.

Esechie, A., Kiss, L., Olah, G., Horvath, E. M., Hawkins, H., Szabo, C., et al. (2008). Protective effect of hydrogen sulfide in a murine model of acute lung injury induced by combined burn and smoke inhalation. *Clinical Science*, *115*, 91–97.

Furne, J., Saeed, A., & Levitt, M. D. (2008). Whole tissue hydrogen sulfide concentrations are orders of magnitude lower than presently accepted values. *American Journal of Physiology. Regulatory, Integrative and Comparative Physiology*, *295*, R1479–R1485.

Gagliardi, R. J. (2000). Neuroprotection, excitotoxicity and NMDA antagonists. *Arquivos de Neuro-Psiquiatria, 58*, 583–588.
Garcia-Bereguiain, M. A., Samhan-Arias, A. K., Martin-Romero, F. J., & Gutierrez-Merino, C. (2008). Hydrogen sulfide raises cytosolic calcium in neurons through activation of L-type Ca^{2+} channels. *Antioxidants & Redox Signaling, 10*, 31–42.
Geng, B., Yang, J., Qi, Y., Zhao, J., Pang, Y., Du, J., et al. (2004). H_2S generated by heart in rat and its effects on cardiac function. *Biochemical and Biophysical Research Communications, 313*, 362–368.
Halassa, M. M., Fellin, T., & Haydon, P. G. (2007). The tripartite synapse: Roles for gliotransmission in health and disease. *Trends in Molecular Medicine, 13*, 54–63.
Halassa, M. M., Fellin, T., Takano, H., Dong, J. H., & Haydon, P. G. (2007). Synaptic islands defined by the territory of a single astrocyte. *The Journal of Neuroscience: The Official Journal of the Society for Neuroscience, 27*, 6473–6477.
Han, Y., Qin, J., Chang, X., Yang, Z., Bu, D., & Du, J. (2005). Modulating effect of hydrogen sulfide on gamma-aminobutyric acid B receptor in recurrent febrile seizures in rats. *Neuroscience Research, 53*, 216–219.
Harris, E. W., Ganong, A. H., & Cotman, C. W. (1984). Long-term potentiation in the hippocampus involves activation of N-methyl-D-aspartate receptors. *Brain Research, 323*, 132–137.
Haydon, P. G. (2001). GLIA: Listening and talking to the synapse. *Nature Reviews. Neuroscience, 2*, 185–193.
Hudspith, M. J. (1997). Glutamate: A role in normal brain function, anaesthesia, analgesia and CNS injury. *British Journal of Anaesthesia, 78*, 731–747.
Hudspith, M., & Munglani, R. (1998). A role for presynaptic NMDA receptors in central sensitization in the spinal cord dorsal horn? *British Journal of Anaesthesia, 81*, 294–295.
Ichinohe, A., Kanaumi, T., Takashima, S., Enokido, Y., Nagai, Y., & Kimura, H. (2005). Cystathionine beta-synthase is enriched in the brains of Down's patients. *Biochemical and Biophysical Research Communications, 338*, 1547–1550.
Ishigami, M., Hiraki, K., Umemura, K., Ogasawara, Y., Ishii, K., & Kimura, H. (2009). A source of hydrogen sulfide and a mechanism of its release in the brain. *Antioxidants & Redox Signaling, 11*, 205–214.
Jantzen, P. T., Connor, K. E., DiCarlo, G., Wenk, G. L., Wallace, J. L., Rojiani, A. M., et al. (2002). Microglial activation and beta-amyloid deposit reduction caused by a nitric oxide-releasing nonsteroidal anti-inflammatory drug in amyloid precursor protein plus presenilin-1 transgenic mice. *The Journal of Neuroscience: The Official Journal of the Society for Neuroscience, 22*, 2246–2254.
Kaila, K. (1994). Ionic basis of GABAA receptor channel function in the nervous system. *Progress in Neurobiology, 42*, 489–537.
Kalani, A., Kamat, P. K., Chaturvedi, P., Tyagi, S. C., & Tyagi, N. (2014). Curcumin-primed exosomes mitigate endothelial cell dysfunction during hyperhomocysteinemia. *Life Sciences, 107*, 1–7.
Kalani, A., Kamat, P. K., Givvimani, S., Brown, K., Metreveli, N., Tyagi, S. C., et al. (2014). Nutri-epigenetics ameliorates blood-brain barrier damage and neurodegeneration in hyperhomocysteinemia: Role of folic acid. *Journal of Molecular Neuroscience: MN, 52*, 202–215.
Kamat, P. K., Kalani, A., Givvimani, S., Sathnur, P. B., Tyagi, S. C., & Tyagi, N. (2013). Hydrogen sulfide attenuates neurodegeneration and neurovascular dysfunction induced by intracerebral-administered homocysteine in mice. *Neuroscience, 252*, 302–319.
Kamat, P. K., Rai, S., Swarnkar, S., Shukla, R., & Nath, C. (2014). Mechanism of synapse redox stress in Okadaic acid (ICV) induced memory impairment: Role of NMDA receptor. *Neurochemistry International, 76*, 32–41.

Kamat, P. K., Tota, S., Saxena, G., Shukla, R., & Nath, C. (2010). Okadaic acid (ICV) induced memory impairment in rats: A suitable experimental model to test anti-dementia activity. *Brain Research, 1309*, 66–74.

Kim, B., Lee, J., Jang, J., Han, D., & Kim, K. H. (2011). Prediction on the seasonal behavior of hydrogen sulfide using a neural network model. *TheScientificWorldJournal, 11*, 992–1004.

Kimura, H. (2000). Hydrogen sulfide induces cyclic AMP and modulates the NMDA receptor. *Biochemical and Biophysical Research Communications, 267*, 129–133.

Kimura, Y., & Kimura, H. (2004). Hydrogen sulfide protects neurons from oxidative stress. *FASEB Journal: Official Publication of the Federation of American Societies for Experimental Biology, 18*, 1165–1167.

Klegeris, A., & McGeer, P. L. (2002). Cyclooxygenase and 5-lipoxygenase inhibitors protect against mononuclear phagocyte neurotoxicity. *Neurobiology of Aging, 23*, 787–794.

Ko, T. H., & Chu, H. (2005). Spectroscopic study on sorption of hydrogen sulfide by means of red soil. *Spectrochimica Acta. Part A, Molecular and Biomolecular Spectroscopy, 61*, 2253–2259.

Kubo, S., Kajiwara, M., & Kawabata, A. (2007). Dual modulation of the tension of isolated gastric artery and gastric mucosal circulation by hydrogen sulfide in rats. *Inflammopharmacology, 15*, 288–292.

Laezza, F., Doherty, J. J., & Dingledine, R. (1999). Long-term depression in hippocampal interneurons: Joint requirement for pre- and postsynaptic events. *Science, 285*, 1411–1414.

Lee, J., Kim, C. H., Kim, D. G., & Ahn, Y. S. (2009). Zinc inhibits amyloid beta production from Alzheimer's amyloid precursor protein in SH-SY5Y cells. *The Korean Journal of Physiology & Pharmacology: Official Journal of the Korean Physiological Society and the Korean Society of Pharmacology, 13*, 195–200.

Li, L., Bhatia, M., Zhu, Y. Z., Zhu, Y. C., Ramnath, R. D., Wang, Z. J., et al. (2005). Hydrogen sulfide is a novel mediator of lipopolysaccharide-induced inflammation in the mouse. *FASEB Journal: Official Publication of the Federation of American Societies for Experimental Biology, 19*, 1196–1198.

Li, F., & Tsien, J. Z. (2009). Memory and the NMDA receptors. *The New England Journal of Medicine, 361*, 302–303.

Lipscombe, D., Helton, T. D., & Xu, W. (2004). L-type calcium channels: The low down. *Journal of Neurophysiology, 92*, 2633–2641.

Lopez, A., Prior, M. G., Reiffenstein, R. J., & Goodwin, L. R. (1989). Peracute toxic effects of inhaled hydrogen sulfide and injected sodium hydrosulfide on the lungs of rats. *Fundamental and Applied Toxicology: Official Journal of the Society of Toxicology, 12*, 367–373.

Lowicka, E., & Beltowski, J. (2007). Hydrogen sulfide (H_2S)—The third gas of interest for pharmacologists. *Pharmacological Reports: PR, 59*, 4–24.

McGeer, P. L., Yasojima, K., & McGeer, E. G. (2002). Association of interleukin-1 beta polymorphisms with idiopathic Parkinson's disease. *Neuroscience Letters, 326*, 67–69.

Michaels, R. L., & Rothman, S. M. (1990). Glutamate neurotoxicity in vitro: Antagonist pharmacology and intracellular calcium concentrations. *The Journal of Neuroscience: The Official Journal of the Society for Neuroscience, 10*, 283–292.

Mishra, P. K., Tyagi, N., Sen, U., Givvimani, S., & Tyagi, S. C. (2010). H_2S ameliorates oxidative and proteolytic stresses and protects the heart against adverse remodeling in chronic heart failure. *American Journal of Physiology. Heart and Circulatory Physiology, 298*, H451–H456.

Monks, T. J., Ghersi-Egea, J. F., Philbert, M., Cooper, A. J., & Lock, E. A. (1999). Symposium overview: The role of glutathione in neuroprotection and neurotoxicity. *Toxicological Sciences: An Official Journal of the Society of Toxicology, 51*, 161–177.

Murai, K. K., Nguyen, L. N., Irie, F., Yamaguchi, Y., & Pasquale, E. B. (2003). Control of hippocampal dendritic spine morphology through ephrin-A3/EphA4 signaling. *Nature Neuroscience*, *6*, 153–160.

Myers, S. J., Dingledine, R., & Borges, K. (1999). Genetic regulation of glutamate receptor ion channels. *Annual Review of Pharmacology and Toxicology*, *39*, 221–241.

Nagai, Y., Tsugane, M., Oka, J., & Kimura, H. (2004). Hydrogen sulfide induces calcium waves in astrocytes. *FASEB Journal: Official Publication of the Federation of American Societies for Experimental Biology*, *18*, 557–559.

Nedergaard, M., Ransom, B., & Goldman, S. A. (2003). New roles for astrocytes: Redefining the functional architecture of the brain. *Trends in Neurosciences*, *26*, 523–530.

Oh, G. S., Pae, H. O., Lee, B. S., Kim, B. N., Kim, J. M., Kim, H. R., et al. (2006). Hydrogen sulfide inhibits nitric oxide production and nuclear factor-kappaB via heme oxygenase-1 expression in RAW264.7 macrophages stimulated with lipopolysaccharide. *Free Radical Biology & Medicine*, *41*, 106–119.

Olson, K. R., Healy, M. J., Qin, Z., Skovgaard, N., Vulesevic, B., Duff, D. W., et al. (2008). Hydrogen sulfide as an oxygen sensor in trout gill chemoreceptors. *American Journal of Physiology. Regulatory, Integrative and Comparative Physiology*, *295*, R669–R680.

Perea, G., & Araque, A. (2010). GLIA modulates synaptic transmission. *Brain Research Reviews*, *63*, 93–102.

Perea, G., Navarrete, M., & Araque, A. (2009). Tripartite synapses: Astrocytes process and control synaptic information. *Trends in Neurosciences*, *32*, 421–431.

Pietrobon, D. (2002). Calcium channels and channelopathies of the central nervous system. *Molecular Neurobiology*, *25*, 31–50.

Qipshidze, N., Metreveli, N., Mishra, P. K., Lominadze, D., & Tyagi, S. C. (2012). Hydrogen sulfide mitigates cardiac remodeling during myocardial infarction via improvement of angiogenesis. *International Journal of Biological Sciences*, *8*, 430–441.

Qu, K., Chen, C. P., Halliwell, B., Moore, P. K., & Wong, P. T. (2006). Hydrogen sulfide is a mediator of cerebral ischemic damage. *Stroke: A Journal of Cerebral Circulation*, *37*, 889–893.

Rai, S., Kamat, P. K., Nath, C., & Shukla, R. (2013). A study on neuroinflammation and NMDA receptor function in STZ (ICV) induced memory impaired rats. *Journal of Neuroimmunology*, *254*, 1–9.

Rai, S., Kamat, P. K., Nath, C., & Shukla, R. (2014). Glial activation and post-synaptic neurotoxicity: The key events in streptozotocin (ICV) induced memory impairment in rats. *Pharmacology, Biochemistry, and Behavior*, *117*, 104–117.

Ransom, B., Behar, T., & Nedergaard, M. (2003). New roles for astrocytes (stars at last). *Trends in Neurosciences*, *26*, 520–522.

Rezessy-Szabo, J. M., Nguyen, Q. D., Hoschke, A., Braet, C., Hajos, G., & Claeyssens, M. (2007). A novel thermostable alpha-galactosidase from the thermophilic fungus Thermomyces lanuginosus CBS 395.62/b: Purification and characterization. *Biochimica et Biophysica Acta*, *1770*, 55–62.

Sen, U., Sathnur, P. B., Kundu, S., Givvimani, S., Coley, D. M., Mishra, P. K., et al. (2012). Increased endogenous H_2S generation by CBS, CSE, and 3MST gene therapy improves ex vivo renovascular relaxation in hyperhomocysteinemia. *American Journal of Physiology. Cell Physiology*, *303*, C41–C51.

Shibuya, N., Tanaka, M., Yoshida, M., Ogasawara, Y., Togawa, T., Ishii, K., et al. (2009). 3-Mercaptopyruvate sulfurtransferase produces hydrogen sulfide and bound sulfane sulfur in the brain. *Antioxidants & Redox Signaling*, *11*, 703–714.

Shigetomi, E., Bowser, D. N., Sofroniew, M. V., & Khakh, B. S. (2008). Two forms of astrocyte calcium excitability have distinct effects on NMDA receptor-mediated slow inward currents in pyramidal neurons. *The Journal of Neuroscience: The Official Journal of the Society for Neuroscience*, *28*, 6659–6663.

Simon, F., Giudici, R., Duy, C. N., Schelzig, H., Oter, S., Groger, M., et al. (2008). Hemodynamic and metabolic effects of hydrogen sulfide during porcine ischemia/reperfusion injury. *Shock, 30*, 359–364.

Sodha, N. R., Clements, R. T., Feng, J., Liu, Y., Bianchi, C., Horvath, E. M., et al. (2008). The effects of therapeutic sulfide on myocardial apoptosis in response to ischemia-reperfusion injury. *European Journal of Cardio-Thoracic Surgery: Official Journal of the European Association for Cardio-thoracic Surgery, 33*, 906–913.

Tang, X. Q., Chen, R. Q., Dong, L., Ren, Y. K., Del Soldato, P., Sparatore, A., et al. (2013). Role of paraoxonase-1 in the protection of hydrogen sulfide-donating sildenafil (ACS6) against homocysteine-induced neurotoxicity. *Journal of Molecular Neuroscience: MN, 50*, 70–77.

Tang, X. Q., Chen, R. Q., Ren, Y. K., Soldato, P. D., Sparatore, A., Zhuang, Y. Y., et al. (2011). ACS6, a hydrogen sulfide-donating derivative of sildenafil, inhibits homocysteine-induced apoptosis by preservation of mitochondrial function. *Medical Gas Research, 1*, 20.

Tanizawa, K. (2011). Production of H_2S by 3-mercaptopyruvate sulphurtransferase. *Journal of Biochemistry, 149*, 357–359.

Tsugane, M., Nagai, Y., Kimura, Y., Oka, J., & Kimura, H. (2007). Differentiated astrocytes acquire sensitivity to hydrogen sulfide that is diminished by the transformation into reactive astrocytes. *Antioxidants & Redox Signaling, 9*, 257–269.

Tyagi, N., Mishra, P. K., & Tyagi, S. C. (2009). Homocysteine, hydrogen sulfide (H_2S) and NMDA-receptor in heart failure. *Indian Journal of Biochemistry & Biophysics, 46*, 441–446.

Tyagi, N., Vacek, J. C., Givvimani, S., Sen, U., & Tyagi, S. C. (2010). Cardiac specific deletion of N-methyl-d-aspartate receptor 1 ameliorates mtMMP-9 mediated autophagy/mitophagy in hyperhomocysteinemia. *Journal of Receptor and Signal Transduction Research, 30*, 78–87.

Umemura, K., & Kimura, H. (2007). Hydrogen sulfide enhances reducing activity in neurons: Neurotrophic role of H_2S in the brain? *Antioxidants & Redox Signaling, 9*, 2035–2041.

Wallace, J. L. (2007). Hydrogen sulfide-releasing anti-inflammatory drugs. *Trends in Pharmacological Sciences, 10*, 501–505.

Walsh, J. J., Friedman, A. K., Sun, H., Heller, E. A., Ku, S. M., Juarez, B., et al. (2014). Stress and CRF gate neural activation of BDNF in the mesolimbic reward pathway. *Nature Neuroscience, 17*, 27–29.

Wang, M., Guo, Z., & Wang, S. (2012). Cystathionine gamma-lyase expression is regulated by exogenous hydrogen peroxide in the mammalian cells. *Gene Expression, 15*, 235–241.

Wang, X., Han, A., Wen, C., Chen, M., Chen, X., Yang, X., et al. (2013). The effects of H_2S on the activities of CYP2B6, CYP2D6, CYP3A4, CYP2C19 and CYP2C9 in vivo in rat. *International Journal of Molecular Sciences, 14*, 24055–24063.

Warenycia, M. W., Goodwin, L. R., Benishin, C. G., Reiffenstein, R. J., Francom, D. M., Taylor, J. D., et al. (1989). Acute hydrogen sulfide poisoning. Demonstration of selective uptake of sulfide by the brainstem by measurement of brain sulfide levels. *Biochemical Pharmacology, 38*, 973–981.

Whiteman, M., Armstrong, J. S., Chu, S. H., Jia-Ling, S., Wong, B. S., Cheung, N. S., et al. (2004). The novel neuromodulator hydrogen sulfide: An endogenous peroxynitrite 'scavenger'? *Journal of Neurochemistry, 90*, 765–768.

Whitfield, N. L., Kreimier, E. L., Verdial, F. C., Skovgaard, N., & Olson, K. R. (2008). Reappraisal of H_2S/sulfide concentration in vertebrate blood and its potential significance in ischemic preconditioning and vascular signaling. *American Journal of Physiology. Regulatory, Integrative and Comparative Physiology, 294*, R1930–R1937.

Yang, Z., Yang, C., Xiao, L., Liao, X., Lan, A., Wang, X., et al. (2011). Novel insights into the role of HSP90 in cytoprotection of H_2S against chemical hypoxia-induced injury in H9c2 cardiac myocytes. *International Journal of Molecular Medicine*, *28*, 397–403.
Zhao, H., Chan, S. J., Ng, Y. K., & Wong, P. T. (2013). Brain 3-mercaptopyruvate sulfurtransferase (3MST): Cellular localization and downregulation after acute stroke. *PLoS One*, *8*, e67322.
Zhao, H., Chen, M. H., Shen, Z. M., Kahn, P. C., & Lipke, P. N. (2001). Environmentally induced reversible conformational switching in the yeast cell adhesion protein alpha-agglutinin. *Protein Science: A Publication of the Protein Society*, *10*, 1113–1123.
Zhou, W., & Freed, C. R. (2005). DJ-1 up-regulates glutathione synthesis during oxidative stress and inhibits A53T alpha-synuclein toxicity. *The Journal of Biological Chemistry*, *280*, 43150–43158.
Zhu, W., Song, X., Li, M., & Dao, J. (2008). [CBS gene variations and serum homocysteine level associated with congenital heart defects]. *Wei Sheng Yan Jiu = Journal of Hygiene Research*, *37*, 463–467.

SECTION IV

H_2S in Plants

CHAPTER TWELVE

Detection of Thiol Modifications by Hydrogen Sulfide

E. Williams*, S. Pead*, M. Whiteman†, M.E. Wood‡, I.D. Wilson*, M.R. Ladomery*, T. Teklic§, M. Lisjak§, J.T. Hancock*,[1]

*Faculty of Health and Applied Sciences, University of the West of England, Bristol, United Kingdom
†University of Exeter Medical School, University of Exeter, Exeter, United Kingdom
‡Biosciences, University of Exeter, Exeter, United Kingdom
§Faculty of Agriculture, University of Osijek, Osijek, Croatia
[1]Corresponding author: e-mail address: john.hancock@uwe.ac.uk

Contents

1. Introduction 234
2. Hydrogen Sulfide Acts as a Signal in Cells 235
3. Modification of Thiols by Signaling Molecules 236
4. Identification of Modified Thiols by Other Methods 238
5. Experimental Protocols 239
6. *Caenorhabditis elegans* as a Model Organism 239
7. Growth of *C. elegans* 240
8. Treatment of Samples with H_2S 240
9. Estimation of Toxicity of H_2S Compounds 241
10. Treatment of Samples with Thiol Tag 242
11. Isolation and Analysis of Modified Proteins 244
12. Estimation of Protein Concentrations in Samples 245
13. Further Analysis and Identification of Modified Proteins 246
14. Concluding Remarks 247
References 248

Abstract

Hydrogen sulfide (H_2S) is an important gasotransmitter in both animals and plants. Many physiological events, including responses to stress, have been suggested to involve H_2S, at least in part. On the other hand, numerous responses have been reported following treatment with H_2S, including changes in the levels of antioxidants and the activities of transcription factors. Therefore, it is important to understand and unravel the events that are taking place downstream of H_2S in signaling pathways. H_2S is known to interact with other reactive signaling molecules such as reactive oxygen species (ROS) and nitric oxide (NO). One of the mechanisms by which ROS and NO have effects in a cell is the modification of thiol groups on proteins, by oxidation or S-nitrosylation, respectively. Recently, it has been reported that H_2S can also modify thiols. Here we report a method for the determination of thiol modifications on proteins following

Methods in Enzymology, Volume 555
ISSN 0076-6879
http://dx.doi.org/10.1016/bs.mie.2014.11.026

the treatment with biological samples with H_2S donors. Here, the nematode *Caenorhabditis elegans* is used as a model system but this method can be used for samples from other animals or plants.

1. INTRODUCTION

Hydrogen sulfide (H_2S) is now known to be an important metabolite involved in the control of cellular function in a range of organisms, including both animals and plants. It has been suggested that it is considered as part of the suite of gasotransmitters in organisms (Wang, 2002, 2003), despite the fact that is very toxic and a potent inhibitor of mitochondrial function through inhibition of Complex IV (Dorman et al., 2002). In mammals, H_2S has been shown to be involved in the control of vascular tone (Liu et al., 2012) and to have a role in related diseases, while in plants H_2S has been shown to be involved in alleviation of numerous stress responses as well as being involved in the control of normal physiological functions (Calderwood and Kopriva, 2014), such as the regulation of stomatal apertures (García-Mata & Lamattina, 2010; Lisjak et al., 2010), flower senescence (Zhang et al., 2011), and root development (Lin, Li, Cui, Lu, & Shen, 2012). Therefore, the downstream events following the exposure of cells to H_2S, either from the generation of H_2S by cells themselves or from the environment, are important to determine.

Cells can be exposed to H_2S from a variety of sources. From the environment, cells from both animals and plants can come into contact with H_2S which has been released from volcanoes, for example, while evolution may have been driven by the H_2S from thermal vents (Martin, Baross, Kelley, & Russell, 2008). Bacteria can make H_2S (Clarke, 1953), so exposing higher organisms to this gas. Even during warfare, humans have been exposed to this gas (Szinicz, 2005), highlighting its toxic nature. Endogenously, in plants H_2S can be generated by desulfhydrases (Alvarez, Calo, Romero, Garcia, & Gotor, 2010), while in animals the gas may be made by 3-mercaptopyruvate sulfurtransferase (3-MST), cystathionine-γ-lyase, or cystathionine-β-synthase (Prabhakar, 2012). It is important to note here that H_2S can be removed by cells too. In plants, the enzyme *O*-acetylserine(thiol) lyase (Tai & Cook, 2000; Youssefian, Nakamura, & Sano, 1993) can reduce H_2S levels, and in animals it has been shown that H_2S can be used by mitochondria as a source of reducing equivalents, leading to the generation of

ATP (Bouillaud, Ransy, & Andriamihaja, 2013), and the removal of H_2S from the cells. Therefore, with enzymes able to generate and remove H_2S, its use as a signal does not seem a surprise.

2. HYDROGEN SULFIDE ACTS AS A SIGNAL IN CELLS

In humans, H_2S is known to be involved in a range of physiological events, often centered around the regulation of the cardiovascular system (Liu et al., 2012). Furthermore, work in humans has shown that treatment with H_2S has been shown to alleviate the onset and symptoms of several diseases (Ahmed, 2013; Al-Magableh, Ng, Kemp-Harper, Miller, & Hart, 2013; Nagpure & Bian, 2013; Wang, 2013), among which are diabetes (Whiteman et al., 2010), atherosclerosis (Mani et al., 2013; Xu, Liu, & Liu, 2014), and vascular inflammation (Liu et al., 2013).

In plants, H_2S has been shown to alleviate the stress responses to a range of challenges, including to heavy metals (Ali et al., 2014; Li, Wang, & Shen, 2012; Zhang et al., 2008, 2010), temperature (Li, Ding, & Du, 2013; Li, Gong, Xie, Yang, & Li, 2012; Stuiver, De Kok, & Kuiper, 1992), and osmotic (Zhang et al., 2009) and oxidative stress (Shan et al., 2011; Zhang et al., 2008).

The interaction of H_2S with oxidative stress is particularly pertinent here, as both are involved in a range of stress responses. H_2S has been shown to have interactions with both the metabolism of reactive oxygen species (ROS) and nitric oxide (NO), as discussed previously (Hancock & Whiteman, 2014). Treatment with H_2S has an influence on the enzymes which produce ROS and NO, as well as the mechanisms to remove these compounds, that is, on antioxidants. For example, in plants, treatment with H_2S can cause alterations in antioxidant levels following salt stress (Lisjak, Teklic, Wilson, Whiteman, & Hancock, 2013), and postharvest storage can be enhanced, again mediated by the alteration of levels of antioxidants (Hu et al., 2012). Glutathione levels too have been reported to be increased in plant cells following treatment with H_2S (De Kok, Bosma, Maas, & Kuiper, 1985), so increasing the capacity of a cell to tolerate oxidative stress conditions.

Therefore, it can be seen that organisms from bacteria to humans are exposed to H_2S and indeed make it for positive reasons. H_2S can be used as a signaling molecule, although at high concentrations it is toxic. It can be used for mitochondria ATP generation (Bouillaud et al., 2013), and it is known to have effects on other signaling pathways, especially those involving ROS and NO (Hancock & Whiteman, 2014). Determining the biomolecules that H_2S interacts with is therefore important.

3. MODIFICATION OF THIOLS BY SIGNALING MOLECULES

Therefore, as can be seen from above, H_2S can impinge on the signaling which involves either ROS or NO, or indeed both (Hancock & Whiteman, 2014). One of the convergence points of this type of signaling is the modification of thiol groups in proteins (Chouchani, James, Fearnley, Lilley, & Murphy, 2011). It has been known for some time that thiol groups can undergo a range of chemistries, as indicated in Fig. 1. The –SH group may interact to form disulfides, an important element in the folding and three-dimensional structure of proteins. Alternatively, the –SH group may be sequentially oxidized to the sulfenic acid group, the sulfinic acid group to the sulfonic acid group. The formation of the sulfenic acid, which is relatively unstable, has been shown to be reversible, or to partake in other reactions (Biteau, Labarre, & Toledano, 2003) and so this could be a good way to control the activity of a protein, akin to phosphorylation and dephosphorylation. Furthermore, the next higher oxidation state of the thiol, that is, the sulfinic acid is also thought to be able to be reversed (Biteau et al., 2003), again suggesting that its involvement in signaling is suitable. The highest oxidation state, the sulfonic acid, is thought to be irreversible. However, in tyrosine phosphatases a further chemistry of the thiol has been suggested, that is, a ring structure has been proposed, allowing the reversion of the thiol back to its original state (Salmeen et al., 2003).

In the presence of NO, the thiol will partake in a reaction which leads to what is referred to as S-nitrosylation, which again has been shown to be

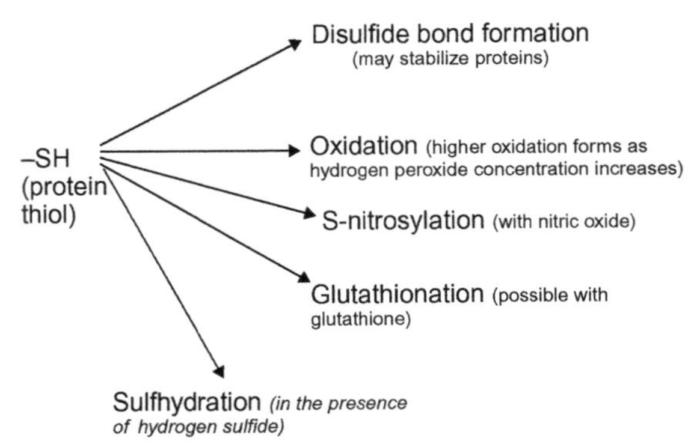

Figure 1 Some of the thiol modifications which can take place in proteins in the presence of a range of chemicals, including H_2S.

reversible and so a good way to control the function of a protein (Hess, Matsumoto, Kim, Marshall, & Stamler, 2005). To assess the extent of S-nitrosylation of proteins, assays such as that dubbed the "biotin switch" assay have been developed (Jaffrey, Erdjument-Bromage, Ferris, Tempst, & Snyder, 2001) and extensively used (Forrester, Foster, Benhar, & Stamler, 2009; Haldar & Stamler, 2013; Nakamura et al., 2013; Zhang, Keszler, Broniowska, & Hogg, 2005). This assay is further discussed below.

A different approach was used to assess the proteins which may be oxidized by hydrogen peroxide (H_2O_2) in plants (Hancock et al., 2005), and it is this rationale which is proposed here (see Fig. 2). A similar approach has also been used with animal cells (Baty, Hampton, & Winterbourn, 2005). In this assay, free thiols are reacted with a labeled iodoacetamide, in this case 5′-iodoacetamide fluorescein (IAF). Once separated on an acrylamide gel by electrophoresis, the reacted proteins can be visualized by the fluorescent nature of the tag. A separate, but identical sample, is pretreated with H_2O_2 and then subsequently reacted with IAF, separated by acrylamide electrophoresis and visualized. Any protein spots deemed to have disappeared when compared to the original sample are assumed to have reacted with the H_2O_2 before the treatment with IAF; hence, they appear to be missing in the electrophoretic analysis. When this approach was used

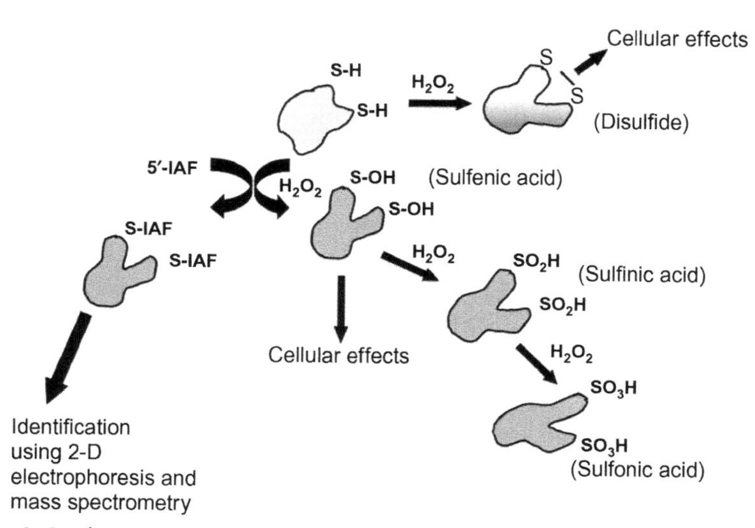

Figure 2 A schematic explaining the rationale behind the approach used to identify thiol groups in proteins which may react with H_2O_2 or H_2S.

previously, several proteins were identified as potentially reacting with H_2O_2 in cells, including glyceraldehyde 3-phosphate dehydrogenase (GAPDH), alcohol dehydrogenase, and SAM synthase. With a subsequent focus on GAPDH, it could be shown *in vitro* that this enzyme did indeed react in a reversible manner with H_2O_2 and interestingly also NO (Hancock et al., 2005), showing the validity of this approach. Further work on the reaction of GAPDH with H_2O_2 or NO showed that on modification the enzyme moved to the nucleus where it appeared to have a function in controlling gene expression. Hence, the identification of GAPDH as a target for H_2O_2 facilitated the elucidation of a whole new role for this enzyme, which was previously thought to be well characterized (Tristan, Shahani, Sedlak, & Sawa, 2011). Clearly, therefore, the identification of thiol modifications of proteins, by ROS, NO, or H_2S, can lead to a further understanding of both the regulation and roles of proteins in cells.

4. IDENTIFICATION OF MODIFIED THIOLS BY OTHER METHODS

One of the most well-used methods for identification of thiol modifications of proteins is the so-called biotin switch assay as mentioned above (Forrester et al., 2009; Zhang et al., 2005). This assay was first suggested over 10 years ago (Jaffrey et al., 2001) and has been used extensively since that time to determine the modification of thiol groups by NO. The rationale behind the assay is thiol groups may have reacted with NO, perhaps in a physiological setting, but not all thiols would have reacted. So remaining free thiols are reacted with methyl-methane thiosulfonate (MMTS), which will readily react with protein thiols. This leaves all the thiols either S-nitrosylated, because they had previously reacted with NO, or S-methylthionated, as they had reacted with the MMTS. Any thiols which had been S-nitrosylated are then reduced with ascorbate, removing the NO group and reforming the –SH group. This reaction is assumed to be specific as thiols which had previously been oxidized to sulfenic acid or sulfinic acid are not rereduced, and any which have previously been glutathionated are also unaffected. So the previously nitrosylated thiols are then reacted with biotin-HPDP, so labeling the thiols for identification, using avidin-horseradish peroxidase. The use of this assay, the pitfalls, and suggested modifications and controls have been previously discussed (Forrester et al., 2009), but it is an assay which has caused some controversy and needs to be undertaken with care (Kovacs & Lindermayr, 2013).

In the biotin switch assay, there is a requirement for the –SNO (S-nitrosylated thiol) to be converted back to the thiol to be identified. However, the assay here for H_2S modification of the thiol has no such requirement. It is easy to undertake and has previously been used successfully to identify H_2O_2-sensitive proteins. Therefore, it is suggested here that with a modification of that assay, H_2S-sensitive proteins may also be identified.

5. EXPERIMENTAL PROTOCOLS

To identify H_2S-sensitive proteins in cell samples, it is suggested that a modification of the methods used to previously identify H_2O_2-sensitive proteins is used (Hancock et al., 2005). However, previously this was used with plant samples, specifically from *Arabidopsis thaliana*. Here, to show that the method is applicable to a wider range of samples, *Caenorhabditis elegans* is used as a model system. Therefore, the methods below will outline the growth of the *C. elegans*, their treatment with H_2S, the tagging of the thiol groups, and the separation of the subsequent proteins with polyacrylamide electrophoresis.

6. *CAENORHABDITIS ELEGANS* AS A MODEL ORGANISM

Caenorhabditis elegans has been used as a model organism for many years, since Brenner suggested its use in 1965 (Brenner, 1974). It is a small nematode which is easy to grow, cheap to maintain, and has a short life span. But more importantly, the genome of *C. elegans* has been very well characterized, as have its developmental stages. Each worm is known to contain exactly 959 somatic cells, which to the large part have been characterized. Furthermore, because of the large amount of homology between genes in *C. elegans* and those in humans, these nematodes are very important in research on human disease and for drug discovery (Kaletta & Hengartner, 2006). For a full in-depth discussion on the use of C. *elegans* in laboratory work, refer to the WormBook (http://www.wormbook.org/).

Interestingly, and of relevance here because the effects of H_2S are being considered, it has been shown that treatment of *C. elegans* with H_2S bestows on the worms a degree of thermal tolerance (Miller & Roth, 2007), and furthermore, the life span of the worms has been shown to be increased. More recently, it has been shown that *C. elegans* are capable of the generation of H_2S, with the presence of three enzymes being reported (Qabazard et al., 2014), and alteration of the levels of H_2S synthesizing enzymes altered the organism's life span. H_2S metabolism, therefore, appears to be part of the normal physiology of the worms. Clearly, therefore, there must be downstream responses to H_2S in *C. elegans*, which may involve the modification of thiol groups on proteins.

However, here *C. elegans* is used simply as a model system in which to show that the tagging and identification of proteins which may react with H_2S can be achieved.

7. GROWTH OF *C. ELEGANS*

C. elegans were grown on NGM plates (Nematode Growth Media plates) with *E. coli* OP50 as a food source, or alternatively they were grown in liquid culture (again with *E. coli* as a source of food). However, here sufficient worms could be collected from NGM plates, dispensing with the need for the liquid cultures (see the WormBook (http://www.wormbook.org/)).

- For NGM plates, 3 g NaCl, 17 g agar, 2.5 g peptone, and 975 mL H_2O were mixed, and subsequently autoclaved and cooled. To this, the following additions were made: 1 mL 1 *M* $CaCl_2$, 1 mL 5 mg/mL cholesterol in ethanol, 1 mL 1 *M* $MgSO_4$, 25 mL 1 *M* KPO_4 buffer [108.3 g KH_2PO_4, 35.6 g K_2HPO_4, H_2O to 1 L]. Large square petri dishes ($120 \times 120 \times 17$ mm) are ideal for growing worms. Plates were poured, and when ready were spread with *E. coli* using a glass rod. After incubation at 37 °C for at least 6 h or overnight at room temperature, the plates were cooled to room temperature and the worms added. Basically, worms are added by "chunking," that is, chunks of NGM plate containing worms and *E. coli* are removed from one plate and added to a new plate, allowing the worms to multiple on a new food source. Alternatively, worms and *E. coli* were washed from a well-populated plate, briefly centrifuged at $1500 \times g$, and then after removal of some of the supernatant the remaining sample can be pipetted onto a fresh *E. coli* plate. *E. coli* ideally should be grown at 37 °C but a 20 °C incubator should be used for growing the worms.

8. TREATMENT OF SAMPLES WITH H_2S

H_2S is a gas and so samples can be treated directly with H_2S from the gaseous phase (gas tanks of H_2S can be purchased, e.g., from Praxair). If whole organisms are to be treated, or parts of an organism, such as the leaves of the plant, this may be a viable option, although strict safety measures will be needed due to the toxicity of H_2S. However, the use of gas is not always convenient and suitable for treating biological samples, such as cells and cell cultures, or protein samples. Therefore, more often than not compounds which release H_2S are used—so-called donor molecules. Common donors include sodium hydrosulfide (NaHS) and sodium sulfide (Na_2S). These are cheap, easy to obtain and store, and easy to use. An example of such as study is one looking at apoptosis in animal cells, mediated by mitogen-activated protein kinases (Adhikari & Bhatia, 2008). But the caution that needs to be exercised here is that once in solution both these compounds will release H_2S quickly, and therefore, samples will be exposed to relatively high H_2S for a short period of time. To overcome this, new donor compounds have been designed and made available such as GYY4137 (Li et al., 2008), which release compounds in a more physiological manner, that is, more slowly for longer periods. Other donors have been generated that are targeted to organelles and are commercially available (from VIVA Biosciences). Very recently, a new H_2S-releasing compound that is designed to be targeted to the mitochondria, AP-39, has been reported, with the view that it may lead to a new therapeutic strategy for a range of diseases (Le Trionnaire et al., 2014).

A further discussion of such compounds can be found in this issue of *Methods in Enzymology*. However, here NaHS was used as a demonstration of the effects which can be seen. If other, and more recent, donors are used, it will be necessary to change the concentrations and time periods of treatment used.

9. ESTIMATION OF TOXICITY OF H_2S COMPOUNDS

H_2S is extremely toxic, so care needs to be taken to ensure the effects of treatment with H_2S donors are what they appear to be. For example, if investigating signaling aspects of H_2S physiology, then it is important to be sure that there is a physiological response to H_2S but that the organisms or cells are not simply being killed. Even though low concentrations of H_2S can be used by cells as a source of reducing power for mitochondria (Bouillaud et al., 2013), H_2S is also a potent inhibitor of mitochondrial function (Dorman et al., 2002) and so treatments need to be of concentrations which do not have this latter effect. Therefore, when using new H_2S donors, or new organisms or tissues, investigations into the effects of H_2S need to be carried out before isolation and analysis of protein samples. Here, the toxicity and physiological responses to the H_2S donor NaHS are used as an example.

- To assess the toxicity of a H_2S donor on the nematodes, the worms were washed from NGM plates with M9 buffer (3 g KH_2PO_4, 6 g Na_2HPO_4, 5 g NaCl, 1 mL 1 *M* $MgSO_4$ in 1 L of water, sterilized by autoclaving).
- A specified amount of this liquid culture was used in each treatment, giving a final volume of 2 mL for each experiment. NaHS was added at the following concentrations: 0 *M* (M9 buffer control), 10 μ*M*, 50 μ*M*. Three repeats were carried out for each concentration of NaHS used. Nematodes were observed using a light microscope, and numbers showing slowed, normal, and no movement (presumed dead: that is, they appeared to be both not moving and stiff) were counted using click counters at 10-min intervals up to 30 min. It could be argued that assessing movement of the worms is quite subjective, but it becomes obvious with experience. Furthermore, slowed movement is often accompanied by jerky movements, making it easier to assess.
- The data for NaHS-treated nematodes are shown in Fig. 3. Here, the percentage of the worms showing reduced movement is shown. There was no significant toxicity during the period of this experiment (data not shown), but with other organisms, or H_2S donors, toxicity may well be seen with this range of concentrations and the H_2S donor may well be needed to be significantly diluted before use.
- With several H_2S donors, there will also be by-products formed on the release of H_2S. Therefore, toxicity experiments, as well as protein analysis, should be repeated with time-depleted donors. That is, donor compounds should be incubated in the absence of biological material for the same periods of time as the experiment and then subsequently used. Such depletion experiments are routinely used when NO donors are employed for example, and the practice should be repeated here.

10. TREATMENT OF SAMPLES WITH THIOL TAG

Samples which can be treated can be either whole organisms, whole tissues, cell cultures, or protein extracts. Clearly, if proteins are to be tagged and identified in whole organisms or tissues, issues around the penetration of the tag have to be considered. It may be preferable to label proteins while they are in their physiological environment, but this may not always be possible. Therefore, often proteins are isolated and treated and labeled *in vitro*, allowing the possible modifications of thiol groups on proteins to be

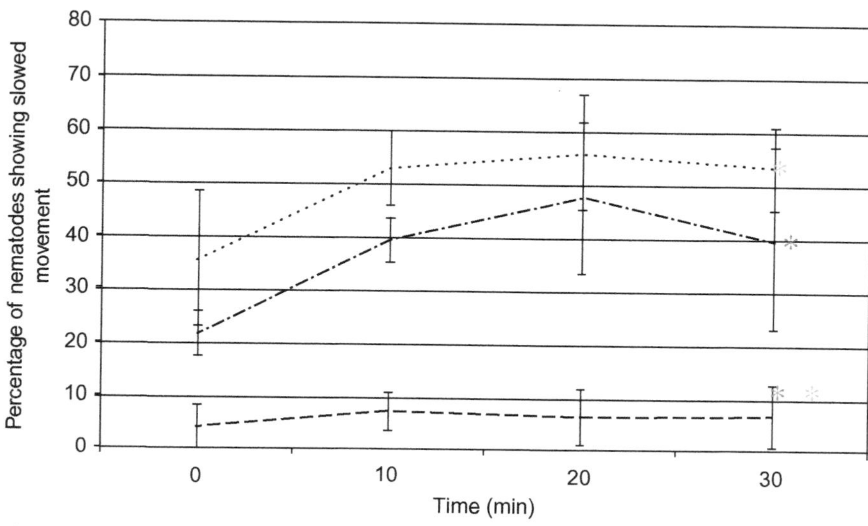

Figure 3 Changes in *C. elegans* movement after treatment with fast-release H_2S donor NaHS. Nematodes were washed from NGM plates with M9 buffer. A specified amount of this liquid culture was used in each treatment, giving a total volume of 2 mL for each repeat. Three repeats were carried out for each concentration of NaHS as well as an M9 buffer control. After addition of NaHS nematodes were observed using a light microscope, and numbers showing slowed, normal, and no movement (presumed dead) were counted using click counters at 10 min intervals up to 30 min. ——— 0 μ*M* (buffer control); —•—•— 10 μ*M* NaHS; • • • • 50 μ*M* NaHS. Error bars show the 95% confidence interval around the mean for each sample. *Significant differences between samples ($p < 0.05$ by *t*-test).

estimated. Subsequently, such work would need to be repeated with the treatment, i.e., H_2S, with the proteins *in situ*. This is the approach used here. Basically, worms were treated, protein extracted, and then tagged with 5′-IAF for identification (see the scheme in Fig. 2).

- To label protein samples, 5′-iodoacetamide fluorescein (IAF) was used. A stock solution needs to be made fresh before each use (5.15 mg in 500 μL PBS, pH ≥7.5). Stock solutions should be kept in the dark (wrapped in foil) before use.
- Labeling the proteins is carried out by the addition of up to 200 μ*M* IAF (final concentration) for 10 min before analysis. Again, samples need to be kept in the dark before analysis (covered in foil or placed in a dark place). Longer incubation times may be required if using whole cells or tissues. Here, to 200 μL of sample, 2 μL of 20 m*M* IAF was added and incubated at room temperature in the dark, before analysis by electrophoresis.

11. ISOLATION AND ANALYSIS OF MODIFIED PROTEINS

The final treated protein samples need to be analyzed by SDS-polyacrylamide gel electrophoresis. Samples are either treated with IAF or treated with H_2S donors and then IAF needs to be denatured and separated.

- For analysis of proteins capable of binding IAF in the example given here, proteins were isolated from the nematodes before treatment. However, it would be possible to treat whole worms or tissues with IAF before analysis.
- For protein isolation, 10 large square petri dishes (120 × 120 × 17 mm) of worms were washed with 5 mL S basal buffer per plate (S basal buffer consists of 5.85 g NaCl, 1 g K_2HPO_4, 6 g K_2PO_4, 1 mL cholesterol (5 mg/mL in ethanol) in a liter of H_2O). This was dispensed into 2 × 50-mL Falcon tubes, and the worms were pelleted at 3000 rpm for 2 min. The supernatant was discarded. Worms were washed in buffer and repelleted as before.
- Samples were treated as required, with concentrations of H_2S donors dictated by the results of the toxicity assay. Here 0–50 μ*M* NaHS was used as a demonstration. Samples were treated for 20 min.
- Worms were removed from the H_2S donor by centrifugation (max 4.4 × *g*) for 5 min, and the supernatant discarded. Worms were washed in S basal buffer and were subsequently resuspended in enough iced triethylammonium bicarbonate buffer (TEAB buffer, 100 m*M* (diluted from 1 *M* stock, pH 8.5)) to bring the volume up to approx. 0.4 mL. This was agitated for 20 s to aid lysis (use a "tissue ruptor" if available), followed by 30 s on ice. This was repeated three times and then centrifuged at 13.3 × *g* for 2 min to remove unlysed material.
- 200 μL supernatant was removed to new 1.5-mL Eppendorf tubes (50 μL was saved for BCA assay: this can be frozen for later analysis if needed for convenience).
- 5′-IAF treatment was carried out as this point. Discussion and details can be found above.
- The proteins were precipitated with acetone for further analysis. 1 mL ice cold acetone was added to the sample, which was vortexed and then incubated at −20 °C for 10 min. This was then pelleted at 13.3 × *g* for 5 min and the acetone discarded. The sample was air dried for approximately 10 min. The sample was still kept covered as much as possible to reduce the amount of light it is exposed to.

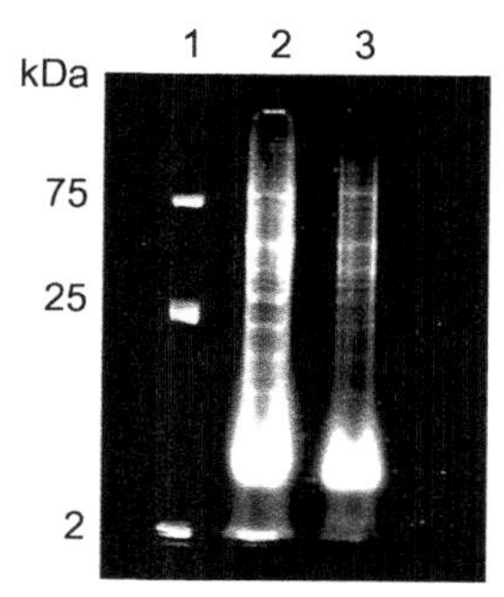

Figure 4 5-IAF labeling of control and treated *C. elegans* protein samples. Protein was extracted from nematode samples either untreated or treated with 50 μ*M* NaHS for 20 min. Protein samples were labeled with 5-IAF and separated using SDS-PAGE. Marker sizes are shown. Lane 1: Precision Plus Protein Standards (Bioline, Hertfordshire, UK); Lane 2: Control—protein from untreated nematodes [2.39 mg/mL]; Lane 3: Protein from nematodes treated with 50 μ*M* NaHS [2 mg/mL]. A representative example of a typical gel is shown.

- To prepare the sample for electrophoresis, the pellet was resuspended in 20 μL sample loading buffer. 2 μL 10 × DTT was added and mixed. The sample was denatured at 70 °C for 10 min and then separated on an acrylamide gel. The electrophoresis was subsequently run at 200 V, 100–125 mA for 30 min to 1 h (checking the progress of the dye front).
- IAF-tagged proteins were then be visualized under UV light. The gel needs to be removed from between the plates and can then be viewed and photographed. An example can be seen in Fig. 4.
- To visualize all the proteins on the gel, and not just those that have bound to the IAF, the gel needs to be stained for the presence of all proteins using Coomassie blue (approximately 30 min), with a methanol-based destain (30% for approximately 6 h). The gel can then once again be photographed, this time using visible light. This will indicate both the presence of non-IAF labeled proteins and allow comparison of the amount of protein present in different lanes on the same gel or across gels.

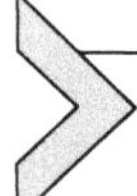

12. ESTIMATION OF PROTEIN CONCENTRATIONS IN SAMPLES

Once samples have been isolated and are ready for treatment, it will be important to carry out a protein assay, as samples will need to be compared, either as different lanes on the same gel or across different gels.

- Protein concentrations were estimated using a Pierce BCA Assay Protein Assay Kit (Thermo Scientific). This is a detergent-compatible assay which is based on bicinchoninic acid (BCA). It has long been known that there is a reduction of Cu^{2+} to Cu^{+} by protein in an alkaline medium (the so-called Biuret reaction). BCA can then be used to create a sensitive colorimetric detection of the cuprous cations (Cu^{+}). Two molecules of BCA react with one Cu^{+} resulting in a purple reaction product. This can then be measured at 492 nm, either with a spectrophotometer or a plate reader.
- Bovine serum albumin (BSA) should be used to create a standard curve against which unknown samples can be compared. BSA should be used as a serial dilution from 0.1 to 1 mg/mL. BSA was dissolved in TEAB buffer so that the buffer is consistent with the samples being assayed.

13. FURTHER ANALYSIS AND IDENTIFICATION OF MODIFIED PROTEINS

Here, the labeling of protein samples from *C. elegans* was used as an example, and they were separated by 1-dimensional SDS-PAGE. It is clear from Fig. 4 that many proteins were labeled with IAF, showing that in a crude protein extract from the nematodes there are numerous proteins which have thiol groups accessible under the conditions used in this experiment. However, on pretreatment with H_2S then many of these proteins are no longer visible under UV light, suggesting that they have bound to H_2S rendering them incapable of subsequently binding to the IAF. However, this does not lead to the identification of these proteins. There are a variety approaches which may subsequently be used.

- To determine which proteins have bound IAF, if the identification of the protein is either already known, or suspected, then an antibody approach may be used. Here the proteins can either be identified using Western blot analysis or an immunoprecipitation approach.
- Alternately, an antibody against fluorescein can be used. This was the rationale used by Wu, Kwon, and Rhee (1998), for example. Here they treated samples with 5-IAF and then probed for the presence of the fluorescein tag with an antifluorescein antibody. One of the H_2O_2-sensitive proteins they identified was protein tyrosine phosphatase 1B, verifying the approach that they used.
- Regardless of the alternatively method chosen the best way to identify the proteins is by separating them by 2-dimensional electrophoresis. The first dimension will be a separation due to the proteins' isoelectric points, and

different gradient and ranges can be used for this. If a complete set of proteins is to be analyzed a wide pH range will need to be employed but often to get better separate of proteins that appear close together a narrow pH range is used. The second dimension will be separation by molecular weight, and again the exact separation seen will depend on the percentage of acrylamide used in the gel. A standard of 10–12% is often used, but acrylamide gels may be more (up to about 15%) or less (down to approximately 5%) or indeed a gradient gel may be used, where the acrylamide gets progressively more concentrated down the length of the gel.

- Once the proteins have been separated by 2-D electrophoresis, then the proteins need to be identified. This is often achieved using mass spectrometry. Protein spots are removed, often manually with a scalpel, and then the proteins are cleaved into peptides. This is usually done with trypsin. Peptides can be identified using matrix-assisted laser desorption ionization time-of-flight spectrophotometry. This yields a peptide mass fingerprint which can be compared to theoretical peptides derived from nucleotide databases, as the cut sites of trypsin can be determined. Alternately, if this does not yield a match, then MS/MS can be used to break the peptides and do further analysis for their identification.
- Even once identified it cannot be assumed that the protein had reacted with H_2S. Ideally further analysis needs to be carried out. This may include the identification of a mass shift in a peptide containing a cysteine residue, to show the exact product of the H_2S reacting with the protein. Further enzyme analysis may also include the assay of enzyme activity (or protein function) in the presence and absence of varying concentrations of H_2S or H_2S donors.

14. CONCLUDING REMARKS

The ability to measure the modifications of thiols is very important to get a full understanding of how reactive compounds such as H_2S interact with proteins to alter their activities and functions. However, other compounds such as ROS, NO, and glutathione also have the ability to reactive with thiols to give altered proteins and peptides. Therefore, it has to be considered that H_2S is not working on its own but rather in concert with a host of other reactive chemicals (Hancock & Whiteman, 2014). Therefore, similar investigations of thiol modifications needs to be carried out with other compounds, perhaps with other assays too, such as the biotin switch assay (Forrester et al., 2009; Haldar & Stamler, 2013; Nakamura et al., 2013;

Zhang et al., 2005), as it is under the physiological conditions in which a protein resides that will determine the exact end result of the thiol alteration. If NO is the predominant signal, then perhaps S-nitrosylation will be the result, but if H_2O_2 is predominant then the thiol may be oxidized—to varying degrees. Getting a full picture of the possible thiol modifications that may take place, considering some thiols may not be solvent accessible, or be used for disulfide formation, and how such thiols are modified in a cell is therefore important. Such discussion also highlights that doing these assays *in vitro* is not always very representative of the true picture, so such experiments of treatments with NO donors, H_2S donors, H_2O_2, etc. need to be carried out on whole cells or whole tissues/organisms too. Lastly, these experiments need to be carried out with combinations of reactive compounds too, so that any competition between such compounds is unraveled.

It has been suggested that H_2S, or H_2S donor molecules, may be used as a future therapeutic agent (Xu et al., 2014) with specific examples including during inflammation (Fox et al., 2012). Furthermore, with plants the storage of fruit has been shown to be affected by the presence of H_2S (Hu et al., 2012), the mechanism being suggested as alteration of antioxidant concentrations within the fruits. Therefore, the interest in the biology of H_2S is surely going to increase, and the direct reacts that H_2S has with biomolecules is going to be important to determine in a range of organisms, from bacteria and plants to animals, and in human tissues.

REFERENCES

Adhikari, S., & Bhatia, M. (2008). H_2S-induced pancreatic acinar cell apoptosis is mediated via JNK and p38 MAP kinase. *Journal of Cellular and Molecular Medicine, 12*, 1374–1383.

Ahmed, A. (2013). Can hydrogen sulfide prevent preeclampsia and fetal growth restriction? *Nitric Oxide, 31*, S17.

Ali, B., Song, W. J., Hu, W. Z., Luo, X. N., Gill, R. A., Wang, J., et al. (2014). Hydrogen sulfide alleviates lead-induced photosynthetic and ultrastructural changes in oilseed rape. *Ecotoxicology and Environmental Safety, 102*, 25–33.

Al-Magableh, M., Ng, H., Kemp-Harper, B., Miller, A., & Hart, J. (2013). Hydrogen sulfide protects endothelial function under conditions of oxidative stress. *Nitric Oxide, 31*, S25.

Alvarez, C., Calo, L., Romero, L. C., Garcia, I., & Gotor, C. (2010). An *O*-acetylserine(-thiol)lyase homolog with L-cysteine desulfhydrase activity regulates cysteine homeostasis in Arabidopsis. *Plant Physiology, 152*, 656–669.

Baty, J. W., Hampton, M. B., & Winterbourn, C. C. (2005). Proteomic detection of hydrogen peroxide-sensitive thiol proteins in Jurkat cells. *Biochemical Journal, 389*, 785–795.

Biteau, B., Labarre, J., & Toledano, M. B. (2003). ATP-dependent reduction of cysteine-sulphinic acid by *S. cerevisiae* sulphiredoxin. *Nature, 425*, 980–984.

Bouillaud, F., Ransy, C., & Andriamihaja, M. (2013). Sulfide and mitochondrial bioenergetics. *Nitric Oxide, 31*, S15.

Brenner, S. (1974). The genetics of *Caenorhabditis elegans*. *Genetics*, 77, 71–94.

Calderwood, A., & Kopriva, S. (2014). Hydrogen sulfide in plants: From dissipation of excess sulfur to signaling molecule. *Nitric Oxide, 41*, 72–78.

Chouchani, E. T., James, A. M., Fearnley, I. M., Lilley, K. S., & Murphy, M. P. (2011). Proteomic approaches to the characterization of protein thiol modification. *Current Opinion in Chemical Biology, 15*, 120–128.

Clarke, P. H. (1953). Hydrogen sulphide production by bacteria. *Journal of General Microbiology, 8*, 397–407.

De Kok, J. L., Bosma, W., Maas, F. M., & Kuiper, P. J. C. (1985). The effect of short-term H_2S fumigation on water-soluble sulphydryl and glutathione levels in spinach. *Plant, Cell and Environment, 8*, 189–194.

Dorman, D. C., Moulin, F. J., McManus, B. E., Mahle, K. C., James, R. A., & Struve, M. F. (2002). Cytochrome oxidase inhibition induced by acute hydrogen sulfide inhalation: Correlation with tissue sulfide concentrations in the rat brain, liver, lung, and nasal epithelium. *Toxicological Sciences, 65*, 18–25.

Forrester, M. T., Foster, M. W., Benhar, M., & Stamler, J. S. (2009). Detection of protein S-nitrosylation with the biotin-switch technique. *Free Radical Biology and Medicine, 46*, 119–126.

Fox, B., Schantz, J.-T., Haigh, R., Wood, M. E., Moore, P. K., Viner, N., et al. (2012). Inducible hydrogen sulfide synthesis in chondrocytes and mesenchymal progenitor cells: Is H_2S a novel cytoprotective mediator in the inflamed joint? *Journal of Cellular and Molecular Medicine, 16*, 896–910.

García-Mata, C., & Lamattina, L. (2010). Hydrogen sulfide, a novel gasotransmitter involved in guard cell signalling. *New Phytologist, 188*, 977–984.

Haldar, S. M., & Stamler, J. S. (2013). S-nitrosylation: Integrator of cardiovascular performance and oxygen delivery. *Journal of Clinical Investigation, 123*, 101–110.

Hancock, J. T., Henson, D., Nyirenda, M., Desikan, R., Harrison, J., Lewis, M., et al. (2005). Proteomic identification of glyceraldehyde 3-phosphate dehydrogenase as an inhibitory target of hydrogen peroxide in *Arabidopsis*. *Plant Physiology and Biochemistry, 43*, 828–835.

Hancock, J. T., & Whiteman, M. (2014). Hydrogen sulfide and cell signaling: Team player or referee? *Plant Physiology and Biochemistry, 78*, 37–42.

Hess, D. T., Matsumoto, A., Kim, S.-O., Marshall, H. E., & Stamler, J. S. (2005). Protein S-nitrosylation: Purview and parameters. *Nature Reviews. Molecular Cell Biology, 6*, 150–166.

Hu, L.-Y., Hu, S.-L., Wu, J., Li, Y.-H., Zheng, J.-L., Wei, Z.-J., et al. (2012). Hydrogen sulfide prolongs postharvest shelf life of strawberry and plays an antioxidative role in fruits. *Journal of Agricultural and Food Chemistry, 60*, 8684–8693.

Jaffrey, S. R., Erdjument-Bromage, H., Ferris, C. D., Tempst, P., & Snyder, S. H. (2001). Protein S-nitrosylation: A physiological signal for neuronal nitric oxide. *Nature Cell Biology, 3*, 193–197.

Kaletta, T., & Hengartner, M. O. (2006). Finding function in novel targets: *C. elegans* as a model organism. *Nature Reviews. Drug Discovery, 5*, 387–399.

Kovacs, I., & Lindermayr, C. (2013). Nitric oxide-based protein modification: Formation and site-specificity of protein *S*-nitrosylation. *Frontiers in plant science, 4*, 137.

Le Trionnaire, S., Perry, A., Szczesny, B., Szabo, C., Winyard, P. G., Whatmore, J. L., et al. (2014). The synthesis and functional evaluation of a mitochondria-targeted hydrogen sulfide donor, (10-oxo-10-(4-(3-thioxo-3H-1,2-dithiol-5-yl)phenoxy)decyl) triphenylphosphonium bromide (AP39). *Medicinal Chemistry Communications, 5*, 728–736.

Li, Z. G., Ding, X. J., & Du, P. F. (2013). Hydrogen sulfide donor sodium hydrosulfide-improved heat tolerance in maize and involvement of proline. *Journal of Plant Physiology, 170*, 741–747.

Li, Z. G., Gong, M., Xie, H., Yang, L., & Li, J. (2012). Hydrogen sulfide donor sodium hydrosulfide-induced heat tolerance in tobacco (Nicotianatabacum L.) suspension cultured cells and involvement of Ca2+ and calmodulin. *Plant Science, 185/186*, 185–189.

Li, L., Wang, Y., & Shen, W. (2012). Roles of hydrogen sulfide and nitric oxide in the alleviation of cadmium-induced oxidative damage in alfalfa seedling roots. *BioMetals, 25*, 617–631.

Li, L., Whiteman, M., Guan, Y. Y., Neo, K. L., Cheng, Y., Lee, S. W., et al. (2008). Characterization of a novel, water-soluble hydrogen sulfide releasing molecule (GYY4137): New insights into the biology of hydrogen sulphide. *Circulation, 7*, 2351–2360.

Lin, Y.-T., Li, M.-Y., Cui, W.-T., Lu, W., & Shen, W.-B. (2012). Haem oxygenase-1 is involved in hydrogen sulfide induced cucumber adventitious root formation. *Journal of Plant Growth Regulation, 31*, 519–528.

Lisjak, M., Srivastava, N., Teklic, T., Civale, L., Lewandowski, K., Wilson, I., et al. (2010). A novel hydrogen sulfide donor causes stomatal opening and reduces nitric oxide accumulation. *Plant Physiology and Biochemistry, 48*, 931–935.

Lisjak, M., Teklic, T., Wilson, I. D., Whiteman, M., & Hancock, J. T. (2013). Hydrogen sulfide: Environmental factor or signaling molecule? *Plant, Cell & Environment, 36*, 1607–1616.

Liu, Z., Han, Y., Li, L., Lu, H., Meng, G., Li, X., et al. (2013). The hydrogen sulfide donor, GYY4137, exhibits anti-atherosclerotic activity in high fat fed apolipoprotein E(−/−) mice. *British Journal of Pharmacology, 169*, 1795–1809.

Liu, Y. H., Lu, M., Hu, L. F., Wong, P. T., Webb, G. D., & Bian, J. S. (2012). Hydrogen sulfide in mammalian cardiovascular system. *Antioxidants & Redox Signaling, 17*, 141–185.

Mani, S., Li, H., Untereiner, A., Wu, L., Yang, G., Austin, R. C., et al. (2013). Decreased endogenous production of hydrogen sulfide accelerates atherosclerosis. *Circulation, 127*, 2523–2534.

Martin, W., Baross, J., Kelley, D., & Russell, M. J. (2008). Hydrothermal vents and the origin of life. *Nature Reviews. Microbiology, 6*, 805–814.

Miller, D. L., & Roth, M. B. (2007). Hydrogen sulfide increases thermotolerance and lifespan in *Caenorhabditis elegans*. *Proceedings of the National Academy of Sciences of the United States of America, 104*, 20618–20622.

Nagpure, B., & Bian, J.-S. (2013). Neuroprotective effect of hydrogen sulfide: Regulation of amyloidosis and inflammation in SH-SY5Y neuroblastoma and BV-2 micro glia cells. *Nitric Oxide, 31*, S22.

Nakamura, T., Tu, S., Waseem Akhtar, M., Sunico, C. R., Okamoto, S.-I., & Lipton, S. A. (2013). Aberrant protein S-nitrosylation in neurodegenerative diseases. *Neuron, 78*, 596–614.

Prabhakar, N. R. (2012). Carbon monoxide (CO) &hydrogen sulfide (H_2S) in hypoxic sensing by the carotid body. *Respiratory Physiology and Neurobiology, 184*, 165–169.

Qabazard, B., Li, L., Gruber, J., Peh, M. T., Ng, L. F., Dinesh Kumar, S., et al. (2014). Hydrogen sulfide is an endogenous regulator of aging in *Caenorhabditis elegans*. *Antioxidants & Redox Signaling, 20*, 2621–2630.

Salmeen, A., Andersen, J. N., Myers, M. P., Meng, T. C., Hinks, J. A., Tonks, N. K., et al. (2003). Redox regulation of protein tyrosine phosphatase 1B involves a sulphenyl-amide intermediate. *Nature, 423*, 769–773.

Shan, C. J., Zhang, S. L., Li, D. F., Zhao, Y. Z., Tian, X. L., Zhao, X. L., et al. (2011). Effects of exogenous hydrogen sulfide on the ascorbate and glutathione metabolism in wheat seedlings leaves under water stress. *Acta Physiologiae Plantarum, 33*, 2533–2540.

Stuiver, C. E. E., De Kok, L. J., & Kuiper, P. J. C. (1992). Freezing tolerance and biochemical changes in wheat shoots as affected by H_2S fumigation. *Plant Physiology and Biochemistry, 30*, 47–55.

Szinicz, L. (2005). History of chemical and biological warfare agents. *Toxicology, 214*, 167–181.

Tai, C. H., & Cook, P. F. (2000). O-acetylserinesulfhydrylase. *Advances in Enzymology and Related Areas of Molecular Biology, 74*, 185–234.

Tristan, C., Shahani, N., Sedlak, T. W., & Sawa, A. (2011). The diverse functions of GAPDH: Views from different subcellular compartments. *Cellular Signalling, 23*, 317–323.

Wang, R. (2002). Two's company, three's a crowd: Can H_2S be the third endogenous gaseous transmitter? *The FASEB Journal, 16*, 1792–1798.

Wang, R. (2003). The gasotransmitter role of hydrogen sulfide. *Antioxidants & Redox Signaling, 5*, 493–501.

Wang, R. (2013). A stinky remedy for atherosclerosis. *Nitric Oxide, 31*, S17–S18.

Whiteman, M., Gooding, K. M., Whatmore, J. L., Ball, C. I., Mawson, D., Skinner, K., et al. (2010). Adiposity is a major determinant of plasma levels of the novel vasodilator hydrogen sulphide. *Diabetologia, 53*, 1722–1726.

Wu, J. F., Kwon, K. S., & Rhee, S. G. (1998). Probing cellular protein targets of H_2O_2 with fluorescein-conjugated iodoacetamide and antibodies to fluorescein. *FEBS Letters, 440*, 111–115.

Xu, S., Liu, Z., & Liu, P. (2014). Targeting hydrogen sulfide as a promising therapeutic strategy for atherosclerosis. *International Journal of Cardiology, 172*, 313–317.

Youssefian, S., Nakamura, M., & Sano, H. (1993). Tobacco plants transformed with the O-acetylserine (thiol) lyase gene of wheat are resistant to toxic levels of hydrogen sulphide gas. *The Plant Journal, 4*, 759–769.

Zhang, H., Hu, L. Y., Hu, K. D., He, Y. D., Wang, S. H., & Luo, J. P. (2008). Hydrogen sulfide promotes wheat seed germination and alleviates the oxidative damage against copper stress. *Journal of Integrative Plant Biology, 50*, 1518–1529.

Zhang, H., Hu, L. Y., Li, P., Hu, K. D., Jiang, C. X., & Luo, J. P. (2010). Hydrogen sulfide alleviated chromium toxicity in wheat. *Biologiae Plantarum, 54*, 743–747.

Zhang, H., Hua, S. L., Zhang, Z. J., Hua, L. Y., Jiang, C. X., Wei, Z. J., et al. (2011). Hydrogen sulfide acts as a regulator of flower senescence in plants. *Postharvest Biology and Technology, 60*, 251–257.

Zhang, Y., Keszler, A., Broniowska, K. A., & Hogg, N. (2005). Characterization and application of the biotin-switch assay for the identification of S-nitrosated proteins. *Free Radical Biology and Medicine, 38*, 871–881.

Zhang, H., Ye, Y. K., Wang, S. H., Luo, J. P., Tang, J., & Ma, D. F. (2009). Hydrogen sulfide counteracts chlorophyll loss in sweet potato seedling leaves and alleviates oxidative damage against osmotic stress. *Plant Growth Regulation, 58*, 243–250.

CHAPTER THIRTEEN

Analysis of Some Enzymes Activities of Hydrogen Sulfide Metabolism in Plants

Zhong-Guang Li[1]
School of Life Sciences, Engineering Research Center of Sustainable Development, Utilization of Biomass Energy Ministry of Education, Key Laboratory of Biomass Energy, Environmental Biotechnology Yunnan Province, Yunnan Normal University, Kunming, Yunnan, PR China
[1]Corresponding author: e-mail address: zhongguang_li@163.com

Contents

1. Theory 254
2. Equipment 256
3. Materials 256
   3.1 Solution and buffer 257
4. Protocol 1 260
   4.1 Duration 260
   4.2 Preparation 260
5. Step 1: Analyze of L-/D-Cysteine Desulfhydrase Activity 261
   5.1 Overview 261
   5.2 Duration 261
   5.3 Tip 261
   5.4 Tip 262
   5.5 Tip 262
   5.6 Tip 262
   5.7 Tip 262
   5.8 Tip 262
6. Protocol 2 263
   6.1 Duration 263
   6.2 Preparation 263
7. Step 1: Analyze of Sulfite Reductase Activity 263
   7.1 Overview 263
   7.2 Duration 264
   7.3 Tip 264
   7.4 Tip 264
   7.5 Tip 264
   7.6 Tip 265
   7.7 Tip 265
   7.8 Tip 265

Methods in Enzymology, Volume 555
ISSN 0076-6879
http://dx.doi.org/10.1016/bs.mie.2014.11.035

8. Protocol 3 265
8.1 Duration 265
8.2 Preparation 265
9. Step 1: Analyze of β-Cyano-L-Alanine Synthase Activity 266
9.1 Overview 266
9.2 Duration 266
9.3 Tip 266
9.4 Tip 266
9.5 Tip 266
9.6 Tip 267
9.7 Tip 267
10. Protocol 4 267
10.1 Duration 267
10.2 Preparation 267
11. Step 1: Analyze of L-Cysteine Synthase Activity 268
11.1 Overview 268
11.2 Duration 268
11.3 Tip 268
11.4 Tip 268
11.5 Tip 269
11.6 Tip 269
Acknowledgment 269
References 269

Abstract

Hydrogen sulfide (H_2S) which is considered as a novel gasotransmitter after reactive oxygen species and nitric oxide in plants has dual character, that is, toxicity that inhibits cytochrome oxidase at high concentration and as signal molecule which is involved in plant growth, development, and the acquisition of tolerance to adverse environments such as extreme temperature, drought, salt, and heavy metal stress at low concentration. Therefore, H_2S homeostasis is very important in plant cells. The level of H_2S in plant cells is regulated by its synthetic and degradative enzymes, L-/D-cysteine desulfhydrase (L-/D-DES), sulfite reductase (SiR), and cyanoalanine synthase (CAS), which are responsible for H_2S synthesis, while cysteine synthase (CS) takes charge of the degradation of H_2S, but its reverse reaction also can produce H_2S. Here, after crude enzyme is extracted from plant tissues, the activities of L-/D-DES, SiR, CAS, and CS are measured by spectrophotometry, the aim is to further understand homeostasis of H_2S in plant cells and its potential mechanisms.

1. THEORY

Sulfur (S) is an essential macronutrient in plants. It is taken up as sulfate and is assimilated into cysteine, an amino acid at the cross roads of primary metabolism, protein synthesis, and the formation of low molecular weight sulfur-containing defense compounds including hydrogen sulfide (H_2S; Fig. 1). H_2S has been considered to be the third gasotransmitter after reactive oxygen species and nitric oxide in plant system, it is thought to be a key signaling molecule. Enzymes which generate H_2S, and remove it, have been characterized in both plants and animals, it is inherently toxic to cells at high concentration, for example, inhibiting cytochrome oxidase. But H_2S is now being thought as a signal molecule, part of signal transduction pathways. H_2S has been implicated in the control of many plant developmental stages and in the responses to a range of stress challenges, such as seed germination, root development, stomatal movement, osmotic stress, salt stress, oxidative stress, heavy metal stress, pathogen challenge, temperature stress, water stress, and organ senescence (Hancock & Whiteman, 2014; Li, 2013).

In plants, there are at least five enzymes involved in H_2S biosynthesis in plants, that is, ① L-cysteine desulfhydrase (L-DES, EC 4.4.1.1), which catalyzes the degradation of L-cysteine to produce H_2S, amine, and pyruvate; ② D-cysteine desulfhydrase (D-DES, EC 4.4.1.15) decomposes D-cysteine to H_2S, similar to L-DES; ③ sulfite is reduced by sulfite reductase (SiR, EC 1.8.7.1) to H_2S using ferredoxin as electron donor; ④ H_2S can be

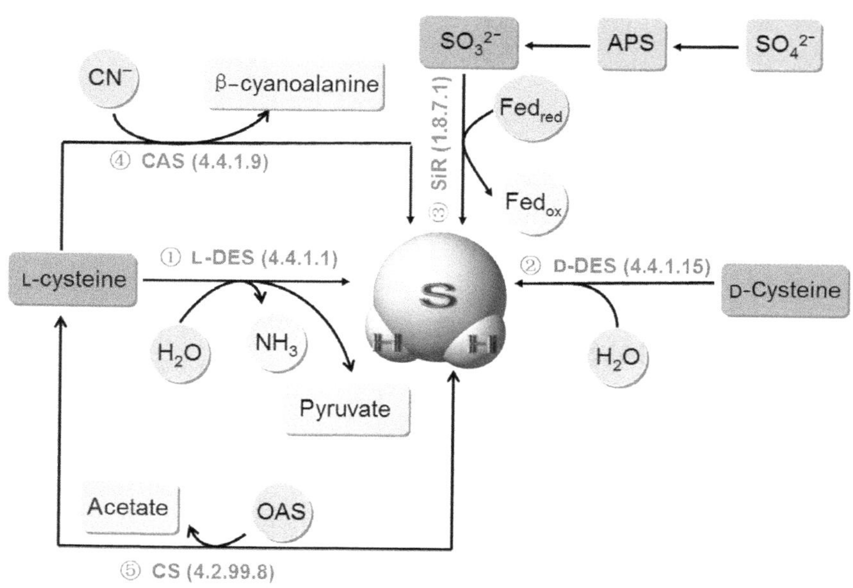

Figure 1 H_2S metabolism pathways in plants.

released from cysteine via cyanoalanine synthase (CAS, EC 4.4.1.9) in the present of hydrogen cyanide; ⑤ H_2S is incorporated into *O*-acetyl-L-serine via cysteine synthase (CS, EC 4.2.99.8) to form cysteine, and its reverse reaction can release H_2S (Li, 2013; Fig. 1). Many results illustrated that endogenous H_2S accumulation in plants is closely related to change in the activities of cysteine-synthesizing and degrading enzymes, such as L-/D-DES, SiR, CAS, and CS. L-DES is a plant L-cysteine-specific desulfhydrase responsible for H_2S release in plants, when plants are subjected to environmental stress such as extreme temperature, drought, salt, heavy-metal stress, and infection by pathogen, its activity and expression can be upregulated (Hancock & Whiteman, 2014; Li, 2013).

2. EQUIPMENT

Electronic balance
Centrifuge
1.5- and 5-ml microcentrifuge tubes
Spectrophotometer
Disposable polystyrene cuvettes (1 cm path length)
Micropipettors
Micropipettor tips
Voter mixer
Mortar and pestle
Volumetric flask
Water bath
Beakers
Test tube

3. MATERIALS

Tris hydroxy methyl aminomethan (Tris)
Sodium phosphate dibasic (Na_2HPO_4)
Sodium phosphate monobasic (NaH_2PO_4)
Concentrated hydrochloride acid
Ninhydrin
Glacial acetic acid
Ferric chloride ($FeCl_3$)

N,*N*-Dimethyl-*p*-phenylenediamine dihydrochloride (DMPD)
Dithiothreitol (DTT)
L-/D-Cysteine (Cys)
O-Acetyl-L-serine (OAS)
Sodium sulfite (Na_2SO_3)
Reduced ferredoxin (Fed_{red})
Potassium cyanide (KCN)
Sodium sulfide (Na_2S)
Zinc acetate ($Za(OAc)_2$)
2-Hydroxyethyl (HEPES)
Potassium hydroxide (KOH)
Potassium chloride (KCl)
Ethylene diamine tetraacetic acid (EDTA)
Ethylene glycol tetraacetic acid (EGTA)
Glycerin
Phenylmethanesulfonyl fluoride (PMSF)
Triton X-100

3.1. Solution and buffer

① L-/D-DES and CAS extraction buffer: 20 m*M* Tris–HCl, pH 8.0

Component	Final concentration	Stock	Amount/100 ml
Tris	20 m*M*	200 m*M*	10 ml
HCl	10.72 m*M*	200 *M*	5.36 ml
Add water to 100 ml			

② L-/D-DES reaction mixture: 100 m*M* Tris–HCl, pH 9.0

Component	Final concentration	Stock	Amount/100 ml
Tris	100 m*M*	200 m*M*	50 ml
HCl	10 m*M*	200 m*M*	5 ml
L- or D-Cys	0.8 m*M*	16 m*M*	5 ml
DTT	2.5 m*M*	100 m*M*	2.5 ml
Add water to 100 ml			

③ 30 m*M* $FeCl_3$

Component	Final concentration	Stock	Amount/100 ml
$FeCl_3$	30 m*M*	100 m*M*	30 ml
HCl	1.2 *M*	12 *M* (concentrated HCl)	10 ml
Add water to 100 ml			

④ 20 m*M* DMPD

Component	Final concentration	Stock	Amount/100 ml
DMPD	20 m*M*	100 m*M*	20 ml
HCl	7.2 *M*	12 *M* (concentrated HCl)	60 ml
Add water to 100 ml			

⑤ SiR extraction buffer: 50 m*M* HEPES–KOH, pH 7.4

Component	Final concentration	Stock	Amount/100 ml
HEPES	50 m*M*	200 m*M*	25 ml
KOH	–	1 *M*	Adjust pH to 7.4
KCl	10 m*M*	100 m*M*	10 ml
EDTA	1 m*M*	100 m*M*	1 ml
EGTA	1 m*M*	100 m*M*	1 ml
Glycerin	10% (v/v)	–	10 ml
DTT	10 m*M*	100 m*M*	10 ml
PMSF	0.5 m*M*	50 m*M*	1 ml
Add water to 100 ml			

⑥ SiR reaction mixture: 50 m*M* HEPES–KOH, pH 7.8

Component	Final concentration	Stock	Amount/100 ml
HEPES	50 m*M*	200 m*M*	25 ml
KOH	–	1 *M*	Adjust pH to 7.8
Na_2SO_3	1 m*M*	10 m*M*	10 ml

Fed_{red}	5 mM	100 mM	5 ml
Add water to 100 ml			

⑦ CAS reaction mixture, pH 9.0

Component	Final concentration	Stock	Amount/100 ml
Tris	100 mM	200 mM	50 ml
HCl	10 mM	200 mM	5 ml
L-Cys	0.8 mM	16 mM	5 ml
KCN	10 mM	1 M	1 ml
Add water to 100 ml			

⑧ CS extraction buffer: 50 mM PBS, pH 7.5

Component	Final concentration	Stock	Amount/200 ml
NaH_2PO_4	8 mM	200 mM	8 ml
Na_2HPO_4	42 mM	200 mM	42 ml
EDTA	1 mM	100 mM	2 ml
$MgCl_2$	5 mM	100 mM	10 ml
DTT	2 mM	100 mM	4 ml
Triton X-100	0.1% (v/v)	–	0.2 ml
PMSF	0.5 mM	50 mM	2 ml
Add water to 200 ml			

⑨ CS reaction mixture: 50 mM PBS, pH 8.0

Component	Final concentration	Stock	Amount/200 ml
NaH_2PO_4	2.65 mM	200 mM	2.65 ml
Na_2HPO_4	47.35 mM	200 mM	47.35 ml
Na_2S	4 mM	100 mM	8 ml
OAS	12.5 mM	100 mM	25 ml
Add water to 200 ml			

⑩ Acidic ninhydrin reagent

Component	Final concentration	Amount/100 ml
Ninhydrin	1.25% (w/v)	1.25 g
Concentrated HCl	2.34 *M*	20 ml
Glacial acetic acid	14.08 *M*	80 ml
Add water to 100 ml		

4. PROTOCOL 1

4.1. Duration

Preparation	2 h
Protocol	About 2 h

4.2. Preparation

Prepare solution and buffer ①, ②, ③, and ④. Sampling and extract crude enzyme from plant tissues in accordance with the flowchart of the complete Protocol 1 (see Fig. 2).

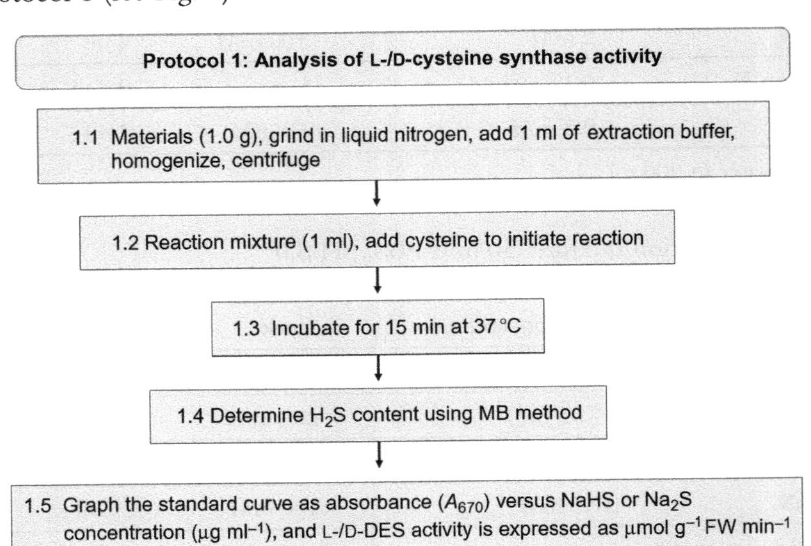

Figure 2 Flowchart for Protocol 1.

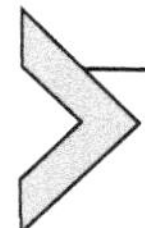

5. STEP 1: ANALYZE OF L-/D-CYSTEINE DESULFHYDRASE ACTIVITY

5.1. Overview

Crude enzyme of L-/D-DES is extracted from plant tissues, after homogenate is centrifuged, L-/D-DES activity is determined using spectrophotometry, and then calculate enzyme activity from a standard curve or extinction coefficient of MB ($15 \times 10^6\ M^{-1}\ \text{cm}^{-1}$) at 670 nm and L-/D-DES activity is expressed as μmol g^{-1} FW (fresh weight) min^{-1}.

5.2. Duration

2 h

1. One gram of plant materials is ground into fine powder with a mortar and pestle under liquid nitrogen and then is homogenized in 1 ml of extraction buffer (pH 8.0).
2. The homogenate is centrifuged at 15,000 × *g* for 15 min at 4 °C, and the supernatant is used to assay L-/D-DES activity.
3. The reaction mixture contained in a total volume of 1 ml: 0.8 m*M*L- (for L-DES assay) or D-cysteine (for D-DES assay), 2.5 m*M* DTT, 100 m*M* Tris–HCl (pH 9.0), and 0.1 mL supernatant (~10 μg protein). The reaction is initiated by the addition of L- or D-cysteine.
4. Incubate for 15 min at 37 °C.
5. The reaction is terminated by adding 100 μl of 30 m*M* $FeCl_3$ dissolved in 1.2 *M* HCl and 100 μl of 20 m*M* DMPD dissolved in 7.2 *M* HCl. The amount of H_2S is determined colorimetrically at 670 nm after incubation for 15 min at room temperature.
6. Blanks are prepared by the same procedures and known concentrations of NaHS or Na_2S are used to make a standard curve, and the activity of L-/D-DES is expressed as μmol g^{-1} FW min^{-1}.

5.3. Tip

Spectrophotometer is warmed up for 15–20 min before use; zero the spectrophotometer using assay mixture without the supernatant (crude enzyme), which is replaced by the equal volume of extraction buffer. After reading the samples, rezero to determine any instrument drift. Ideally, the absorbance should fall between 0.1 and 1 units, depending on the performance of the spectrophotometer.

5.4. Tip

Graph the standard curve as absorbance (A_{670}) versus concentration (μg ml^{-1}). Determine the concentrations of the unknown samples from the graph. Alternatively, the concentration can be determined using MB molar extinction coefficient $15 \times 10^6\ M^{-1}\ cm^{-1}$, and the Beer–Lambert equation

$$A = \varepsilon cl$$

where ε is the molar extinction coefficient, c the concentration of MB, and l the path length in cm (Papenbrock & Schmidt, 2000).

5.5. Tip

In this protocol, $FeCl_3$ and DMPD are soluted in 1.2 and 7.2 m*M* HCl, respectively. In addition, both can be soluted in 3.5 and 100 m*M* H_2SO_4 as well as form 5 m*M* DMPD and 50 m*M* $FeCl_3$, respectively (Sekiya, Schmidt, Wilson, & Filner, 1982).

5.6. Tip

The activity of L-/D-DES also is expressed as the amount of emission of H_2S from reaction mixture. Briefly, the homogenate is mixed with solution containing 100 m*M* phosphate-buffered saline (pH 7.0), 10 m*M* L-/D-cysteine, and 2 m*M* phosphopyridoxal in a test tube at room temperature, and the released H_2S is absorbed in a $Za(OAc)_2$ trap, a small glass tube containing $Za(OAc)_2$ (20 m*M*), fixed in the bottom of the reaction vial, and then add the equal volume of $FeCl_3$ and DMPD solution, respectively, to determine H_2S content as above-mentioned methods (Riemenschneider, Nikiforova, Hoefgen, DeKok, & Papenbrock, 2005).

5.7. Tip

The activity of L-/D-DES also is expressed as μmol mg^{-1} protein min^{-1}, and the content of soluble proteins is determined using Bradford assay (Noble, 2014).

5.8. Tip

To accurately determine enzyme activity, the volume of supernatant should be measured or brought up to 1 ml with extraction buffer.

6. PROTOCOL 2

6.1. Duration

Preparation	2 h
Protocol	About 2 h

6.2. Preparation

Prepare solution and buffer ⑤ and ⑥. Sampling and extract crude enzyme from plant tissues in accordance with the flowchart of the complete Protocol 2 (see Fig. 3).

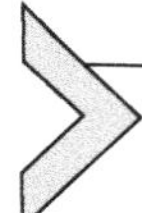

7. STEP 1: ANALYZE OF SULFITE REDUCTASE ACTIVITY

7.1. Overview

Crude enzyme of SiR is extracted from plant tissues, after homogenate is centrifuged, SiR activity is determined using spectrophotometry, and then calculate enzyme activity from a standard curve or extinction coefficient of MB

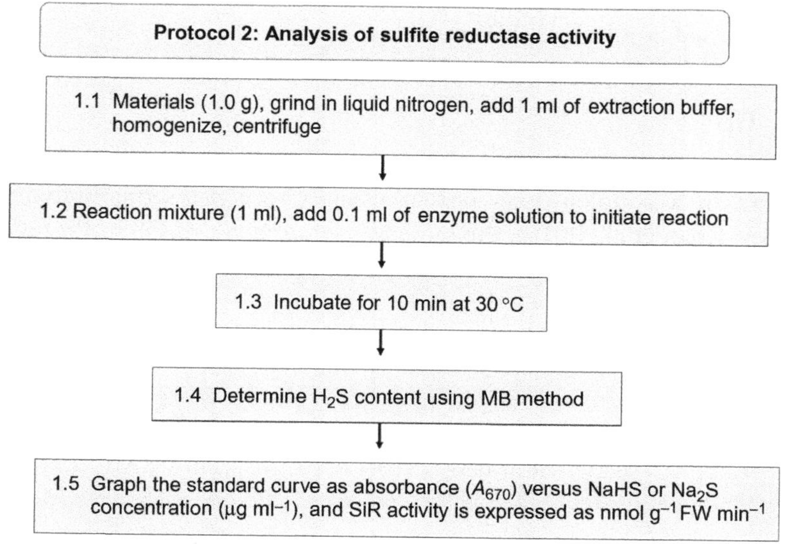

Figure 3 Flowchart for Protocol 2.

($15 \times 10^6\ M^{-1}\ cm^{-1}$) at 670 nm and SiR activity is expressed as nmol g^{-1} FW min^{-1}.

7.2. Duration

2 h

1. One gram of plant materials is ground into fine powder with a mortar and pestle under liquid nitrogen and then is homogenized in 1 ml of extraction buffer (pH 7.4) containing 10 m*M* KCl, 1 m*M* EDTA, 1 m*M* EGTA, 10% (v/v) glycerin, 10 m*M* DTT, and 0.5 m*M* PMSF.
2. The homogenate is centrifuged at $14{,}000 \times g$ for 15 min at 4 °C, and the supernatant is measured the enzymatic activity of SiR.
3. The enzymatic activity of SiR is measured in reaction mixture contained in a total volume of 1 ml, containing 25 m*M* HEPES, pH 7.8, 1 m*M* Na_2SO_3, and 5 m*M* Fed_{red}, as well as the reaction is initiated by the addition of 0.1 ml supernatant (~10 μg protein).
4. Incubate for 10 min at 30 °C.
5. The reaction is terminated by adding 100 μl of 30 m*M* $FeCl_3$ dissolved in 1.2 *M* HCl and 100 μl of 20 m*M* DMPD dissolved in 7.2 *M* HCl, and the amount of H_2S is determined colorimetrically at 670 nm after incubation for 15 min at room temperature.
6. Blanks are prepared by the same procedures and known concentrations of NaHS or Na_2S are used to make a standard curve, and SiR activity is expressed as nmol g^{-1} FW min^{-1}.

7.3. Tip

SiR activity also is expressed as the amount of cysteine by measuring cysteine content in a coupled assay system using *O*-acetyl-L-serine sulfydrylase (OSSA; Khan et al., 2010), the same as measurement of CS in Protocol 4.

7.4. Tip

Due to reduced ferredoxin (Fed_{red}) has significant absorbance at 422 nm, but oxidized ferredoxin does not, SiR activity also is expressed as the decrease in absorbance of Fed_{red} at 422 nm and calculate SiR activity according to the extinction coefficient of Fed_{red} ($10.4 \times 10^3\ M^{-1}\ cm^{-1}$; Arb, 1990).

7.5. Tip

The same as Section 5.3.

7.6. Tip

The same as Section 5.4.

7.7. Tip

The same as Section 5.7.

7.8. Tip

The same as Section 5.8.

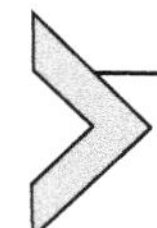

8. PROTOCOL 3

8.1. Duration

Preparation	2 h
Protocol	About 2 h

8.2. Preparation

Prepare solution and buffer ① and ⑦. Sampling and extract crude enzyme from plant tissues in accordance with the flowchart of the complete Protocol 3 (see Fig. 4).

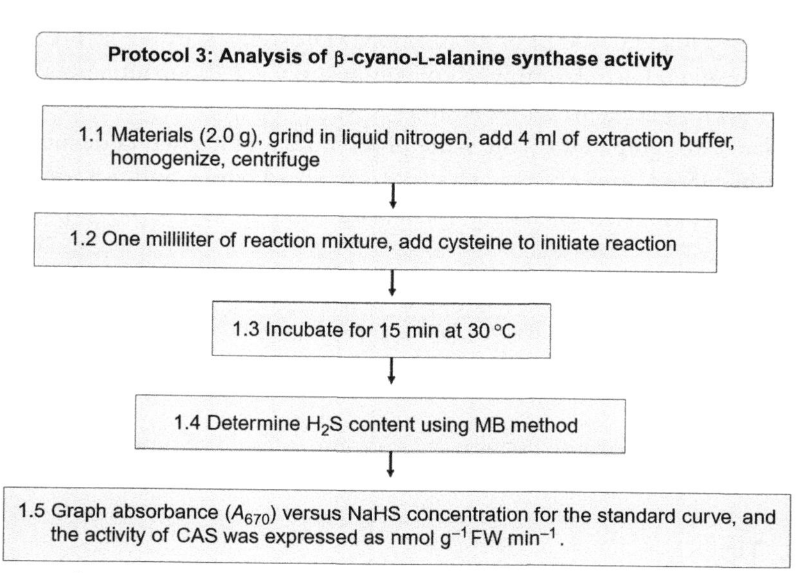

Figure 4 Flowchart for Protocol 3.

9. STEP 1: ANALYZE OF β-CYANO-L-ALANINE SYNTHASE ACTIVITY

9.1. Overview

Crude enzyme of CAS is extracted from plant tissues, after homogenate is centrifuged, CAS activity is determined using spectrophotometry, and then calculate enzyme activity from a standard curve or extinction coefficient of MB ($15 \times 10^6\ M^{-1}\ \mathrm{cm}^{-1}$) at 670 nm and CAS activity is expressed as nmol g^{-1} FW min^{-1}.

9.2. Duration

2 h

1. Two gram of plant materials are ground into fine powder with a mortar and pestle under liquid nitrogen and then are homogenized in 4 ml of extraction buffer: 20 m*M* Tris–HCl buffer (pH 8.0).
2. The homogenate is centrifuged at 14,000 × *g* for 15 min at 4 °C, and the supernatant is used for CAS activity assay.
3. The reaction mixture contained in a total volume of 1 ml: 0.8 m*M* L-cysteine, 10 m*M* KCN, 100 m*M* Tris–HCl (pH 9.0), and 50 µl supernatant. The reaction is initiated by the addition of L-cysteine.
4. Incubate for 15 min at 30 °C.
5. The reaction is terminated by adding 100 µl of 30 m*M* $FeCl_3$ dissolved in 1.2 *M* HCl and 100 µl of 20 m*M* DMPD dissolved in 7.2 *M* HCl. The amount of H_2S is determined colorimetrically at 670 nm after incubation for 15 min at room temperature.
6. Blanks are prepared by the same procedures and known concentrations of NaHS or Na_2S are used to make a standard curve, and the activity of CAS is expressed as nmol g^{-1} FW min^{-1}.

9.3. Tip

KCN is a toxicity, use carefully, and dispose properly.

9.4. Tip

To accurately determine enzyme activity, the volume of supernatant should be measured or brought up to 4 ml with extraction buffer.

9.5. Tip

The same as Section 5.3.

9.6. Tip

The same as Section 5.4.

9.7. Tip

The same as Section 5.7.

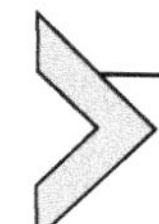

10. PROTOCOL 4

10.1. Duration

Preparation	2 h
Protocol	About 2 h

10.2. Preparation

Prepare solution and buffer ⑧, ⑨, and ⑩. Sampling and extract crude enzyme from plant tissues in accordance with the flowchart of the complete Protocol 4 (see Fig. 5).

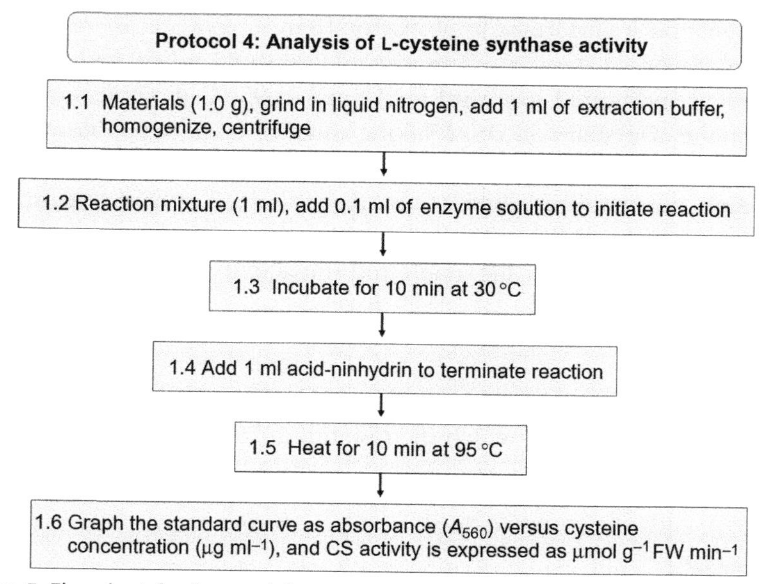

Figure 5 Flowchart for Protocol 4.

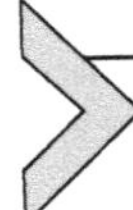

11. STEP 1: ANALYZE OF L-CYSTEINE SYNTHASE ACTIVITY

11.1. Overview

Crude enzyme of CS is extracted from plant tissues, after homogenate is centrifuged, CS activity is determined using spectrophotometry, and then calculate enzyme activity from a standard curve or extinction coefficient of $3.24 \times 10^6\ M^{-1}\ cm^{-1}$ and CS activity is expressed as μmol g^{-1} FW min^{-1}.

11.2. Duration

2 h

1. The plant tissues (1 g) are ground into fine powder with a mortar and pestle under liquid nitrogen and then are homogenized in 1 ml of extraction buffer: 50 m*M* phosphate buffer (pH 7.5) containing 1 m*M* EDTA, 5 m*M* $MgCl_2$, 2 m*M* DTT, 0.1% Triton X-100, and 0.5 m*M* PMSF.
2. The homogenate is centrifuged at 14,000 × *g* for 15 min at 4 °C, and the resulting supernatant is used for CS activity assay.
3. The enzymatic activity of CS is determined in 1 ml of the reaction mixture, containing 50 m*M* potassium phosphate, pH 8.0, 4 m*M* Na_2S, 12.5 m*M* *O*-acetyl-L-Ser, and the reaction is initiated by the addition of 0.1 ml of the enzyme solution.
4. Incubation is performed at 30 °C for 10 min.
5. The reaction is terminated by the addition of 1 ml of acid-ninhydrin reagent and heated afterward for 10 min at 95 °C (boiling water). The amount of cysteine in the resulting mixture is spectrophotometrically determined at 560 nm.
6. Graph the standard curve as absorbance (A_{560}) versus cysteine concentration (μg ml^{-1}). Determine the concentrations of the unknown samples from the graph and the activity of CS is expressed as μmol g^{-1} FW min^{-1}.

11.3. Tip

In color reaction, timing which the resulting mixture is boiling at 95 °C is started.

11.4. Tip

The same as Section 5.3.

11.5. Tip

The same as Section 5.7.

11.6. Tip

The same as Section 5.8.

ACKNOWLEDGMENT

This work is supported by the National Natural Science Foundation of China (31360057).

REFERENCES

Arb, C. V. (1990). Sulphur metabolism. C. Ssulphite reductase. *Methods in Plant Biochemistry*, *3*, 345–348.

Hancock, J. T., & Whiteman, M. (2014). Hydrogen sulfide and cell signaling: Team player or referee? *Plant Physiology and Biochemistry*, *78*, 37–42.

Khan, M. S., Haas, F. H., Samami, A. A., Gholami, A. M., Bauer, A., Fellenberg, K., et al. (2010). Sulfite reductase defines a newly discovered bottleneck for assimilatory sulfate reduction and is essential for growth and development in *Arabidopsis thaliana*. *The Plant Cell*, *22*, 1216–1231.

Li, Z. G. (2013). Hydrogen sulfide: A multifunctional gaseous molecule in plants. *Russian Journal of Plant Physiology*, *60*, 733–740.

Noble, J. E. (2014). Quantification of protein concentration using UV absorbance and coomassie dyes. *Methods in Enzymology*, *536*, 17–26.

Papenbrock, J., & Schmidt, A. (2000). Characterization of a sulfurtransferase from *Arabidopsis thaliana*. *European Journal of Biochemistry*, *267*, 145–154.

Riemenschneider, A., Nikiforova, V., Hoefgen, R., DeKok, L. J., & Papenbrock, J. (2005). Impact of elevated H_2S on metabolite levels, activity of enzymes and expression of genes involved in cysteine metabolism. *Plant Physiology and Biochemistry*, *43*, 473–483.

Sekiya, J., Schmidt, A., Wilson, L. G., & Filner, P. (1982). Emission of hydrogen sulfide by leaf tissue in response to L-cysteine. *Plant Physiology*, *70*, 430–436.

CHAPTER FOURTEEN

Sulfide Detoxification in Plant Mitochondria

Hannah Birke*[,2], Tatjana M. Hildebrandt†, Markus Wirtz*, Rüdiger Hell*[,1]

*Centre for Organismal Studies Heidelberg, University of Heidelberg, Heidelberg, Germany
†Institute for Plant Genetics, Leibniz University Hannover, Hannover, Germany
[1]Corresponding author: e-mail address: ruediger.hell@cos.uni-heidelberg.de

Contents

1. Introduction 272
   1.1 Formation of sulfide in plant cells 272
   1.2 Sulfide toxicity and detoxification mechanisms in plant mitochondria 274
2. Methods 275
   2.1 Determination of CAS activity 275
   2.2 SAT affinity purification of OAS-TL from plant tissue and determination of enzymatic activity 276
   2.3 Discrimination between CAS and OAS-TL 280
   2.4 Determination of SDO activity 281
3. Summary 283
Acknowledgments 283
References 284

Abstract

In contrast to animals, which release the signal molecule sulfide in small amounts from cysteine and its derivates, phototrophic eukaryotes generate sulfide as an essential intermediate of the sulfur assimilation pathway. Additionally, iron–sulfur cluster turnover and cyanide detoxification might contribute to the release of sulfide in mitochondria. However, sulfide is a potent inhibitor of cytochrome *c* oxidase in mitochondria. Thus, efficient sulfide detoxification mechanisms are required in mitochondria to ensure adequate energy production and consequently survival of the plant cell. Two enzymes have been recently described to catalyze sulfide detoxification in mitochondria of *Arabidopsis thaliana*, *O*-acetylserine(thiol)lyase C (OAS-TL C), and the sulfur dioxygenase (SDO) ethylmalonic encephalopathy protein 1 (ETHE1). Biochemical characterization of sulfide producing and consuming enzymes in mitochondria of plants is fundamental to understand the regulatory network that enables mitochondrial sulfide homeostasis under nonstressed and stressed conditions. In this chapter, we provide established protocols

[2] Current address: CSIRO Agriculture Flagship, Black Mountain Laboratories, Canberra, Australia.

Methods in Enzymology, Volume 555
ISSN 0076-6879
http://dx.doi.org/10.1016/bs.mie.2014.11.027

to determine the activity of the sulfide releasing enzyme β-cyanoalanine synthase as well as sulfide-consuming enzymes OAS-TL and SDO. Additionally, we describe a reliable and efficient method to purify OAS-TL proteins from plant material.

1. INTRODUCTION

Hydrogen sulfide (H_2S) is considered to be toxic for animals and humans due to its inhibitory effect on cytochrome *c* oxidase (COX) in mitochondria (Li, Rose, & Moore, 2011). However, there is increasing evidence that H_2S also acts as signaling molecule, for example, during neuromodulation in the brain and relaxation of smooth muscles in the vascular system (Li et al., 2011). In animals, sulfide is released from cysteine by cystathionine-β-synthase, cystathionine-γ-lyase, or β-mercaptopyruvate sulfurtransferase in combination with cysteine aminotransferase (Li et al., 2011). The toxic effect of H_2S is also well described for plants (Birke, De Kok, Wirtz, & Hell, 2014; Birke, Haas, et al., 2012; Maas, de Kok, Hoffmann, & Kuiper, 1987). Recent work indicates that sulfide also acts as signaling molecule in plants, for example, during stomata closure (Garcia-Mata & Lamattina, 2010) and autophagy in *Arabidopsis thaliana* (Gotor, Garcia, Crespo, & Romero, 2013) as well as chloroplast biogenesis in spinach seedlings (Chen et al., 2011). However, in contrast to animals, sulfide is not only generated by release from cysteine and its derivates in plants.

1.1. Formation of sulfide in plant cells

Cysteine degradation, cyanide detoxification, and presumably also iron–sulfur cluster turnover can contribute to total sulfide generation in plant cells (Fig. 1; Frazzon et al., 2007; Hatzfeld et al., 2000; Hell & Wirtz, 2011). However, the bulk of sulfide is produced by the assimilatory sulfate reduction pathway: Activated sulfate, adenosine-5′-phosphosulfate (APS), is reduced in a two-step process to sulfide by APS reductase and sulfite reductase, which are both exclusively localized in plastids (Hell & Wirtz, 2011; Takahashi, Kopriva, Giordano, Saito, & Hell, 2011). Sulfide is then incorporated into *O*-acetylserine (OAS) by OAS-(thiol)lyase (OAS-TL, EC 2.5.1.47) to form cysteine. The cysteine precursor OAS is produced by serine acetyltransferase (SAT, EC 2.3.1.30) that can reversibly interact with OAS-TL to form the hetero-oligomeric cysteine synthase complex (Fig. 2) (Hell & Wirtz, 2011). The interaction is stabilized in the presence of sulfide

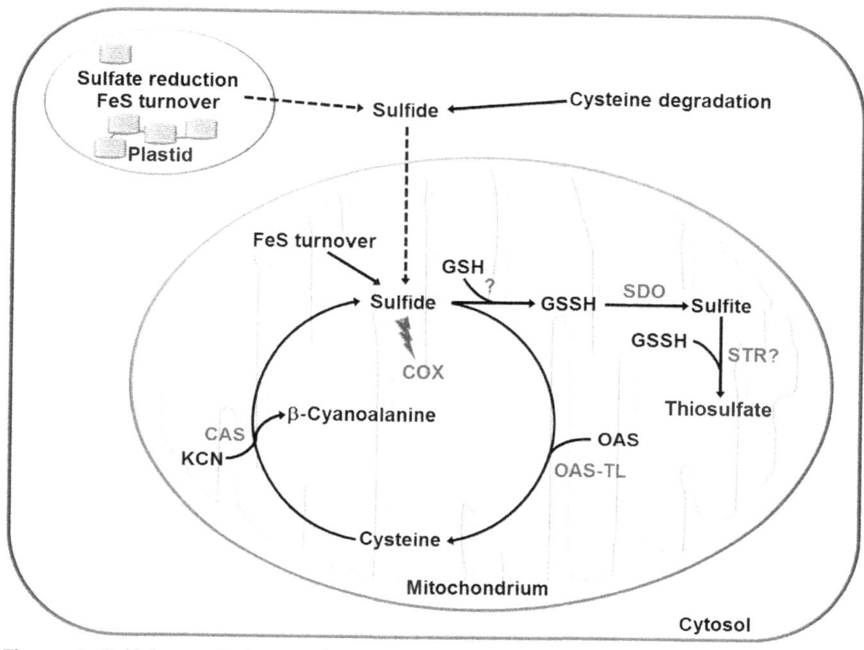

Figure 1 Sulfide-producing and -consuming processes in plant cells. For details, see text. CAS: β-cyanoalanine synthase; COX: cytochrome *c* oxidase; OAS-TL: *O*-acetylserine (thiol)lyase; SDO: sulfur dioxygenase; STR?: unknown sulfur transferase; ?: unknown protein or nonenzymatically.

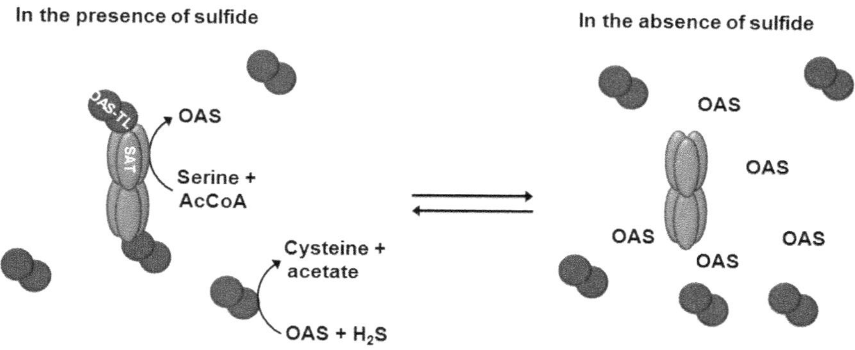

Figure 2 Cysteine synthase complex formation is regulated by sulfide and OAS steady-state levels. For details, see text. OAS-TL: *O*-acetylserine(thiol)lyase; SAT: serine acetyltransferase.

and mediated by the SAT C-terminus that binds to the active site of OAS-TL. As a consequence, OAS-TL is inactive within the complex and OAS is released. However, a high OAS-TL:SAT ratio within the cell ensures metabolization of OAS by free and thus active OAS-TL. In the absence of sulfide, OAS accumulates and is able to dissociate the complex due to competitive binding with the SAT C-terminus in the active site of OAS-TL. Free SAT is less active than complex-bound SAT, since it is more sensitive to enzymatic inhibition by cysteine (Wirtz et al., 2012; Wirtz, Birke, et al., 2010). Consequently, production of OAS is decreased upon limitation of sulfide. Formation of the cysteine synthase complex thus contributes to regulation of cysteine synthesis and might act as sulfide sensor (Hell & Wirtz, 2011). Noteworthy, SAT and OAS-TL isoforms are present in the plastids as well as cytosol and mitochondria and reverse genetic approaches showed that the cytosolic OAS-TL A isoform rather than plastidic OAS-TL B incorporates the majority of sulfide into cysteine (Birke, Heeg, Wirtz, & Hell, 2013; Heeg et al., 2008; Watanabe et al., 2008). This indicates that sulfide moves from the plastid into the cytosol and mitochondria which is in agreement with permeability of the membranes for sulfide (Jacques, 1936; Mathai et al., 2009).

In cytosol and mitochondria, sulfide is formed as by-product of cysteine degradation or cysteine consuming processes. For example, the cytosolic cysteine desulfhydrase DES1 (EC 4.4.1.1) catalyzes the desulfuration of cysteine in *Arabidopsis* and contributes to maintenance of cysteine homeostasis, mainly at late developmental stages and under environmental stress conditions (Alvarez, Calo, Romero, Garcia, & Gotor, 2010). Detoxification of cyanide, which is produced during ethylene synthesis (Peiser et al., 1984), requires mitochondrial β-cyanoalanine synthase (CAS, EC 4.4.1.9) that catalyzes the substitution of the thiol group in cysteine with cyanide to form β-cyanoalanine while sulfide is released (Miller & Conn, 1980). Additionally, NSF1, which is part of the iron–sulfur cluster assembly machinery in *Arabidopsis* mitochondria, has cysteine desulfurase activity (EC 2.8.1.7) and releases sulfide *in vitro* (Frazzon et al., 2007), making iron–sulfur cluster biosynthesis a potential sulfide source in mitochondria.

1.2. Sulfide toxicity and detoxification mechanisms in plant mitochondria

Main target of sulfide toxicity in *Arabidopsis* is COX of the mitochondrial respiratory chain (Fig. 1) which is strongly inhibited by sulfide ($IC_{50} \leq 10$ n*M*; Birke, Haas, et al., 2012). Maintenance of low sulfide levels in mitochondria is therefore crucial for energy production of the cell and survival of the plant. Recently, two enzymes have been shown to be involved in local sulfide detoxification in *Arabidopsis*, namely mitochondrial OAS-TL C (Birke, Haas, et al., 2012) and the sulfur dioxygenase (SDO) ethylmalonic encephalopathy protein 1 (ETHE1) (Krüßel et al., 2014). Although OAS-TL C hardly contributes to net cysteine synthesis of the cell (Heeg et al., 2008; Watanabe et al., 2008), lack of OAS-TL C in mitochondria results in partial inhibition of COX activity in seedlings (Birke, Haas, et al., 2012), indicating a significant contribution of OAS-TL C to sulfide detoxification by incorporation of sulfide into OAS. However, spinach and the moss *Physcomitrella patens* lack OAS-TL activity in mitochondria (Birke, Müller, et al., 2012; Warrilow & Hawkesford, 2000), which provides further evidence for the significance of additional sulfide detoxification mechanisms. *Arabidopsis* ETHE1 has been shown to be an essential mitochondrial SDO (EC 1.13.11.18) that can detoxify sulfide by thiosulfate formation (Holdorf et al., 2012; Krüßel et al., 2014). According to the current model, ETHE1 oxidizes glutathione persulfide (GSSH), which is formed from sulfide and glutathione either spontaneously or by a so far unknown enzyme, to sulfite. Sulfite and a second GSSH are then used by a sulfur transferase to form thiosulfate (Birke et al., 2014; Krüßel et al., 2014).

In this chapter, we describe established methods to determine activities of enzymes that actively contribute to keep sulfide concentration in plant mitochondria tolerable. Such biochemical tools are necessary to understand the network of sulfide-consuming and -generating metabolic reactions under nonstressed and stressed condition in plants and will help to uncover the regulatory circuits within plant cells that allow sulfide homeostasis. Due to the striking connection between cyanide and sulfide toxicity, we focus on CAS (Section 2.1) and OAS-TL (Section 2.2) in order to define the different biochemical properties of these structurally related enzymes (Section 2.3). Finally, we describe a method to determine sulfide detoxification capacity of ETHE1 (Section 2.4).

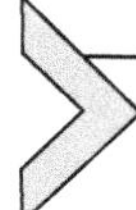

2. METHODS

2.1. Determination of CAS activity

CAS catalyzes the reaction L-cysteine + KCN → β-cyanoalanine + H_2S. To determine CAS activity of proteins *in vitro*, quantification of bisulfide formation using the methylene blue assay is the most commonly used method. The assay was first described by Hasegawa, Tada, Torii, and Esashi (1994) to determine CAS activity in seed protein extracts and has since been applied to soluble protein fractions from spinach and *Arabidopsis* tissue as well as recombinantly expressed proteins (e.g., Hatzfeld et al., 2000; Warrilow & Hawkesford, 2000; Yamaguchi, Nakamura, Kusano, & Sano, 2000). A standard reaction according to Warrilow and Hawkesford (1998, 2000) contains 2–20 μg of purified CAS, 1 m*M* L-cysteine, 0.75 m*M* KCN, and 160 m*M* 2-amino-2-methyl-1-propanol buffer, pH 9.8, in a final volume of 1 ml. Upon incubation at 26 °C for 20 min, the reaction is terminated by addition of 0.5 ml acidic dye precursor solution (15 m*M* *N,N*-dimethyl-1,4-phenylenediamine dihydrochloride (DMPD), 3 m*M* $FeCl_3$, 4.2 *M* HCl) and methylene blue is allowed to form from the formed bisulfide and DMPD as well as Fe(III) for 20 min at 26 °C. After centrifugation at 1500 × *g* for 20 min at room temperature, the amount of methylene blue is determined by measuring the absorbance of the obtained supernatant at 745 nm against Na_2S standards. Specific activity of CAS is defined as μmol bisulfide formed per min per mg of protein.

It is worth mentioning that the above indicated concentration of KCN in the assay is optimized to quantify activity of a true CAS and might have to be adjusted when testing other members of the β-substituted alanine synthase family (see Section 2.3).

2.2. SAT affinity purification of OAS-TL from plant tissue and determination of enzymatic activity

The high affinity interaction of SAT and OAS-TL in the cysteine synthase complex (K_D = 2–40 n*M*, Berkowitz, Wirtz, Wolf, Kuhlmann, & Hell, 2002; Campanini et al., 2005; Droux, Ruffet, Douce, & Job, 1998; Wirtz, Birke, et al., 2010; Zhao et al., 2006) is mediated by the SAT C-terminus which is conserved among bacteria and plant species (Droux et al., 1998; Mino et al., 1999; Wirtz, Birke, et al., 2010). In fact, it has been demonstrated that SAT and OAS-TL from different species are able to interact (Birke, Müller, et al., 2012; Droux et al., 1998; Wirtz, Birke, et al., 2010). The specific interaction between SAT and OAS-TL has been used successfully to purify OAS-TL proteins from various *Arabidopsis* tissues (Heeg et al., 2008; Wirtz, Heeg, Samami, Ruppert, & Hell, 2010), the alga Chlorella (Salbitani, Wirtz, Hell, Carfagna, 2014), and *P. patens* protonemata (Birke, Müller, et al., 2012). The SAT affinity purification protocol can be completed within 2 days and is subdivided into three parts: (1) recombinant expression and immobilization of His_6-tagged *Arabidopsis* SAT5 on a nickel-loaded chromatography column, (2) large-scale extraction of soluble proteins from plant tissue, and (3) purification of OAS-TL proteins from the soluble protein extract.

2.2.1 Recombinant expression and immobilization of His_6:AtSAT5 on a nickel-loaded chromatography column

We here describe the method using *Arabidopsis* SAT5 (At5g56760) as bait for OAS-TL purification. However, in theory AtSAT5 can be replaced by any *bona fide* SAT harboring the canonical SAT C-terminus, which ends with an Ile (Francois, Kumaran, & Jez, 2006). The full-length coding sequence of AtSAT5 is cloned into pET28a(+) (Novagen) or a comparable vector to allow expression of His_6:AtSAT5. The N-terminal position of the His_6 tag is crucial since a tag at the C-terminus of SAT would prevent interaction with OAS-TL. Expression of the protein in *Escherichia coli* HMS174 (DE3) (Novagen) or a comparable strain is induced with isopropyl-β-D-thiogalactopyranosid according to the manufacturer's instructions. After harvesting of cells by centrifugation for 10 min at 20,000 × *g* and 4 °C, cells can be stored at −80 °C until used for extraction of the expressed protein. Prior to extraction, cells are disintegrated on ice by 10 min ultrasonic treatment with a Sonopuls GM70 (Bandelin). The protein is extracted in binding buffer (50 m*M* Tris–HCl, pH 8.0, 250 m*M* NaCl, 20 m*M* imidazole, 0.5 m*M* phenylmethylsulfonyl fluoride (PMSF)) and cell debris are removed by centrifugation at 4 °C for 10 min at maximum speed. The supernatant is filtered through a 0.45-μm filter and circulated at room temperature with a flow rate of 1 ml min^{-1} for 30–60 min over a Ni^{2+}-loaded HiTrap Chelating HP column (GE Healthcare Life Science) that is equilibrated with binding buffer. Bacterial proteins are removed by application of 10 column volumes (CV) washing buffer (50 m*M* Tris–HCl, pH 8.0, 250 m*M* NaCl, 80 m*M* imidazole, 0.5 m*M* PMSF). The bacterial OAS-TL is released from His_6:AtSAT5 with 10 CV OAS-TL elution buffer (washing buffer supplemented with 10 m*M* OAS). To allow binding of the plant OAS-TL proteins in the next step, OAS has to be removed from the column by washing with 10 CV washing buffer. The His_6:AtSAT5-loaded column can be stored in washing buffer at room temperature for approximately 16 h so that the following steps can be performed the next day, if necessary.

2.2.2 Large-scale extraction of soluble proteins from plant tissue

For extraction of soluble proteins from *Arabidopsis* leaves, 100 g are ground to a fine powder in liquid nitrogen. The powder is then stirred for 30 min in

350 ml ice-cold extraction buffer (50 m*M* Tris–HCl, pH 8.0, 250 m*M* NaCl, 80 m*M* imidazole, 0.5 m*M* PMSF, 1 m*M* dithiothreitol (DTT)). The resulting crude extract is filtered through Miracloth tissue (Calbiochem) and the flow through is used for centrifugation at 27,500 × *g* and 4 °C for 20 min to collect residual cell debris. The obtained supernatant is then fractionated by successive precipitation of proteins with 20% and 75% ammonium sulfate. In the first step, the respective amount of ammonium sulfate to obtain a saturation of 20% (Sambrook, Fritsch, & Maniatis, 1989) is added to the supernatant and the mixture is stirred on ice for 30 min followed by centrifugation as described above to collect proteins. The amount of ammonium sulfate in the obtained supernatant is then increased to a final saturation of 75% (Sambrook et al., 1989) and the mixture is again stirred on ice for 30 min followed by centrifugation as described above. The pellet is resuspended in 12 ml extraction buffer and the obtained protein extract is desalted by size exclusion chromatography using PD-10 columns (GE Healthcare Life Sciences) according to the manufacturer's instructions in the same buffer. This step removes ammonium sulfate that might interfere with CSC formation in the last step of the purification. The concentration of proteins can then be determined with the method of choice, for example, according to Bradford (1976) using bovine albumin as reference. The desalted protein extract should be used for purification of OAS-TL proteins at the same day.

The amount of starting material varies when extracting proteins from different tissues. When using *Arabidopsis* stem, root, or heterotrophic cell culture, 4–5 g of starting material is sufficient to purify μg amounts of OAS-TL protein (Wirtz, Heeg, et al., 2010), whereas 150 g starting material is recommended for Physcomitrella protonemata (Birke, Müller, et al., 2012). The volume of extraction buffer is adjusted accordingly. The yield of purified OAS-TL protein is about 40% from *Arabidopsis* leaf extracts. The purified OAS-TL is typically more than 150-fold enriched in activity, apparently free of contaminations and has a specific activity of more than 300 μmol min^{-1} mg^{-1} protein (Heeg et al., 2008).

2.2.3 Purification of OAS-TL proteins from soluble protein extract

The soluble protein extract obtained in Section 2.2.2 is circulated for 60 min over the His_6:AtSAT5-loaded column (Section 2.2.1) with a flow rate of 1 ml min^{-1} and at room temperature. After a washing step with 10 CV washing buffer, OAS-TL proteins are eluted with 10 CV OAS-TL elution buffer (Section 2.2.1) and collected in 0.5-ml fractions. The concentration of

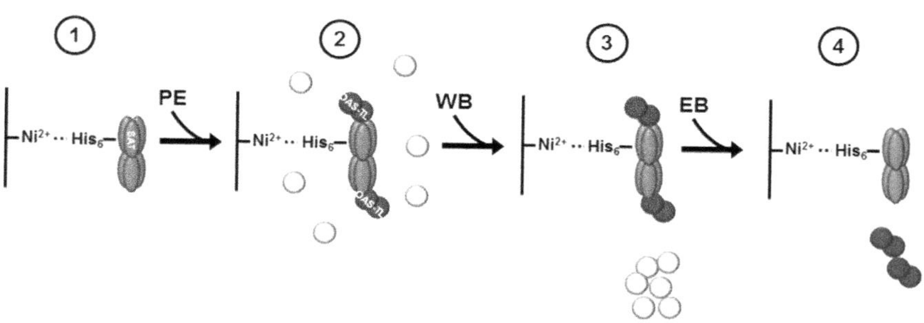

Figure 3 SAT affinity purification of OAS-TL proteins. His_6-tagged recombinant SAT is immobilized on a Ni^{2+}-loaded HiTrap column (1) and serves as bait for OAS-TL in crude plant protein extracts (PE). After application of PE (2), the column is washed with washing buffer (WB) to remove nonspecifically bound plant proteins (3). Finally, OAS-TL proteins are eluted (4) by application of washing buffer supplemented with OAS (EB).

protein in the obtained fractions can then be determined with the method of choice, for example, according to Bradford (1976) using bovine albumin as reference. OAS-TL proteins are stable at room temperature for several hours and can be stored in elution buffer at −20 °C without losing activity for approximately 1 year. The principle of the purification is illustrated in Fig. 3.

2.2.4 Determination of OAS-TL activity

OAS-TL catalyzes the reaction OAS + $H_2S \rightarrow$ L-cysteine + acetate. To determine OAS-TL activity of purified OAS-TL proteins *in vitro*, a modified protocol from Gaitonde (1967) is used, in which the product cysteine is quantified using ninhydrin. A standard reaction contains 2–5 ng of purified OAS-TL protein, 50 m*M* HEPES, pH 7.5, 10 m*M* OAS, 5 m*M* DTT, and 5 m*M* Na_2S in a total volume of 0.1 ml. The reaction is started by addition of Na_2S, incubated for 5–10 min at 25 °C, and terminated by addition of 0.05 ml trichloroacetic acid. Upon centrifugation for 5 min at room temperature and maximum speed, the supernatant is mixed with 0.3 ml freshly prepared ninhydrin–acetic acid mix (1 part 100% (v/v) acetic acid and 2 parts ninhydrin solution (1.25% (w/v) ninhydrin, 25% (v/v) HCl, 75% (v/v) acetic acid; stable at room temperature for 1 year when protected from light)) and incubated for 10 min at 100 °C. Under these conditions, ninhydrin forms a pink colored complex with cysteine that is stable for up to 20 min. The sample is cooled down on ice for 1 min and diluted with 100% (v/v) ethanol to a final volume of 1 ml. The absorbance at 560 nm is determined against a blank containing no enzyme and the amount of cysteine is quantified using a cysteine standard curve.

The linear range of the here described cysteine detection method is between 50 and 500 μ*M* cysteine in the enzymatic assay (0.1 ml). This corresponds to 5–50 nmol cysteine in the final test volume (1 ml). If less than 5 nmol cysteine is produced by the enzyme, for example, when substrate supply is decreased for determination of the Michaelis–Menten constant, the ninhydrin-based cysteine quantification must be replaced by a more sensitive detection method. Detection of cysteine using the dye monobromobimane has been applied successfully and allows detection of 0.1 pmol of cysteine–bimane after separation of bimane derivates by reversed-phased high-performance liquid chromatography (HPLC) (Wirtz, Droux, & Hell, 2004). The disadvantages of the bimane-based OAS-TL assay are necessity of analytical instrumentation (fluorescence detector coupled to an HPLC) and longer execution time.

Noteworthy, the ninhydrin-based OAS-TL assay can be used successfully to determine *in vitro* OAS-TL activity in crude protein or mitochondrial extracts of plants (Birke et al., 2013; Heeg et al., 2008), when availability of substrates is not limiting. Typically 2 μg of crude protein extract is used in the assay, which is allowed to proceed for 2–10 min at 25 °C.

Specific activity of OAS-TL is defined as μmol cysteine formed per min per mg protein.

It is also worth mentioning that the above indicated concentration of OAS in the assay is optimized to quantify activity of a true OAS-TL and might have to be adjusted when testing other members of the β-substituted alanine synthase family (see Section 2.3).

2.3. Discrimination between CAS and OAS-TL

Both CAS and OAS-TL belong to the family of β-substituted alanine synthases and show high similarity in amino acid sequence (Hatzfeld et al., 2000; Heeg et al., 2008). In fact, CAS from various plant species has been shown to catalyze the OAS-TL reaction and some OAS-TL proteins possess CAS activity (Hatzfeld et al., 2000; Warrilow & Hawkesford, 2000). However, CAS does not interact with SAT to form the CSC (Heeg et al., 2008). Additionally, all CAS proteins identified so far are characterized by substitution of the highly conserved Thr residue in the OAS binding pocket of OAS-TL (Thr72 in OAS-TL from *Salmonella typhimurium*; Burkhard et al., 1998). The absence of Thr and its hydroxyl group in the binding pocket probably results in weaker binding of OAS (Warrilow & Hawkesford, 2002), which would provide a structural explanation for the higher K_m(OAS) of CAS compared to OAS-TL that is observed in spinach and *Arabidopsis* (Hatzfeld et al., 2000; Warrilow & Hawkesford, 2000). Determination of kinetic parameters therefore helps to classify β-substituted alanine synthases. The following criteria for discrimination between CAS and OAS-TL are derived from spinach enzymes (Warrilow & Hawkesford, 1998, 2000, 2002). However, high amino acid sequence similarity between β-substituted alanine synthases from different plant species (Hatzfeld et al., 2000; Jost et al., 2000; Maruyama, Ishizawa, Takagi, & Esashi, 1998; Warrilow & Hawkesford, 2000; Yi, Juergens, & Jez, 2012) make it reasonable to believe that these criteria can be applied to proteins from other plant species. (a) A true CAS has higher specific enzymatic activity in the CAS assay than in the OAS-TL assay under substrate-saturated conditions, whereas a true OAS-TL shows the contrary. (b) A true CAS shows a higher specific activity than a true OAS-TL in the CAS assay under substrate-saturated conditions, whereas the contrary is true in the OAS-TL assay. (c) The Michaelis–Menten constant for KCN is significantly lower than 0.1 m*M*. In contrast, affinity of a true OAS-TL for cyanide is very low, the K_m(KCN) is typically higher than 0.1 m*M*. (d) A true CAS is almost entirely inhibited by nonphysiologically high cyanide concentrations (≥0.2 m*M*), while OAS-TL displays highest enzymatic activity under this artificial condition. As a consequence of (c) and (d), true CAS detoxifies cyanide at low cyanide concentrations, whereas OAS-TL is almost inactive under these conditions. In planta, OAS-TL would gain significant CAS activity only at cyanide concentrations that are extremely toxic for plant cells (Birke, Haas, et al., 2012).

2.4. Determination of SDO activity

The SDO activity assay was originally described to determine SDO activity in chemolithotrophic sulfur bacteria (Suzuki, 1965) but has been adapted recently to animals (Hildebrandt & Grieshaber, 2008) and plants (Holdorf et al., 2012; Krüßel et al., 2014). In eukaryotes, the enzyme catalyzing the SDO reaction is called ETHE1, since mutations in the human gene cause the fatal metabolic disease ethylmalonic encephalopathy (Tiranti et al., 2009). ETHE1 is localized in the mitochondrial matrix and acts as dimer using Fe^{2+} as a cofactor (Tiranti et al., 2009).

SDO oxidizes persulfide groups, for example, of GSSH, to sulfite while incorporating molecular oxygen: $GSSH + O_2 + H_2O \rightarrow GSH + SO_3^{2-} + 2H^+$. SDO activity is measured as oxygen consumption in the

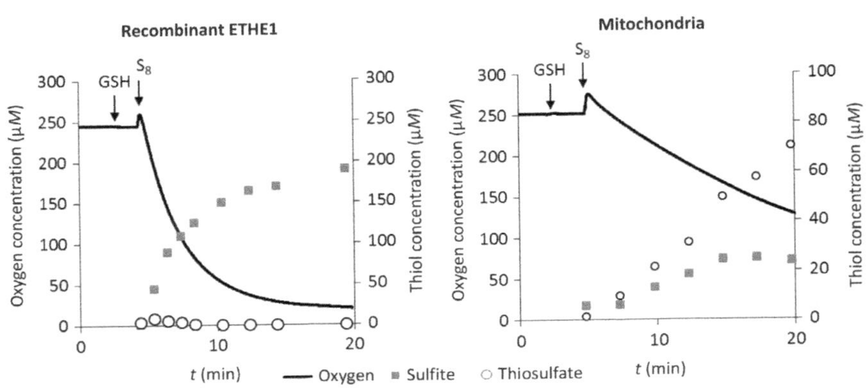

Figure 4 Original traces of SDO activity tests. Recombinant AtETHE1 (1.5 μg protein ml^{-1}) and wild-type *Arabidopsis* mitochondria (0.25 mg protein ml^{-1}) were used in the assay. 1 m*M* GSH and 15 μl ml^{-1} saturated solution of elemental sulfur (S_8) in acetone were added (indicated by arrows) to produce the substrate GSSH nonenzymatically. The oxygen concentration was measured using an oxygen electrode and concentrations of sulfite and thiosulfate were quantified by HPLC. *Reproduced from Krüßel et al. (2014), Copyright American Society of Plant Biologists, www.plantphysiol.org; with permission from Plant Physiology.*

presence of GSSH as a substrate. However, GSSH is unstable and cannot be provided as a stock solution but has to be produced directly in the reaction mixture. The standard assay therefore includes reduced glutathione (GSH) and elemental sulfur (S_8) (Fig. 4) which react nonenzymatically to form polysulfane compounds (GSS_nH), including GSSH (Rohwerder & Sand, 2003). It is worth mentioning that GSH acts as a catalyst for elemental sulfur activation and is not consumed during the reaction. Alternatively, GSSH can be produced by preincubation of hydrogen sulfide with oxidized glutathione (GSSG; e.g., 3 m*M* Na_2S + 0.5 m*M* GSSG, 10 min at 30 °C). However, in some samples results can be compromised by the presence of sulfide-oxidizing enzymes.

The SDO activity assay according to Hildebrandt and Grieshaber (2008) is performed using a Clark-type oxygen electrode at 25 °C. The reaction mixture contains 1–2 μg purified enzyme or 150–300 μg mitochondrial protein in 0.1 *M* potassium phosphate buffer, pH 7.4. 1 m*M* GSH (final concentration) is added, followed by 15 μl of a saturated elemental sulfur solution in acetone. The final reaction volume is 1 ml. Oxygen consumption rates are measured during the linear phase of oxygen depletion, which usually occurs in the first 2–3 min, and corrected for oxygen consumption in a blank without enzyme. An example of original data from SDO activity tests is shown in

Fig. 4. In addition to the oxygen consumption as measure for SDO activity, products of the SDO reaction can be analyzed by HPLC (Völkel & Grieshaber, 1994). Noteworthy, sulfite produced by the SDO is immediately converted to thiosulfate in mitochondria of animals and plants (Hildebrandt & Grieshaber, 2008; Krüßel et al., 2014).

The following issues are critical when performing the SDO assay: first, a saturated solution of sulfur in acetone, stored at room temperature, is usually used for the assay due to the low solubility of S_8 in water. Acetone does not have an effect on SDO activity. However, if a system without acetone is required, a solution containing dispersed sulfur in water can be prepared by mixing the saturated sulfur solution in acetone with an equal volume of deionized water and dialyzing against water overnight (Rohwerder & Sand, 2003). Second, when working with plant material, it is necessary to use isolated mitochondria for the SDO assay, since background oxygen consumption with GSH is too high in tissue homogenates.

3. SUMMARY

In plant cells, sulfide is produced in plastids as intermediate of the sulfur assimilation pathway but also as a consequence of thiol catabolism and utilization of cysteine in processes like cyanide detoxification and iron–sulfur cluster turnover in mitochondria. On the other hand, sulfide is a potent inhibitor of COX in mitochondria. To protect COX and ensure energy production of the cell, multiple sulfide detoxification mechanisms are therefore present in mitochondria. So far, two enzymes have been shown to be relevant for sulfide detoxification in mitochondria of *Arabidopsis*. OAS-TL C removes excess sulfide by incorporation of sulfide into OAS, whereas the SDO ETHE1 is one of the key enzyme for sulfide detoxification by thiosulfate formation. Methods to purify OAS-TL proteins from plant tissue as well as to determine OAS-TL and SDO activity have been described in this chapter, providing guidelines to analyze sulfide detoxification in plant mitochondria.

ACKNOWLEDGMENTS

The authors gratefully acknowledge support of H. B. by the Landesgraduiertenförderung Baden-Württemberg, the Schmeil-Stiftung Heidelberg, and the German Research Society (grant no. He1848/13-1/14-1).

REFERENCES

Alvarez, C., Calo, L., Romero, L. C., Garcia, I., & Gotor, C. (2010). An *O*-acetylserine (thiol)lyase homolog with L-cysteine desulfhydrase activity regulates cysteine homeostasis in Arabidopsis. *Plant Physiology*, *152*, 656–669.

Berkowitz, O., Wirtz, M., Wolf, A., Kuhlmann, J., & Hell, R. (2002). Use of biomolecular interaction analysis to elucidate the regulatory mechanism of the cysteine synthase complex from *Arabidopsis thaliana*. *The Journal of Biological Chemistry*, *277*, 30629–30634.

Birke, H., Haas, F. H., De Kok, L. J., Balk, J., Wirtz, M., & Hell, R. (2012). Cysteine biosynthesis, in concert with a novel mechanism, contributes to sulfide detoxification in mitochondria of *Arabidopsis thaliana*. *The Biochemical Journal*, *445*, 275–283.

Birke, H., Heeg, C., Wirtz, M., & Hell, R. (2013). Successful fertilization requires the presence of at least one major *O*-acetylserine(thiol)lyase for cysteine synthesis in pollen of Arabidopsis. *Plant Physiology*, *163*, 959–972.

Birke, H., De Kok, L.J., Wirtz, M., & Hell, R. (2014) The role of compartment-specific cysteine synthesis for sulfur homeostasis during H_2S exposure in Arabidopsis. *Plant Cell Physiology*, in press, http://dx.doi.org/10.1093/pcp/pcu166.

Birke, H., Müller, S. J., Rother, M., Zimmer, A. D., Hoernstein, S. N., Wesenberg, D., et al. (2012). The relevance of compartmentation for cysteine synthesis in phototrophic organisms. *Protoplasma*, *249*(Suppl. 2), 147–155.

Bradford, M. M. (1976). A rapid and sensitive method for the quantitation of microgram quantities of protein utilizing the principle of protein-dye binding. *Analytical Biochemistry*, *72*, 248–254.

Burkhard, P., Rao, G. S., Hohenester, E., Schnackerz, K. D., Cook, P. F., & Jansonius, J. N. (1998). Three-dimensional structure of *O*-acetylserine sulfhydrylase from *Salmonella typhimurium*. *Journal of Molecular Biology*, *283*, 121–133.

Campanini, B., Speroni, F., Salsi, E., Cook, P. F., Roderick, S. L., Huang, B., et al. (2005). Interaction of serine acetyltransferase with *O*-acetylserine sulfhydrylase active site: Evidence from fluorescence spectroscopy. *Protein Science*, *14*, 2115–2124.

Chen, J., Wu, F.-H., Wang, W.-H., Zheng, C.-J., Lin, G.-H., Dong, X.-J., et al. (2011). Hydrogen sulphide enhances photosynthesis through promoting chloroplast biogenesis, photosynthetic enzyme expression, and thiol redox modification in *Spinacia oleracea* seedlings. *Journal of Experimental Botany*, *62*, 4481–4493.

Droux, M., Ruffet, M. L., Douce, R., & Job, D. (1998). Interactions between serine acetyltransferase and *O*-acetylserine (thiol) lyase in higher plants-structural and kinetic properties of the free and bound enzymes. *European Journal of Biochemistry*, *255*, 235–245.

Francois, J. A., Kumaran, S., & Jez, J. M. (2006). Structural basis for interaction of *O*-acetylserine sulfhydrylase and serine acetyltransferase in the Arabidopsis cysteine synthase complex. *Plant Cell*, *18*, 3647–3655.

Frazzon, A. P., Ramirez, M. V., Warek, U., Balk, J., Frazzon, J., Dean, D. R., et al. (2007). Functional analysis of Arabidopsis genes involved in mitochondrial iron-sulfur cluster assembly. *Plant Molecular Biology*, *64*, 225–240.

Gaitonde, M. K. (1967). A spectrophotometric method for the direct determination of cysteine in the presence of other naturally occurring amino acids. *The Biochemical Journal*, *104*, 627–633.

Garcia-Mata, C., & Lamattina, L. (2010). Hydrogen sulphide, a novel gasotransmitter involved in guard cell signalling. *The New Phytologist*, *188*, 977–984.

Gotor, C., Garcia, I., Crespo, J. L., & Romero, L. C. (2013). Sulfide as a signaling molecule in autophagy. *Autophagy*, *9*, 609–611.

Hasegawa, R., Tada, T., Torii, Y., & Esashi, Y. (1994). Presence of β-cyanoalanine synthase in unimbibed dry seeds and its activation by ethylene during pre-germination. *Physiologia Plantarum*, *91*, 141–146.

Hatzfeld, Y., Maruyama, A., Schmidt, A., Noji, M., Ishizawa, K., & Saito, K. (2000). β-Cyanoalanine synthase is a mitochondrial cysteine synthase-like protein in spinach and Arabidopsis. *Plant Physiology, 123*, 1163–1172.

Heeg, C., Kruse, C., Jost, R., Gutensohn, M., Ruppert, T., Wirtz, M., et al. (2008). Analysis of the Arabidopsis *O*-acetylserine(thiol)lyase gene family demonstrates compartment-specific differences in the regulation of cysteine synthesis. *Plant Cell, 20*, 168–185.

Hell, R., & Wirtz, M. (2011). Molecular biology, biochemistry and cellular physiology of cysteine metabolism in *Arabidopsis thaliana. The Arabidopsis Book, 9*, e0154.

Hildebrandt, T. M., & Grieshaber, M. K. (2008). Three enzymatic activities catalyze the oxidation of sulfide to thiosulfate in mammalian and invertebrate mitochondria. *The FEBS Journal, 275*, 3352–3361.

Holdorf, M. M., Owen, H. A., Lieber, S. R., Yuan, L., Adams, N., Dabney-Smith, C., et al. (2012). Arabidopsis ETHE1 encodes a sulfur dioxygenase that is essential for embryo and endosperm development. *Plant Physiology, 160*, 226–236.

Jacques, A. G. (1936). The kinetics of penetration: XII. Hydrogen sulfide. *The Journal of General Physiology, 19*, 397–418.

Jost, R., Berkowitz, O., Wirtz, M., Hopkins, L., Hawkesford, M. J., & Hell, R. (2000). Genomic and functional characterization of the *oas* gene family encoding *O*-acetylserine (thiol) lyases, enzymes catalyzing the final step in cysteine biosynthesis in *Arabidopsis thaliana. Gene, 253*, 237–247.

Krüßel, L., Junemann, J., Wirtz, M., Birke, H., Thornton, J. D., Browning, L. W., et al. (2014). The mitochondrial sulfur dioxygenase ETHYLMALONIC ENCEPHALOPATHY PROTEIN1 is required for amino acid catabolism during carbohydrate starvation and embryo development in Arabidopsis. *Plant Physiology, 165*, 92–104.

Li, L., Rose, P., & Moore, P. K. (2011). Hydrogen sulfide and cell signaling. *Annual Review of Pharmacology and Toxicology, 51*, 169–187.

Maas, F. M., de Kok, L. J., Hoffmann, I., & Kuiper, P. J. C. (1987). Plant responses to H_2S and SO_2 fumigation. I. Effects on growth, transpiration and sulfur content of spinach. *Physiologia Plantarum, 70*, 713–721.

Maruyama, A., Ishizawa, K., Takagi, T., & Esashi, Y. (1998). Cytosolic β-cyanoalanine synthase activity attributed to cysteine synthases in cocklebur seeds. Purification and characterization of cytosolic cysteine synthases. *Plant & Cell Physiology, 39*, 671–680.

Mathai, J. C., Missner, A., Kugler, P., Saparov, S. M., Zeidel, M. L., Lee, J. K., et al. (2009). No facilitator required for membrane transport of hydrogen sulfide. *Proceedings of the National Academy of Sciences of the United States of America, 106*, 16633–16638.

Miller, J. M., & Conn, E. E. (1980). Metabolism of hydrogen cyanide by higher plants. *Plant Physiology, 65*, 1199–1202.

Mino, K., Yamanoue, T., Sakiyama, T., Eisaki, N., Matsuyama, A., & Nakanishi, K. (1999). Purification and characterization of serine acetyltransferase from *Escherichia coli* partially truncated at the C-terminal region. *Bioscience, Biotechnology, and Biochemistry, 63*, 168–179.

Peiser, G. D., Wang, T. T., Hoffman, N. E., Yang, S. F., Liu, H. W., & Walsh, C. T. (1984). Formation of cyanide from carbon 1 of 1-aminocyclopropane-1-carboxylic acid during its conversion to ethylene. *Proceedings of the National Academy of Sciences of the United States of America, 81*, 3059–3063.

Rohwerder, T., & Sand, W. (2003). The sulfane sulfur of persulfides is the actual substrate of the sulfur-oxidizing enzymes from Acidithiobacillus and Acidiphilium spp. *Microbiology, 149*, 1699–1709.

Salbitani, G., Wirtz, M., Hell, R., & Carfagna, S. (2014). Affinity purification of O-acetylserine(thiol)lyase from Chlorella sorokiniana by recombinant proteins from Arabidopsis thaliana. *Metabolites, 4*, 629–639. http://dx.doi.org/10.3390/metabo4030629.

Sambrook, J., Fritsch, E. F., & Maniatis, T. (1989). *Molecular cloning: A laboratory manual.* New York: Cold Spring Harbour Laboratory Press.
Suzuki, I. (1965). Oxidation of elemental sulfur by an enzyme system of *Thiobacillus thiooxidans*. *Biochimica et Biophysica Acta, 104*, 359–371.
Takahashi, H., Kopriva, S., Giordano, M., Saito, K., & Hell, R. (2011). Sulfur assimilation in photosynthetic organisms: Molecular functions and regulations of transporters and assimilatory enzymes. *Annual Review of Plant Biology, 62*, 157–184.
Tiranti, V., Viscomi, C., Hildebrandt, T., Di Meo, I., Mineri, R., Tiveron, C., et al. (2009). Loss of ETHE1, a mitochondrial dioxygenase, causes fatal sulfide toxicity in ethylmalonic encephalopathy. *Nature Medicine, 15*, 200–205.
Völkel, S., & Grieshaber, M. K. (1994). Oxygen dependent sulfide detoxification in the lugworm *Arenicola marina*. *Marine Biology, 118*, 137–147.
Warrilow, A., & Hawkesford, M. (1998). Separation, subcellular location and influence of sulphur nutrition on isoforms of cysteine synthase in spinach. *Journal of Experimental Botany, 49*, 1625–1636.
Warrilow, A. G., & Hawkesford, M. J. (2000). Cysteine synthase (*O*-acetylserine (thiol) lyase) substrate specificities classify the mitochondrial isoform as a cyanoalanine synthase. *Journal of Experimental Botany, 51*, 985–993.
Warrilow, A. G., & Hawkesford, M. J. (2002). Modulation of cyanoalanine synthase and *O*-acetylserine (thiol) lyases A and B activity by beta-substituted alanyl and anion inhibitors. *Journal of Experimental Botany, 53*, 439–445.
Watanabe, M., Kusano, M., Oikawa, A., Fukushima, A., Noji, M., & Saito, K. (2008). Physiological roles of the β-substituted alanine synthase gene family in Arabidopsis. *Plant Physiology, 146*, 310–320.
Wirtz, M., Beard, K. F. M., Lee, C. P., Boltz, A., Schwarzländer, M., Fuchs, C., et al. (2012). Mitochondrial cysteine synthase complex regulates *O*-acetylserine biosynthesis in plants. *The Journal of Biological Chemistry, 287*, 27941–27947.
Wirtz, M., Birke, H., Heeg, C., Mueller, C., Hosp, F., Throm, C., et al. (2010). Structure and function of the hetero-oligomeric cysteine synthase complex in plants. *The Journal of Biological Chemistry, 285*, 32810–32817.
Wirtz, M., Droux, M., & Hell, R. (2004). *O*-acetylserine (thiol) lyase: An enigmatic enzyme of plant cysteine biosynthesis revisited in *Arabidopsis thaliana*. *Journal of Experimental Botany, 55*, 1785–1798.
Wirtz, M., Heeg, C., Samami, A. A., Ruppert, T., & Hell, R. (2010). Enzymes of cysteine synthesis show extensive and conserved modifications patterns that include N^{α}-terminal acetylation. *Amino Acids, 39*, 1077–1086.
Yamaguchi, Y., Nakamura, T., Kusano, T., & Sano, H. (2000). Three Arabidopsis genes encoding proteins with differential activities for cysteine synthase and β-cyanoalanine synthase. *Plant & Cell Physiology, 41*, 465–476.
Yi, H., Juergens, M., & Jez, J. M. (2012). Structure of soybean β-cyanoalanine synthase and the molecular basis for cyanide detoxification in plants. *Plant Cell, 24*, 2696–2706.
Zhao, C., Moriga, Y., Feng, B., Kumada, Y., Imanaka, H., Imamura, K., et al. (2006). On the interaction site of serine acetyltransferase in the cysteine synthase complex from *Escherichia coli*. *Biochemical and Biophysical Research Communications, 341*, 911–916.

Molecular Hydrogen

CHAPTER FIFTEEN

Molecular Hydrogen as a Novel Antioxidant: Overview of the Advantages of Hydrogen for Medical Applications

Shigeo Ohta[1]

Department of Biochemistry and Cell Biology, Institute of Development and Aging Sciences, Graduate School of Medicine, Nippon Medical School, Kawasaki, Japan

[1]Corresponding author: e-mail address: ohta@nms.ac.jp

Contents

1. Introduction 290
2. Comparison of H_2 with Other Medical Gasses 291
3. Oxidative Stress as Pathogenic Sources 292
4. Physiological Roles of H_2O_2 294
5. Measurement of H_2 Gas Concentration 295
6. Advantages of Hydrogen in Medical Applications 296
   6.1 Selective reaction of H_2 with highly reactive ROS 296
   6.2 Rapid diffusion 298
7. Methods of Ingesting Molecular Hydrogen 301
   7.1 Inhalation of hydrogen gas 301
   7.2 Oral ingestion by drinking hydrogen water 301
   7.3 Injection of hydrogen-saline 302
   7.4 Direct incorporation of molecular hydrogen by diffusion: Eye drops, bath, and cosmetics 304
   7.5 Maternal intake of H_2 304
8. Medical Effects of H_2 304
   8.1 Acute oxidative stress by ischemia/reperfusion 304
   8.2 Chronic oxidative stress loading to neurodegeneration 305
   8.3 Stimulatory effects on energy metabolism 305
   8.4 Anti-inflammatory effects 306
9. Possible Molecular Mechanisms Underlying Various Effects of Molecular Hydrogen 306
   9.1 Direct reduction of hydroxyl radicals with molecular hydrogen 306
   9.2 Direct reduction of peroxynitrite with molecular hydrogen to regulate gene expression 307
   9.3 Indirect reduction of oxidative stress by regulating gene expression 308
10. Unresolved Questions and Closing Remarks 309
References 310

Methods in Enzymology, Volume 555
ISSN 0076-6879
http://dx.doi.org/10.1016/bs.mie.2014.11.038

Abstract

Molecular hydrogen (H_2) was believed to be inert and nonfunctional in mammalian cells. We overturned this concept by demonstrating that H_2 reacts with highly reactive oxidants such as hydroxyl radical ($^{\bullet}OH$) and peroxynitrite ($ONOO^-$) inside cells. H_2 has several advantages exhibiting marked effects for medical applications: it is mild enough neither to disturb metabolic redox reactions nor to affect signaling by reactive oxygen species. Therefore, it should have no or little adverse effects. H_2 can be monitored with an H_2-specific electrode or by gas chromatography. H_2 rapidly diffuses into tissues and cells to exhibit efficient effects. Thus, we proposed the potential of H_2 for preventive and therapeutic applications. There are several methods to ingest or consume H_2: inhaling H_2 gas, drinking H_2-dissolved water (H_2-water), injecting H_2-dissolved saline (H_2-saline), taking an H_2 bath, or dropping H_2-saline onto the eyes. Recent publications revealed that, in addition to the direct neutralization of highly reactive oxidants, H_2 indirectly reduces oxidative stress by regulating the expression of various genes. Moreover, by regulating gene expression, H_2 functions as an anti-inflammatory, antiallergic, and anti-apoptotic molecule, and stimulates energy metabolism. In addition to growing evidence obtained by model animal experiments, extensive clinical examinations were performed or are under way. Since most drugs specifically act on their specific targets, H_2 seems to differ from conventional pharmaceutical drugs. Owing to its great efficacy and lack of adverse effects, H_2 has potential for clinical applications for many diseases.

1. INTRODUCTION

Molecular hydrogen with the molecular formula H_2 is a colorless, odorless, tasteless, nonmetallic, and nontoxic gas at room temperature. Hydrogen gas is flammable and will burn in air at a very wide range of concentrations between 4% and 75% by volume. Its autoignition temperature, the temperature of spontaneous ignition in air, is about 500 °C (http://en.wikipedia.org/wiki/Hydrogen). These facts suggest that H_2 is not so dangerous in daily life when its concentration is under 4%.

In turns of biological reactions in several microorganisms, H_2 is a product of certain types of anaerobic metabolism, usually via reactions catalyzed by iron- or nickel-containing enzymes called hydrogenases (Adams, Mortenson, & Chen, 1980; Fritsch, Lenz, & Friedrich, 2013). H_2 is also enzymatically metabolized as an energy source by providing electrons to the electron transport chain. These enzymes catalyze the reversible redox reaction between H_2 and its constituent two protons and two electrons (van Berkel-Arts et al., 1986).

On the other hand, in all photosynthetic organisms, the water-splitting reaction occurs in the light reactions, where water is decomposed into protons, electrons, and oxygen. Some organisms, including the alga *Chlamydomonas reinhardtii* and cyanobacteria, have evolved a second step in the dark reactions in which protons and electrons are reduced to form H_2 gas by specialized hydrogenases in cyanobacteria or chloroplast (Carrieri, Wawrousek, Eckert, Yu, & Maness, 2011). For industrial uses, extensive efforts have also been undertaken with alga in a bioreactor by genetically modifying cyanobacterial hydrogenases to synthesize H_2 gas efficiently (King, 2013; van Berkel-Arts et al., 1986).

In contrast, H_2 was accepted to behave as an inert gas in mammalian cells because of the lack of no hydrogenase genes. Thus, it had been believed that H_2 is nonfunctional in our cells. In fact, H_2 seemed to react with no biological compounds, including oxygen (O_2), in the absence of catalysts at body temperature. Indeed, owing to its characteristics, H_2 gas was used for measuring local blood flow (Aukland, Bower, & Berliner, 1964).

We overturned this concept in a publication in 2007 describing that H_2 acts as a therapeutic and preventive antioxidant by selectively reducing highly active oxidants, such as hydroxyl radical ($^{\bullet}OH$) and peroxynitrite ($ONOO^-$) in cultured cells, and that H_2 has cytoprotective effects against oxidative stress (Ohsawa et al., 2007). Since then, a large number of studies have explored therapeutic and preventive effects of H_2. These publications cover many biological effects against oxidative stress in almost all organs (Ohta, 2011, 2012). Moreover, it has been revealed that H_2 has more roles, including anti-inflammatory, antiapoptotic, and antiallergic effects, in most tissues of model animals, and that H_2 stimulates energy metabolism. In addition to publications on model animal experiments, more than 10 papers on clinical examinations have been published. As of 2013, the number of publications on its biologically or medically beneficial effects had surpassed 300 (Ohta, 2014).

2. COMPARISON OF H_2 WITH OTHER MEDICAL GASSES

Gas inhalation as disease therapy has recently received attention (Kajimura, Fukuda, Bateman, Yamamoto, & Suematsu, 2010; Szabó, 2007). In recent decades, there has been extraordinary and rapid growth in our knowledge of gaseous molecules, including hydrogen sulfide (H_2S), nitric oxide ($NO^{\bullet}$), and carbon monoxide (CO). H_2S, CO, and $NO^{\bullet}$ are extremely toxic molecules; however, they play important roles as signaling molecules in biological systems (Kimura, 2010; Motterlini & Otterbein, 2010).

In contrast, H_2 has advantages in terms of toxicity: it has no cytotoxicity even at high concentration (Abraini, Gardette-Chauffour, Martinez, Rostain, & Lemaire, 1994; Fontanari et al., 2000; Lillo & Parker, 2000; Lillo, Parker, & Porter, 1997). Furthermore, safety standards have been established for high concentrations of hydrogen gas for inhalation since high-pressure hydrogen gas was actually used in deep diving gas mixes to prevent decompression sickness and arterial gas thrombi (Fontanari et al., 2000). The safety of H_2 for humans is demonstrated by its application in Hydreliox, an exotic, breathing gas mixture of 49% H_2, 50% helium, and 1% O_2, which is used to prevent decompression sickness and nitrogen narcosis during very deep technical diving (Abraini et al., 1994; Fontanari et al., 2000; Lillo & Parker, 2000; Lillo et al., 1997).

As the primary target of H_2S, CO, and $NO^{\bullet}$, heme-based proteins play central roles. Integrated approaches revealed the physiological significance of H_2S, CO, and $NO^{\bullet}$ on mitochondrial cytochrome *c* oxidase, a key target and central mediator of mitochondrial respiration (Kajimura et al., 2010). As far as briefly examined (Ohsawa et al., 2007), H_2 does not reduce the oxidized heme of cytochrome *c*. Thus, the primary target of H_2 seems to differ from that of the other medical gaseous molecules.

Moreover, the production of $NO^{\bullet}$, H_2S, or CO is carried out by different enzymes, $NO^{\bullet}$ synthases, cystathionine γ-lyase/cystathionine β-synthase, or hemeoxygenase-1 (HO-1), respectively (Kashfi & Olson, 2013). In contrast, as mentioned above, mammalian cells have no enzyme for producing intracellular H_2.

Regarding the interaction between H_2 and the other toxic medical gasses, combined therapy with H_2 and CO demonstrated additional therapeutic efficacy via both antioxidant and anti-inflammatory mechanisms, and may be a clinically feasible approach for preventing ischemia/reperfusion injury in the myocardium (Nakao et al., 2010). Breathing $NO^{\bullet}$ plus H_2 during ischemia/reperfusion reduced the infarct size and maintained cardiac function, and reduced the generation of myocardial nitro-tyrosine associated with $NO^{\bullet}$ inhalation (Shinbo et al., 2013). These findings suggest that the target of H_2 differs from those of CO and $NO^{\bullet}$.

3. OXIDATIVE STRESS AS PATHOGENIC SOURCES

First, the author would like to introduce how the biological function of H_2 was discovered regarding its contribution to reducing oxidative stress.

Reactive oxygen species (ROS) are generated inside the body during daily life as a by-product of energy metabolism by oxidative phosphorylation in every aerobic organism. Occasionally, excess ROS are produced, such as by smoking or air pollution, exposure to ultraviolet or irradiation rays, intense exercise, and physical or psychological stress (Agarwal, 2005; Grassi et al., 2010; Harma, Harma, & Erel, 2006; Liu et al., 1996; Tanriverdi et al., 2006). When ROS are produced excessively or endogenous antioxidant capacity is diminished, indiscriminate oxidation elicits harmful effects, resulting in "oxidative stress."

Acute oxidative stress arises from various different situations: inflammation, ischemia/reperfusion in cardiac or cerebral infarction, organ transplantation, and cessation of operative bleeding, among others (Ferrari et al., 1991; Reuter, Gupta, Chaturvedi, & Aggarwal, 2010; Vaziri & Rodriguez-Iturbe, 2006). Under normal conditions, ROS induced by strenuous exercise result in muscle fatigue (Westerblad & Allen, 2011). Evidence has established strong links between chronic oxidative stress and a wide variety of pathologies, including malignant diseases, diabetes mellitus, atherosclerosis, and chronic inflammatory processes, as well as many neurodegenerative diseases and the aging process (Andersen, 2004; El Assar, Angulo, & Rodriguez-Manas, 2013; Kim & Byzova, 2014).

As a first step in generating ROS, superoxide anion radicals ($^{\bullet}O_2^-$) are the primary ROS mostly generated by electron leakage from the mitochondrial electron transport chain (Andersen, 2004; Finkel & Holbrook, 2000; Lin & Beal, 2006; Turrens, 2003). Other enzymes, including NADPH oxidases, cytochrome p450s, lipoxygenase, cyclooxygenase, and xanthine oxidase, also participate in ROS generation in the immune- or detoxifying system (Droge, 2002). Superoxide dismutase enzymatically converts $^{\bullet}O_2^-$ to hydrogen peroxide (H_2O_2), which is metabolized to generate water (H_2O). Highly reactive $^{\bullet}OH$ is generated from H_2O_2 or $^{\bullet}O_2^-$ via the Fenton or Weiss reaction in the presence of catalytically active metals, such as Fe^{2+} and Cu^+ (Halliwell & Gutteridge, 1992). Reaction of $^{\bullet}O_2^-$ with $NO^{\bullet}$ generates $ONOO^-$, which is a very active nitrogen species (RNS) (Radi, 2013). $^{\bullet}OH$ is the major cause of the oxidation and destruction of biomolecules by direct reaction or by triggering the chain reaction of free radicals (Lipinski, 2011). Ionizing radiation, including cosmic rays, also generates $^{\bullet}OH$ as a damaging intermediate through the reaction with water, a process termed radiolysis (Schoenfeld, Ansari, Nakao, & Wink, 2012; Schoenfeld et al., 2011).

Although antioxidation therapy or prevention of various diseases is expected owing to the clinical importance of oxidative damage, many antioxidants have been of limited therapeutic success (Steinhubl, 2008). Antioxidant supplements have exhibited little effect on preventing cancer, myocardial infarction, and atherosclerosis, but conversely have increased mortality (Bjelakovic, Nikolova, Gluud, Simonetti, & Gluud, 2007; Brambilla et al., 2008; Hackam, 2007; Hercberg et al., 2010; Steinhubl, 2008).

4. PHYSIOLOGICAL ROLES OF H_2O_2

As mentioned above, ROS had historically been believed to cause cellular damage and to lack physiological functions; however, cellular redox homeostasis is a delicate balance between ROS production and the antioxidant system (Bashan, Kovsan, Kachko, Ovadia, & Rudich, 2009; Brewer, Mustafi, Murray, Rajasekaran, & Benjamin, 2013). Some ROS are now appreciated to function as signaling molecules to regulate a wide variety of physiological process (Bell, Klimova, Eisenbart, Schumacker, & Chandel, 2007; Liu, Colavitti, Rovira, & Finkel, 2005). H_2O_2 was shown to be required for cytokine, insulin, growth factor, AP-1, c-Jun N-terminal kinase 1, p53, and nuclear factor kappa B signaling and to promote phosphatase inactivation by cysteine oxidation (Chandel, Trzyna, McClintock, & Schumacker, 2000; Chandel, Vander Heiden, Thompson, & Schumacker, 2000; Finkel, 1998). These reactions provide a plausible biochemical mechanism by which ROS can impinge on signaling pathways (Collins et al., 2012).

Additionally, oxidative stress caused by H_2O_2 and $NO^{\bullet}$ induces enzymes involved in antioxidation and tolerance to protect cells against oxidative stress (Endo et al., 2009; Ristow & Zarse, 2010). For example, translocation of NF-E2-related factor 2 (Nrf2) into the nucleus leads to the regulation of gene expression involved in defense systems against oxidative stress (Jazwa & Cuadrado, 2010) and other toxic sources including heavy metals (Gan & Johnson, 2014). Moreover, H_2O_2 is a key factor to regulate cellular differentiation (Tormos et al., 2011; Tsukagoshi, Busch, & Benfey, 2010), the immune system (West et al., 2011; Zhou, Yazdi, Menu, & Tschopp, 2011), autophagy (Garg et al., 2013; Li, Ishdorj, & Gibson, 2012), and apoptosis (Mates, Segura, Alonso, & Marquez, 2012). Thus, it is crucial for functional H_2O_2 not to be completely eliminated in order to maintain homeostasis; as such, it is very important to be aware of side effects when developing an effective antioxidant for the prevention of oxidative stress-related diseases.

Unexpectedly, recent notable studies have suggested that excessive antioxidants increased mortality and rates of cancer (Bjelakovic et al., 2007; Bjelakovic, Nikolova, Gluud, Simonetti, & Gluud, 2008; Gray et al., 2008; Hackam, 2007; Hercberg et al., 2010; Walker, 2008) probably because they may interfere with some essential defensive mechanisms (Bjelakovic & Gluud, 2007; Bjelakovic et al., 2008; Carriere et al., 2004; Chandel et al., 1998; Mandal et al., 2010; Miller et al., 2005; Salganik, 2001). Against this background, an ideal antioxidant is expected to mitigate excess oxidative stress, but not disturb redox homeostasis. In other words, an ideal molecule should not reduce signaling molecules, such as H_2O_2 but should effectively reduce strong oxidants, such as $^{\bullet}OH$.

Since H_2 reduces $^{\bullet}OH$ but does not react with $^{\bullet}O_2^-$, H_2O_2, and $NO^{\bullet}$ that have physiological roles (Ohsawa et al., 2007), we propose that the adverse effects of H_2 are very small compared with those of other antioxidants. Thus, we have reached the conclusion that the ideal antioxidant could be H_2.

5. MEASUREMENT OF H_2 GAS CONCENTRATION

H_2 gas concentration is measureable by gas chromatography. Additionally, H_2 concentration dissolved in a solution can be measured by this method. For example, H_2 in blood can be monitored by the following method: Venous or arterial blood (e.g., 5 ml) is collected in a closed aluminum bag with no dead space, followed by the addition of a defined volume of air (e.g., 30 ml) into the bag. After complete transfer of the H_2 gas from the blood to the air in the closed bag, H_2 can be measured by gas chromatography (Fig. 1). The inhalation of H_2 actually increased H_2 dissolved in arterial blood in a hydrogen gas concentration-dependent manner, and the H_2 levels in venous blood were lower than in arterial blood; the different level between arterial and venous blood indicates the amount of H_2 incorporated into and consumed by tissues (Ohsawa et al., 2007). In a clinical examination, Ono et al. also showed a difference in H_2 concentrations between arterial and venous blood (Ono et al., 2012).

H_2 concentration can be measured using an H_2 electrode that specifically detects H_2; however, this sensor is also somewhat sensitive to H_2S. Thus, when H_2S is contaminated in a solution, one must take into consideration its effects.

H_2 can be measured in tissues using a needle-type H_2 sensor (Unisense, Aarhus, Denmark). The electrode current was measured with a picoammeter (Keithley, Cleveland, Ohio) attached to a recorder. The negative

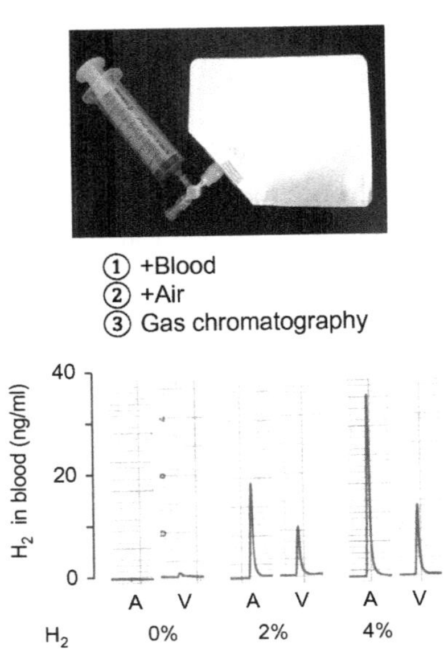

Figure 1 Incorporation of H_2 into blood by inhaling hydrogen gas. Rats inhaled a mixed gas of H_2 (1% or 2%) and O_2 (30%) under anesthetic N_2O and halothane for 1 h, and arterial (indicated by A) or venous blood (indicated by V) was collected into a closed aluminum bag from a three-way stopper (upper panel). After the transfer of H_2 into an accurate volume of air phase from the blood, amounts of H_2 were examined by gas chromatography. Lower panel shows profiles of gas chromatography. The vertical scale indicates the amounts of blood H_2 after calculations. *Adapted from after Ohsawa et al. (2007) modified version of Fig. 5A, with permission from Nature Publishing Group.*

current obtained from the H_2 sensor was converted to regional H_2 concentration using a calibration curve generated from known levels of H_2-saturated saline.

6. ADVANTAGES OF HYDROGEN IN MEDICAL APPLICATIONS

6.1. Selective reaction of H_2 with highly reactive ROS

H_2 dissolved in culture medium did not change the cellular levels of $^{\bullet}O_2^{-}$ and H_2O_2, as judged by the fluorescent signals of MitoSOX and dichlorofluorescein-diacetate (DCF-DA), respectively (Ohsawa et al., 2007). Additionally, H_2 did not decrease the cellular level of $NO^{\bullet}$. In contrast, H_2 treatment significantly decreased levels of $^{\bullet}OH$, as judged by the decrease in the fluorescent signal of hydroxyphenyl fluorescein (HPF) (Setsukinai, Urano, Kakinuma, Majima, & Nagano, 2003).

In terms of the experimental protocol, culture media containing H_2 were prepared as follows: H_2 was dissolved beyond the saturated level into DMEM medium under 0.4 MPa pressure of hydrogen gas for 2 h, O_2 was also dissolved into another medium by bubbling, the third medium contained CO_2, and then fetal bovine serum was supplemented to 1% in all three media. The three media were combined at various ratios to obtain the desired concentration of H_2 and 8.5 mg/l of O_2 at 25 °C. For culture, the combined media were put into a culture flask and immediately examined for H_2 or O_2 concentration with an H_2 or O_2 electrode, and in turn gas composed of the desired ratio of H_2 and N_2 ($H_2 + N_2 = 75\%$), 20% of O_2, and 5% of CO_2 was filled into the culture flask; for example, when the medium contained 0.6 m*M* H_2, the H_2 gas was adjusted to 75%. The mixed gas was obtained by regulating the flow rates of its constituents with connected flow meters. As a control, degassed medium lacking H_2 was prepared by stirring medium that had been saturated with H_2 in an open vessel for 4 h, and the concentration of H_2 was checked with an H_2 electrode.

Then, PC12 cells were incubated in medium with or without 0.6 m*M* H_2, and exposed to antimycin A or L-NAME (N^G-nitro-L-arginine methyl ester) to induce $^{\bullet}O_2^-$, H_2O_2, and $^{\bullet}OH$ or $NO^{\bullet}$. Fluorescent images of MitoSOX-, DCF-DA (2′,7′-dichlorodihydrofluorescein)-, HPF-, and DAF-2 DA (diaminofluorescein-2 diacetate)-treated cells were obtained by laser-scanning confocal microscopy (Olympus FV300) to estimate intracellular $^{\bullet}O_2^-$, H_2O_2, $^{\bullet}OH$, and NO, respectively (Fig. 2).

Alternatively, PC12 cells were exposed to intracellular $^{\bullet}OH$ produced by the Fenton reaction ($H_2O_2 + Cu^+ \rightarrow OH + OH^- + Cu^{2+}$), with or without 0.6 m*M* H_2. Cells were preincubated with 1 m*M* $CuSO_4$, washed, and exposed for 1 h to 0.1 m*M* ascorbate (Vit. C) in order to reduce intracellular Cu^{2+} to Cu^+. In this case, endogenous H_2O_2 would be sufficient to produce $^{\bullet}OH$. H_2 indeed protected the cells against $^{\bullet}OH$.

Moreover, the decrease in the cellular $^{\bullet}OH$ level by H_2 was confirmed by spin-trapping technology (Halliwell & Gutteridge, 1992). Standard electron spin resonance (ESR) signals of the DMPO-$OH^{\bullet}$ radical were obtained by trapping $^{\bullet}OH$ with a spin-trapping reagent (DMPO). PC12 cells were preincubated with 0.1 *M* DMPO and 2 m*M* $CuSO_4$ for 30 min at 37 °C with or without 0.6 m*M* H_2. After removal of this medium, the cells were treated with 0.2 m*M* ascorbate and 0.1 m*M* H_2O_2 for 5 min at 23 °C to produce $^{\bullet}OH$ by the Fenton reaction, and then scraped into a flat cuvette for ESR measurement. Alternatively, PC12 cells were incubated in PBS containing 0.1 *M* DMPO and 30 g/ml antimycin A for 7 min at 23 °C to produce

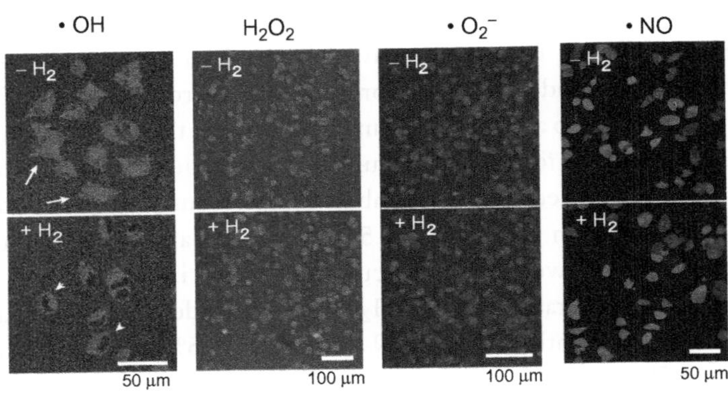

Figure 2 Selective reduction of reactive oxygen or nitrogen species by H_2 in cultured cells. PC12 cells were kept in medium with 0.6 m*M* H_2 (indicated by $+H_2$) or without H_2 (indicated by $-H_2$), and exposed to antimycin A or L-NAME (N^G-nitro-L-arginine methyl ester) to induce $^{\bullet}OH$, H_2O_2, and $^{\bullet}O_2^-$ (ROS) or $NO^{\bullet}$ (RNS) for 30 min. Each ROS or RNS was detected using flowing fluorescent dye; HPF, H_2DCF (2′,7′-dichlorodihydrofluorescein), MitoSOX, and DAF-2 DA (diaminofluorescein-2 diacetate) were used to detect $^{\bullet}OH$, H_2O_2, $^{\bullet}O_2^-$, and $NO^{\bullet}$, respectively. These representative fluorescence images obtained by laser-scanning confocal microscopy demonstrate the selective reduction of $^{\bullet}OH$ by H_2. *Adapted from Ohsawa et al. (2007) modified version of Fig. 1A, B and supplementary Fig. 1A, C, with permission from Nature Publishing Group.* (See the color plate.)

excess $^{\bullet}OH$, with or without 0.6 m*M* H_2, and then scraped into a flat cuvette for ESR measurement.

The selective reduction of ROS can be explained by the marked oxidative strength of $^{\bullet}OH$. In other words, $^{\bullet}OH$ is strong enough to react with even inert H_2, but that $^{\bullet}O_2^-$, H_2O_2, and $NO^{\bullet}$ are insufficient to react with H_2 according to their activities. Namely, H_2 is mild enough neither to disturb metabolic redox reactions nor to affect ROS that function in cellular signaling (Fig. 3).

6.2. Rapid diffusion

Most hydrophilic antioxidants cannot penetrate biomembranes and most hydrophobic antioxidants remain on the membranes. In contrast, H_2 can be infused into lipids as well as aqueous solutions. It has favorable distribution characteristics having the physical ability to penetrate biomembranes and diffuse into the cytosol, as illustrated in Fig. 4.

Despite the clinical importance of overcoming oxidative damage, antioxidants had limited therapeutic success. This may be because most

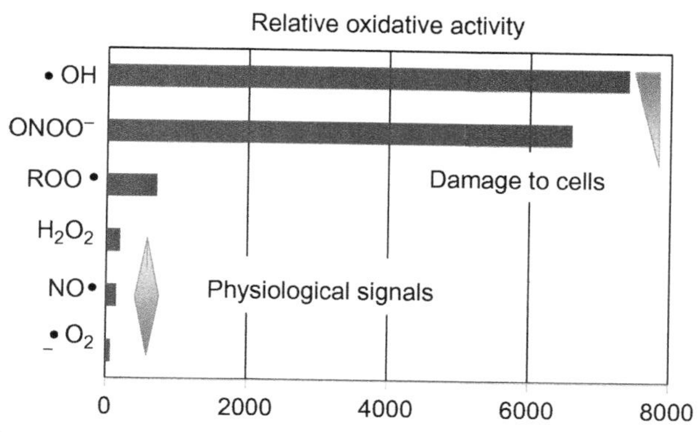

Figure 3 Relative oxidative activities in each reactive oxygen and nitrogen species. $^{\bullet}OH$ and $ONOO^-$ are highly reactive to damaged cells, whereas $^{\bullet}O_2^-$, $NO^{\bullet}$, and H_2O_2 have physiological roles as signaling molecules. *This graph is based on data from a previous publication (Setsukinai et al., 2003).*

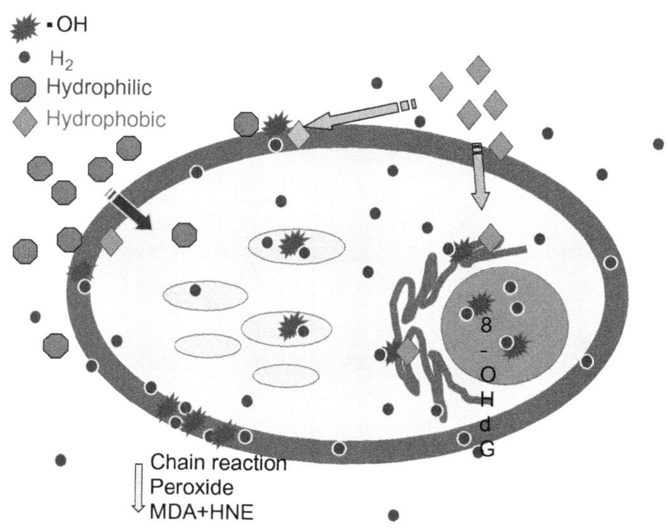

Figure 4 Illustration of gaseous diffusion of H_2 into a cell. Most hydrophilic compounds are retained at membranes and cannot reach the cytosol, whereas most hydrophobic ones cannot penetrate biomembranes in the absence of specific carriers. In contrast, H_2 can be rapidly distributed into cytosol and organelles. On the membrane, $^{\bullet}OH$ triggers the initiation of a free radical chain reaction to generate lipid peroxides, which are converted to some oxidative stress markers, 4-hydroxyl-2-nonenal (4-HNE), and malondialdehyde (MDA). In the nucleus, $^{\bullet}OH$ oxidizes DNA for modification to 8-OHdG (8-hydroxy-deoxyguanine).

antioxidants do not reach specific regions (Murphy, 1997; Murphy & Smith, 2000; Smith & Murphy, 2011). As H_2 effectively reaches the nucleus and mitochondria, the protection of nuclear DNA and mitochondria suggests preventive effects against lifestyle-related diseases, cancer, and the aging process (Ohsawa et al., 2007). Moreover, H_2 passes through the blood brain barrier, although most antioxidant compounds cannot; this is also an advantage of H_2.

The gaseous diffusion of H_2 can be monitored inside various tissues by detection with a specific H_2 electrode. For example, H_2 concentration has been monitored within the rat myocardium. The electrode was inserted into the "at-risk" area for infarction to estimate the diffusion of H_2 into the ischemic myocardium area after coronary artery occlusion. H_2 concentration was increased by its diffusion, even with coronary artery occlusion (Hayashida et al., 2008) (Fig. 5).

Moreover, we devised eye drops with dissolved H_2 to administer H_2 to the retina directly, and monitored the time course of changes in H_2 levels using the needle-shaped hydrogen sensor electrode inserted through the sclera to the vitreous body in rats. H_2 could reach the vitreous body by administering H_2 saturated in normal saline. When H_2 eye drops were

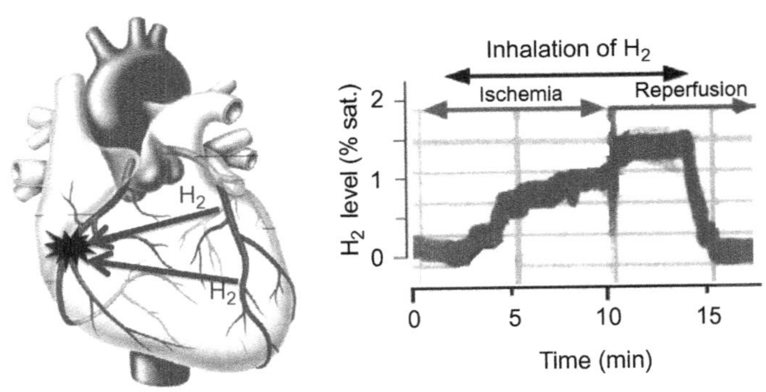

Figure 5 Inhalation of H_2 gas increases the intramyocardial H_2 by diffusion. Left panel illustrates that H_2 can reach an infarct area by diffusion even without blood flow. Right panel indicates an experimental result obtained as follows: Regional myocardial ischemia was induced by transient occlusion of the left anterior descending coronary artery of a rat. A needle-type hydrogen sensor (Unisense, Aarhus, Denmark) was inserted in the LV cavity (arterial blood) and H_2 gas at 2% was administered by respiration to intubated rats receiving mechanical ventilation and the concentration of H_2 in the "at-risk" area for infarction during ischemia and reperfusion was monitored with the needle-type H_2 sensor. *Adapted from Hayashida et al. (2008) modified version of Fig. 2C, with permission from Elsevier.*

administered continuously, approximately 70% H_2 was detected on the ocular surface (Oharazawa et al., 2010).

These experiments indicate that H_2 can rapidly diffuse into tissues even without blood flow.

7. METHODS OF INGESTING MOLECULAR HYDROGEN

7.1. Inhalation of hydrogen gas

Inhalation of H_2 gas is the most straightforward therapeutic method. H_2 gas can be inhaled through a ventilator circuit, facemask, or nasal cannula. Since inhaled H_2 gas acts rapidly, it may be suitable for defense against acute oxidative stress. In particular, inhalation of gas does not affect blood pressure (Ohsawa et al., 2007); on the other hand, drip infusion of drugs increases blood pressure and causes serious obstacles during the treatment of myocardial infarction. In particular, excess oxidative stress gives damages to tissues at the time of the initiation of reperfusion. Notably, most antioxidants cannot reach the at-risk area for infarction before initiating reperfusion. As pointed out above, H_2 can reach the region without blood flow by rapid diffusion (Fig. 5).

By a clinical examination, Ono et al. monitored H_2 and showed that inhalation of 3–4% H_2 gas did not affect any physiological parameters, suggesting no adverse effects (Ono et al., 2012).

7.2. Oral ingestion by drinking hydrogen water

Inhalation of H_2 gas is actually unsuitable or impractical for continuous H_2 consumption in daily life for preventive use. In contrast, solubilized H_2 (H_2-dissolved water; i.e., H_2-water) may be beneficial since it is a portable, easily administered, and safe way to ingest H_2 (Nagata, Nakashima-Kamimura, Mikami, Ohsawa, & Ohta, 2009; Ohsawa, Nishimaki, Yamagata, Ishikawa, & Ohta, 2008). H_2 can be dissolved in water up to 0.8 mM (1.6 mg/l) under atmospheric pressure at room temperature without any change of pH.

H_2-water can be made by several methods: infusing H_2 gas into water under high pressure, electrolyzing water to produce H_2, and reacting magnesium metal or its hydride with water. These methods may be applicable not only to water but also to other solvents. H_2 penetrates glass and plastic walls of any vessel in a short time, while aluminum containers can retain H_2 for a long time.

In brief, for experimental treatments, H_2 was dissolved in water under high pressure (0.4 MPa) to a supersaturated level and the saturated H_2-water was stored under atmospheric pressure in an aluminum bag with no dead volume. Mice were given water freely using closed glass vessels equipped with an outlet line containing two ball bearings, which kept the water from being degassed. The vessel was refilled with fresh H_2-water at the same time (e.g., at 4:00 pm) every day.

When water saturated with H_2 was placed into the stomach of a rat, H_2 was detected at several micromoles in blood (Nagata et al., 2009; Nakashima-Kamimura, Mori, Ohsawa, Asoh, & Ohta, 2009). In addition, a rat received H_2-water (0.8 mmol/l H_2 in water) orally by stomach gavage, for example, at 15 ml/kg. Hepatic H_2 was monitored with a needle-type hydrogen electrode (Kamimura, Nishimaki, Ohsawa, & Ohta, 2011) (Fig. 6).

Furthermore, after seven adult volunteers had drunk H_2-water, the H_2 content of their expired breath was measured by gas chromatography with a semiconductor (Shimouchi, Nose, Shirai, & Kondo, 2012). The ingestion of H_2-water rapidly increased breath H_2 content to its maximal level 10 min after ingestion, which thereafter decreased to the baseline level within 60 min. H_2 lost from the water during the experimental procedures accounted for 3% or less of the total. The rate of H_2 release from the skin surface was estimated as approximately 0.1%. On the basis of the remaining H_2 mass balance, approximately 40% of H_2 that had been drunk was consumed inside the body. This report suggests that exogenous H_2 is at least partially trapped by oxygen radicals, such as $^{\bullet}OH$ (Shimouchi et al., 2012).

7.3. Injection of hydrogen-saline

H_2 is intravenously or intraperitoneally injectable as H_2-saline (H_2-dissolved saline), which allows the delivery of H_2 with great efficacy in model animals (Cai et al., 2009; Li et al., 2013; Sun et al., 2011).

Nagatani et al. performed an open-label, prospective, nonrandomized study of intravenous H_2 administration in 38 patients hospitalized for acute ischemic stroke. All patients received an H_2 intravenous solution immediately after the diagnosis of acute ischemic stroke. Data from this study indicated that an H_2 intravenous solution is safe for patients with acute cerebral infarction, including patients treated with tissue-plasminogen activator (Nagatani et al., 2013).

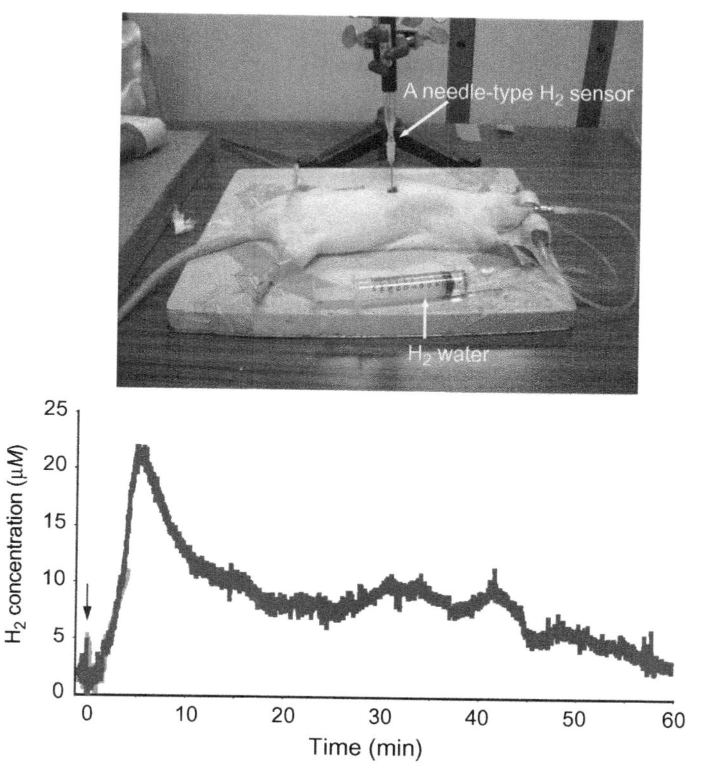

Figure 6 Incorporation of H_2 into the liver from the stomach. H_2 was dissolved in water under high pressure (0.4 MPa) to a supersaturated level. Upper panel: A needle-type hydrogen sensor (Unisense, Aarhus, Denmark) was inserted into rat liver and the rat received saturated H_2-water orally by stomach gavage at 15 ml/kg (upper panel). Lower panel: H_2 concentration was monitored with a picoammeter (Keithley, Cleveland, Ohio). Arrow indicates the time point when rat was administered H_2-water. *Adapted from Kamimura et al. (2011), with permission from Wiley Online Library.*

To rats, H_2-water, H_2-saline, and hydrogen gas were orally administered, intraperitoneally or intravenously injected, and inhaled, respectively. A method for determining the H_2 concentration was applied using high-quality sensor gas chromatography, after which the specimen was prepared via tissue homogenization in airtight tubes. The hydrogen concentration reached a peak at 5 min after oral and intraperitoneal administration, compared with 1 min after intravenous administration. These results indicate that H_2 can reach most organs or blood independently by the three methods (Liu et al., 2014).

7.4. Direct incorporation of molecular hydrogen by diffusion: Eye drops, bath, and cosmetics

Alternatively, H_2-loaded eye drops were prepared by dissolving H_2 in saline and directly administering them to the ocular surface (Kubota et al., 2011; Oharazawa et al., 2010).

H_2 should easily penetrate the skin and is distributed throughout the body via blood flow. Thus, taking a warm water bath with dissolved H_2 is a method of incorporating H_2 into the body in daily life. It takes only 10 min for it to be distributed throughout the whole body, as judged by measuring H_2 gas in expiration (unpublished results). Indeed, powders that can be used to produce H_2 baths are commercially available in Japan.

H_2 delivery to cardiac grafts during cold preservation using a hydrogen-supplemented water bath efficiently ameliorated myocardial injury due to cold ischemia and reperfusion. This device to saturate organs with H_2 during cold storage merits further investigation for possible therapeutic and preventative use during transplantation (Noda et al., 2013).

7.5. Maternal intake of H_2

H_2 intake helps prevent the hippocampal impairment of offspring induced by ischemia/reperfusion during pregnancy (Mano et al., 2014). The effects of H_2 on rat fetal hippocampal damage caused by ischemia and reperfusion in pregnancy were examined with the transient occlusion of bilateral utero-ovarian arteries. Starting 2 days before the operation, the mothers were provided with H_2-saturated water *ad libitum* until vaginal delivery. A significant increase in the concentration of H_2 in the placenta was observed after the oral administration of H_2-saturated water to the mothers, with less placental oxidative damage after ischemia and reperfusion in the presence of H_2. Neonatal growth retardation was observed in the ischemia/reperfusion group, which was alleviated by H_2 administration. Maternal H_2 administration improved oxidative stress and the reference memory of the offspring to the sham level after ischemia and reperfusion injury during pregnancy. Thus, this finding supports the idea that maternal H_2 intake helps prevent the impairment of offspring induced by oxidative stress.

8. MEDICAL EFFECTS OF H_2

8.1. Acute oxidative stress by ischemia/reperfusion

As a type of acute oxidative stress, ischemia/reperfusion induces serious oxidative stress, and its injuries should be considered in many clinical treatments.

Inhalation of H_2 gas improved ischemia/reperfusion injuries in cerebral (Ohsawa et al., 2007) and myocardial infarction (Hayashida et al., 2008; Yoshida et al., 2012). Hydrogen-saline protected against renal ischemia/reperfusion injury (Wang et al., 2011). All clinical manifestations related to post-cardiac arrest (CA) syndrome are attributed to ischemia/reperfusion injury in various organs, including the brain and heart. H_2 gas inhalation yielded great improvement in survival and the neurological deficit score in post-CA syndrome in a rat model (Hayashida et al., 2012). H_2 also mitigated damage during the transplantation of various organs in the form of H_2 gas (Buchholz et al., 2008), H_2-water (Cardinal et al., 2010), and H_2-preservation solution (Noda et al., 2013). A clinical study showed a positive effect of H_2 on patients with acute brain stem infarction (Ono et al., 2011). These acute effects may be due to the direct reduction of oxidative stress by H_2 because no lag time was necessary.

8.2. Chronic oxidative stress loading to neurodegeneration

Chronic oxidative stress is accepted as one of the causes of neurodegeneration, including dementia and Parkinson's disease (PD) (Andersen, 2004; Federico et al., 2012). Experimental oxidative stress in the brain can be induced by chronic physical restraint stress and can impair learning and memory (Abrous, Koehl, & Le Moal, 2005; Liu et al., 1996). Drinking H_2-water suppressed the increase in this oxidative stress and prevented this cognitive impairment (Nagata et al., 2009). In PD, mitochondrial dysfunction and associated oxidative stress are major causes of dopaminergic cell loss in the substantia nigra (Schapira, 2008; Yoritaka et al., 1996). H_2 in drinking water was given before or after stereotactic surgery for 6-hydroxydopamine-induced nigrostriotal degeneration in a rat model of PD. H_2-water prevented both the development and the progression of nigrostriatal degeneration in rats (Fu et al., 2009). Moreover, drinking H_2-water also suppressed dopaminergic neuronal loss in another PD mouse model induced by MPTP (1-methyl-4-phenyl-1,2,3,6-tetrahydropyridine) (Fujita et al., 2009). In a placebo-controlled, randomized, double-blind, parallel-group clinical pilot study, the efficacy of H_2-water in patients with PD was assessed for 48 weeks. Total Unified Parkinson's Disease Rating Scale (UPDRS) scores in the H_2-water group significantly improved, whereas UPDRS scores in the placebo group worsened (Yoritaka et al., 2013).

8.3. Stimulatory effects on energy metabolism

Obesity induces oxidative stress (Matsuda & Shimomura, 2013). H_2-water significantly alleviated fatty liver in *db/db* mice, which are type 2 diabetes model mice with obesity, as well as high-fat diet-induced fatty liver in wild-type mice. Long-term H_2-water drinking significantly decreased fat and body weights, despite no increase in the consumption of diet and water, in *db/db* mice, and decreased levels of plasma glucose, insulin, and triglyceride by stimulating energy metabolism (Kamimura et al., 2011). Analysis of gene expression revealed that a hepatic hormone, fibroblast growth factor 21 (FGF21), showed increased expression upon drinking H_2-water (Kamimura et al., 2011). FGF21 functions to stimulate fatty acid and glucose expenditure. Thus, H_2-water stimulates energy metabolism (Kamimura et al., 2011). Beneficial roles of H_2-water in the prevention of potential metabolic syndrome were also reported by a clinical study (Song et al., 2013).

8.4. Anti-inflammatory effects

Inflammation is closely involved in oxidative stress. H_2-reduced inflammation in experimental model animals induced by concanavalin A and dextran sodium sulfate (Kajiya, Silva, Sato, Ouhara, & Kawai, 2009), lipopolysaccharide (Chen et al., 2013; Xu et al., 2012), Zymosan, an inducer of generalized inflammation and polymicrobial sepsis (Li et al., 2013). H_2 gas, H_2-saline, and H_2-water decreased the levels of proinflammatory cytokines to suppress inflammation. Rheumatoid arthritis (RA) is a chronic inflammatory disease characterized by the destruction of bone and cartilage. The symptoms of RA were significantly improved with H_2-water (Ishibashi et al., 2012).

In terms of the current state of knowledge, H_2 exhibits not only antioxidative effects but also affects many phenotypes in various model animals. H_2 has many beneficial effects on animal models and patients besides its antioxidative effects: anti-inflammation, antiapoptosis, antiallergy, and stimulation of energy metabolism (Ohta, 2011, 2012, 2014). Their mutual relationships are not clear, but the reduction of oxidative stress may primarily lead to various subsequent effects. H_2 seems to exhibit a variety of phenotypic effects toward improving many pathogenic states by regulating the expression of various genes. The molecules encoding by these genes are, probably, not primary responders to H_2, but indirectly act to enable the various effects of H_2. The primary target of H_2 remains unknown.

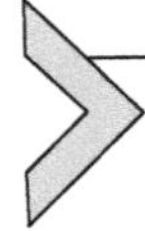

9. POSSIBLE MOLECULAR MECHANISMS UNDERLYING VARIOUS EFFECTS OF MOLECULAR HYDROGEN

9.1. Direct reduction of hydroxyl radicals with molecular hydrogen

H_2 was shown to reduce $^{\bullet}OH$ in an experiment using cultured cells (Ohsawa et al., 2007). Later, it was shown that H_2 eye drops directly decreased $^{\bullet}OH$

induced by ischemia/reperfusion in retinas (Oharazawa et al., 2010). Moreover, it has been demonstrated that, at the tissue level, H_2 neutralized $^{\bullet}OH$ that had been induced by ionizing irradiation in testes, as judged by the decreased HFP signal, and exhibited a radioprotective role (Chuai et al., 2012).

Considering the reaction rate of $^{\bullet}OH$ with H_2 in dilute aqueous solutions, this rate may be too slow to enable fully a decrease in $^{\bullet}OH$ in order to exhibit its beneficial roles (Buxton, Greenstock, Helman, & Ross, 1988). Mammalian cells are, however, highly structured with complicated biomembranes and viscous solutions with multiple concentrated components. Since collision frequency is rate-limiting in a viscous environment, the marked diffusion rate of H_2 could be advantageous to overcome the slow reaction rate constant. $^{\bullet}OH$ is known as a major trigger of the chain reaction of free radicals (Niki, 2009). Once this chain reaction occurs on biomembranes, it continues and expands causing serious damage to cells (Fig. 4). H_2 accumulates in the lipid phase more than in the aqueous phase, especially in unsaturated lipid regions, which are the major target of the initial chain reaction (unpublished results). Thus, H_2 may have an advantage to suppress the chain reaction, which produces lipid peroxide, and leads to the generation of oxidative stress markers, such as 4-hydroxyl-2-nonenal (4-HNE) and malondialdehyde (MDA) (Niki, 2014). Indeed, H_2 decreased these oxidative markers in many studies (Ning et al., 2013; Ohsawa et al., 2008; Zhou et al., 2013). Additionally, $^{\bullet}OH$ can modify deoxyguanine (dG) to 8-hydroxy-deoxyguanine (8-OHdG) (Delaney, Jarem, Volle, & Yennie, 2012; Kawai et al., 2012) (Fig. 4). H_2 decreased the level of 8-OHdG in most of the examined patients and animals (Ishibashi et al., 2012; Kawai et al., 2012)..

These experimental observations suggest that sufficient H_2 can efficiently mitigate tissue oxidation induced by $^{\bullet}OH$. However, when animals or humans drink H_2-water, it is not clear whether H_2-water provides a sufficient amount of H_2 to scavenge $^{\bullet}OH$ efficiently (Fig. 7).

9.2. Direct reduction of peroxynitrite with molecular hydrogen to regulate gene expression

As another molecular mechanism, the scavenging of $ONOO^-$ by H_2 should be considered. $ONOO^-$ is known to modify tyrosine of proteins to generate nitro-tyrosine (Radi, 2013). Several studies have shown that H_2 efficiently decreases nitro-tyrosine in animal models regardless of whether H_2-water (Cardinal et al., 2010), H_2 gas (Shinbo et al., 2013), or H_2-saline

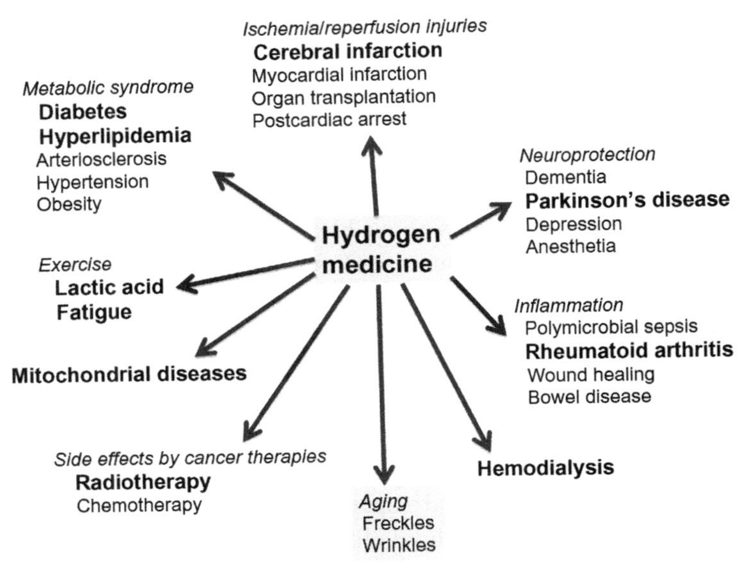

Figure 7 Summary of potential of various preventive and therapeutic effects of H_2. Bold letters indicate published results for clinical examinations (Ohta, 2014). Positive effects obtained by animal experiments for disease models are shown by normal text.

(Chen et al., 2010; Yu et al., 2011; Zhang et al., 2011; Zhu et al., 2011) is used. Moreover, drinking H_2-water decreased nitro-tyrosine in patients with RA (Ishibashi et al., 2012). Thus, at least part of the effect of H_2 can be attributed to the decreased production of nitro-tyrosine in proteins.

Many protein factors involved in transcriptional control are nitrolated (−O-NO_2) or nitrosolated (−S-NO_2). It is possible that the decrease in –O-NO_2 or –S-NO_2 may regulate the expression of various genes (Radi, 2013). However, major targets have not been identified and are under investigation.

9.3. Indirect reduction of oxidative stress by regulating gene expression

H_2 reduces oxidative stress not only directly, but also indirectly, by inducing antioxidation systems, including HO-1 SOD (Zhai et al., 2013), catalase (Cai, Zhang, Yu, & Cai, 2013), and myeloperoxidase (Zhang et al., 2011). Nrf2 is known to function as a defense system against oxidative stress and various poisons by inducing various genes including HO-1. HO-1, a microsomal enzyme degrading heme to carbon monoxide, free iron, and biliverdin, participates in the cell defense against oxidative stress (Jazwa & Cuadrado, 2010).

In Nrf2-deficient mice, mitigating effects by the inhalation of H_2 gas declined in hyperoxic lung injury accompanying by a decrease in HO-1, indicating that H_2 gas can ameliorate hyperoxic lung injury in an Nrf2-dependent manner (Kawamura et al., 2013). Activation of Nrf2 is also required for the amelioration of cerebral ischemia–reperfusion injury in rats by H_2 (Zhai et al., 2013).

H_2 influences some signal transductions as an indirect modulator; however, it is unlikely that H_2 could directly bind to some receptors involved in the signal transductions. The primary target molecule of H_2 has not been identified in these signal transduction pathways. These regulatory molecules are, probably, not primary responders to H_2, but indirectly act to enable the various effects of H_2. The primary target of H_2 remains unknown.

10. UNRESOLVED QUESTIONS AND CLOSING REMARKS

H_2 can be incorporated or ingested into the body by various methods: inhalation of H_2 gas, drinking H_2-infused water (H_2-water), injection of H_2-infused saline, and incorporation through the skin. Drinking H_2-water was efficacious for various disease models and patients; however, H_2 can be infused up to only 0.8 mM under atmospheric pressure and drinking H_2-water provides a blood H_2 concentration up to only ~10 μM with short dwelling time in the body (Nagata et al., 2009; Nakashima-Kamimura et al., 2009). Moreover, inhaling 1–4% (vol/vol) of H_2 gas was effective, by which H_2 should reach 8–32 μM in blood. Under these conditions, H_2 should be insufficient to scavenge $^{\bullet}OH$ for fully exhibiting H_2 benefits because the direct reaction rate of $^{\bullet}OH$ with H_2 in an aqueous solution may be too slow to decrease $^{\bullet}OH$ (Buxton et al., 1988) as pointed out earlier (Wood & Gladwin, 2007). Thus, it remained elusive how such low levels of H_2 with a short dwelling time could effectively compete with the numerous cellular targets in chronic or acute pathogenesis. Unexpectedly, H_2 was shown to regulate the expression of many genes and the phosphorylation of factors involved in various types of signal transduction to exhibit various phenotypes. For example, drinking H_2-water reduces the gene expressions of proinflammatory cytokines to relieve inflammation, as mentioned above, FGF21 to stimulate energy metabolism (Kamimura et al., 2011), and Grelin for neuroprotection (Matsumoto et al., 2013). However, it essentially remains unsolved what the primary target of H_2 is.

Many other mysteries regarding H_2 therapy also remain unresolved. For initiating cellular signals by H_2, H_2 should be too inert to react with most molecules except highly reactive ones, such as $^{\bullet}OH$ or $ONOO^-$. To activate H_2 to react with other molecules, a sufficient level of a putative catalyst must be present; however, it is highly unlikely that such a putative catalyst would be abundant. Moreover, H_2 should be too small to bind a putative H_2-binding receptor because its intramolecular fluctuation should lead to the instability.

H_2 can be easily applied because of a lack of adverse effects and great efficacy for nearly all pathogenic statuses involved in oxidative stress and inflammation. Since most pharmacological drugs specifically act on their targets, H_2 seems to differ from conventional drugs or other medical gasses because of its extensive and varied effects. H_2 has great potential for preventive and therapeutic applications owing to its great efficacy and its "novel" concept.

REFERENCES

Abraini, J. H., Gardette-Chauffour, M. C., Martinez, E., Rostain, J. C., & Lemaire, C. (1994). Psychophysiological reactions in humans during an open sea dive to 500 m with a hydrogen–helium–oxygen mixture. *Journal of Applied Physiology, 76*(3), 1113–1118.

Abrous, D. N., Koehl, M., & Le Moal, M. (2005). Adult neurogenesis: From precursors to network and physiology. *Physiological Reviews, 85*(2), 523–569.

Adams, M. W., Mortenson, L. E., & Chen, J. S. (1980). Hydrogenase. *Biochimica et Biophysica Acta, 594*(2–3), 105–176.

Agarwal, R. (2005). Smoking, oxidative stress and inflammation: Impact on resting energy expenditure in diabetic nephropathy. *BMC Nephrology, 6*, 13.

Andersen, J. K. (2004). Oxidative stress in neurodegeneration: Cause or consequence? *Nature Medicine, 10*(Suppl.), S18–S25.

Aukland, K., Bower, B. F., & Berliner, R. W. (1964). Measurement of local blood flow with hydrogen gas. *Circulation Research, 14*, 164–187.

Bashan, N., Kovsan, J., Kachko, I., Ovadia, H., & Rudich, A. (2009). Positive and negative regulation of insulin signaling by reactive oxygen and nitrogen species. *Physiological Reviews, 89*(1), 27–71.

Bell, E. L., Klimova, T. A., Eisenbart, J., Schumacker, P. T., & Chandel, N. S. (2007). Mitochondrial reactive oxygen species trigger hypoxia-inducible factor-dependent extension of the replicative life span during hypoxia. *Molecular and Cellular Biology, 27*, 5737–5745.

Bjelakovic, G., & Gluud, C. (2007). Surviving antioxidant supplements. *Journal of the National Cancer Institute, 99*(10), 742–743.

Bjelakovic, G., Nikolova, D., Gluud, L. L., Simonetti, R. G., & Gluud, C. (2007). Mortality in randomized trials of antioxidant supplements for primary and secondary prevention: Systematic review and meta-analysis. *JAMA, 297*(8), 842–857.

Bjelakovic, G., Nikolova, D., Gluud, L. L., Simonetti, R. G., & Gluud, C. (2008). Antioxidant supplements for prevention of mortality in healthy participants and patients with various diseases. *The Cochrane Database of Systematic Reviews, 2*, CD007176.

Brambilla, D., Mancuso, C., Scuderi, M. R., Bosco, P., Cantarella, G., Lempereur, L., et al. (2008). The role of antioxidant supplement in immune system, neoplastic, and neurodegenerative disorders: A point of view for an assessment of the risk/benefit profile. *Nutrition Journal*, 7, 29.

Brewer, A. C., Mustafi, S. B., Murray, T. V., Rajasekaran, N. S., & Benjamin, I. J. (2013). Reductive stress linked to small HSPs, G6PD, and Nrf2 pathways in heart disease. *Antioxidants & Redox Signaling*, *18*(9), 1114–1127.

Buchholz, B. M., Kaczorowski, D. J., Sugimoto, R., Yang, R., Wang, Y., Billiar, T. R., et al. (2008). Hydrogen inhalation ameliorates oxidative stress in transplantation induced intestinal graft injury. *American Journal of Transplantation*, *8*(10), 2015–2024.

Buxton, G. V., Greenstock, C. L., Helman, W. P., & Ross, A. B. (1988). Critical view of rate constants for reactions of hydrated electrons, hydrogen atoms and hydroxyl radicals ($^{\bullet}OH/^{\bullet}OH^{-}$) in aqueous solution. *Journal of Physical and Chemical Reference Data*, *17*, 513–886.

Cai, J., Kang, Z., Liu, K., Liu, W., Li, R., Zhang, J. H., et al. (2009). Neuroprotective effects of hydrogen saline in neonatal hypoxia-ischemia rat model. *Brain Research*, *1256*, 129–137.

Cai, W. W., Zhang, M. H., Yu, Y. S., & Cai, J. H. (2013). Treatment with hydrogen molecule alleviates TNFalpha-induced cell injury in osteoblast. *Molecular and Cellular Biochemistry*, *373*(1–2), 1–9.

Cardinal, J. S., Zhan, J., Wang, Y., Sugimoto, R., Tsung, A., McCurry, K. R., et al. (2010). Oral hydrogen water prevents chronic allograft nephropathy in rats. *Kidney International*, 77(2), 101–109.

Carriere, A., Carmona, M. C., Fernandez, Y., Rigoulet, M., Wenger, R. H., Penicaud, L., et al. (2004). Mitochondrial reactive oxygen species control the transcription factor CHOP-10/GADD153 and adipocyte differentiation: A mechanism for hypoxia-dependent effect. *The Journal of Biological Chemistry*, *279*(39), 40462–40469.

Carrieri, D., Wawrousek, K., Eckert, C., Yu, J., & Maness, P. C. (2011). The role of the bidirectional hydrogenase in cyanobacteria. *Bioresource Technology*, *102*(18), 8368–8377.

Chandel, N. S., Maltepe, E., Goldwasser, E., Mathieu, C. E., Simon, M. C., & Schumacker, P. T. (1998). Mitochondrial reactive oxygen species trigger hypoxia-induced transcription. *Proceedings of the National Academy of Sciences of the United States of America*, *95*(20), 11715–11720.

Chandel, N. S., Trzyna, W. C., McClintock, D. S., & Schumacker, P. T. (2000). Role of oxidants in NF-kappa B activation and TNF-alpha gene transcription induced by hypoxia and endotoxin. *Journal of Immunology*, *165*(2), 1013–1021.

Chandel, N. S., Vander Heiden, M. G., Thompson, C. B., & Schumacker, P. T. (2000). Redox regulation of p53 during hypoxia. *Oncogene*, *19*(34), 3840–3848.

Chen, C. H., Manaenko, A., Zhan, Y., Liu, W. W., Ostrowki, R. P., Tang, J., et al. (2010). Hydrogen gas reduced acute hyperglycemia-enhanced hemorrhagic transformation in a focal ischemia rat model. *Neuroscience*, *169*(1), 402–414.

Chen, H. G., Xie, K. L., Han, H. Z., Wang, W. N., Liu, D. Q., Wang, G. L., et al. (2013). Heme oxygenase-1 mediates the anti-inflammatory effect of molecular hydrogen in LPS-stimulated RAW 264.7 macrophages. *International Journal of Surgery*, *11*(10), 1060–1066.

Chuai, Y., Gao, F., Li, B., Zhao, L., Qian, L., Cao, F., et al. (2012). Hydrogen-rich saline attenuates radiation-induced male germ cell loss in mice through reducing hydroxyl radicals. *The Biochemical Journal*, *442*(1), 49–56.

Collins, Y., Chouchani, E. T., James, A. M., Menger, K. E., Cocheme, H. M., & Murphy, M. P. (2012). Mitochondrial redox signalling at a glance. *Journal of Cell Science*, *125*(Pt. 4), 801–806.

Delaney, S., Jarem, D. A., Volle, C. B., & Yennie, C. J. (2012). Chemical and biological consequences of oxidatively damaged guanine in DNA. *Free Radical Research*, *46*(4), 420–441.

Droge, W. (2002). Free radicals in the physiological control of cell function. *Physiological Reviews*, *82*(1), 47–95.

El Assar, M., Angulo, J., & Rodriguez-Manas, L. (2013). Oxidative stress and vascular inflammation in aging. *Free Radical Biology & Medicine*, *65*, 380–401.

Endo, J., Sano, M., Katayama, T., Hishiki, T., Shinmura, K., Morizane, S., et al. (2009). Metabolic remodeling induced by mitochondrial aldehyde stress stimulates tolerance to oxidative stress in the heart. *Circulation Research*, *105*, 1118–1127.

Federico, A., Cardaioli, E., Da Pozzo, P., Formichi, P., Gallus, G. N., & Radi, E. (2012). Mitochondria, oxidative stress and neurodegeneration. *Journal of the Neurological Sciences*, *322*(1–2), 254–262.

Ferrari, R., Ceconi, C., Curello, S., Cargnoni, A., Pasini, E., & Visioli, O. (1991). The occurrence of oxidative stress during reperfusion in experimental animals and men. *Cardiovascular Drugs and Therapy*, *5*(Suppl. 2), 277–287.

Finkel, T. (1998). Oxygen radicals and signaling. *Current Opinion in Cell Biology*, *10*(2), 248–253, S0955-0674(98)80147-6 [pii].

Finkel, T., & Holbrook, N. J. (2000). Oxidants, oxidative stress and the biology of ageing. *Nature*, *408*(6809), 239–247.

Fontanari, P., Badier, M., Guillot, C., Tomei, C., Burnet, H., Gardette, B., et al. (2000). Changes in maximal performance of inspiratory and skeletal muscles during and after the 7.1-MPa Hydra 10 record human dive. *European Journal of Applied Physiology*, *81*(4), 325–328.

Fritsch, J., Lenz, O., & Friedrich, B. (2013). Structure, function and biosynthesis of O(2)-tolerant hydrogenases. *Nature Reviews Microbiology*, *11*(2), 106–114.

Fu, Y., Ito, M., Fujita, Y., Ichihara, M., Masuda, A., Suzuki, Y., et al. (2009). Molecular hydrogen is protective against 6-hydroxydopamine-induced nigrostriatal degeneration in a rat model of Parkinson's disease. *Neuroscience Letters*, *453*(2), 81–85.

Fujita, K., Seike, T., Yutsudo, N., Ohno, M., Yamada, H., Yamaguchi, H., et al. (2009). Hydrogen in drinking water reduces dopaminergic neuronal loss in the 1-methyl-4-phenyl-1,2,3,6-tetrahydropyridine mouse model of Parkinson's disease. *PLoS One*, *4*(9), e7247.

Gan, L., & Johnson, J. A. (2014). Oxidative damage and the Nrf2-ARE pathway in neurodegenerative diseases. *Biochimica et Biophysica Acta*, *1842*(8), 1208–1218.

Garg, A. D., Dudek, A. M., Ferreira, G. B., Verfaillie, T., Vandenabeele, P., Krysko, D. V., et al. (2013). ROS-induced autophagy in cancer cells assists in evasion from determinants of immunogenic cell death. *Autophagy*, *9*(9), 1292–1307.

Grassi, D., Desideri, G., Ferri, L., Aggio, A., Tiberti, S., & Ferri, C. (2010). Oxidative stress and endothelial dysfunction: Say no to cigarette smoking! *Current Pharmaceutical Design*, *16*(23), 2539–2550.

Gray, S. L., Anderson, M. L., Crane, P. K., Breitner, J. C., McCormick, W., Bowen, J. D., et al. (2008). Antioxidant vitamin supplement use and risk of dementia or Alzheimer's disease in older adults. *Journal of the American Geriatrics Society*, *56*(2), 291–295.

Hackam, D. G. (2007). Review: Antioxidant supplements for primary and secondary prevention do not decrease mortality. *ACP Journal Club*, *147*(1), 4.

Halliwell, B., & Gutteridge, J. M. (1992). Biologically relevant metal ion-dependent hydroxyl radical generation. An update. *FEBS Letters*, *307*(1), 108–112.

Harma, M. I., Harma, M., & Erel, O. (2006). Measuring plasma oxidative stress biomarkers in sport medicine. *European Journal of Applied Physiology*, *97*(4), 505, author reply 506–508.

Hayashida, K., Sano, M., Kamimura, N., Yokota, T., Suzuki, M., Maekawa, Y., et al. (2012). H(2) gas improves functional outcome after cardiac arrest to an extent comparable to therapeutic hypothermia in a rat model. *Journal of the American Heart Association, 1*(5), e003459.

Hayashida, K., Sano, M., Ohsawa, I., Shinmura, K., Tamaki, K., Kimura, K., et al. (2008). Inhalation of hydrogen gas reduces infarct size in the rat model of myocardial ischemia-reperfusion injury. *Biochemical and Biophysical Research Communications, 373*(1), 30–35.

Hercberg, S., Kesse-Guyot, E., Druesne-Pecollo, N., Touvier, M., Favier, A., Latino-Martel, P., et al. (2010). Incidence of cancers, ischemic cardiovascular diseases and mortality during 5-year follow-up after stopping antioxidant vitamins and minerals supplements: A postintervention follow-up in the SU.VI.MAX Study. *International Journal of Cancer, 127*(8), 1875–1881.

Ishibashi, T., Sato, B., Rikitake, M., Seo, T., Kurokawa, R., Hara, Y., et al. (2012). Consumption of water containing a high concentration of molecular hydrogen reduces oxidative stress and disease activity in patients with rheumatoid arthritis: An open-label pilot study. *Medical Gas Research, 2*(1), 27.

Jazwa, A., & Cuadrado, A. (2010). Targeting heme oxygenase-1 for neuroprotection and neuroinflammation in neurodegenerative diseases. *Current Drug Targets, 11*(12), 1517–1531.

Kajimura, M., Fukuda, R., Bateman, R. M., Yamamoto, T., & Suematsu, M. (2010). Interactions of multiple gas-transducing systems: Hallmarks and uncertainties of CO, NO, and H2S gas biology. *Antioxidants & Redox Signaling, 13*(2), 157–192.

Kajiya, M., Silva, M. J., Sato, K., Ouhara, K., & Kawai, T. (2009). Hydrogen mediates suppression of colon inflammation induced by dextran sodium sulfate. *Biochemical and Biophysical Research Communications, 386*(1), 11–15.

Kamimura, N., Nishimaki, K., Ohsawa, I., & Ohta, S. (2011). Molecular hydrogen improves obesity and diabetes by inducing hepatic FGF21 and stimulating energy metabolism in db/db mice. *Obesity, 19*(7), 1396–1403.

Kashfi, K., & Olson, K. R. (2013). Biology and therapeutic potential of hydrogen sulfide and hydrogen sulfide-releasing chimeras. *Biochemical Pharmacology, 85*, 689–703.

Kawai, D., Takaki, A., Nakatsuka, A., Wada, J., Tamaki, N., Yasunaka, T., et al. (2012). Hydrogen-rich water prevents progression of nonalcoholic steatohepatitis and accompanying hepatocarcinogenesis in mice. *Hepatology, 56*(3), 912–921.

Kawamura, T., Wakabayashi, N., Shigemura, N., Huang, C. S., Masutani, K., Tanaka, Y., et al. (2013). Hydrogen gas reduces hyperoxic lung injury via the Nrf2 pathway in vivo. *American Journal of Physiology. Lung Cellular and Molecular Physiology, 304*(10), L646–L656.

Kim, Y. W., & Byzova, T. V. (2014). Oxidative stress in angiogenesis and vascular disease. *Blood, 123*(5), 625–631.

Kimura, H. (2010). Hydrogen sulfide: From brain to gut. *Antioxidants & Redox Signaling, 12*(9), 1111–1123.

King, P. W. (2013). Designing interfaces of hydrogenase-nanomaterial hybrids for efficient solar conversion. *Biochimica et Biophysica Acta, 1827*(8–9), 949–957.

Kubota, M., Shimmura, S., Kubota, S., Miyashita, H., Kato, N., Noda, K., et al. (2011). Hydrogen and N-acetyl-L-cysteine rescue oxidative stress-induced angiogenesis in a mouse corneal alkali-burn model. *Investigative Ophthalmology & Visual Science, 52*(1), 427–433.

Li, L., Ishdorj, G., & Gibson, S. B. (2012). Reactive oxygen species regulation of autophagy in cancer: Implications for cancer treatment. *Free Radical Biology & Medicine, 53*(7), 1399–1410.

Li, G. M., Ji, M. H., Sun, X. J., Zeng, Q. T., Tian, M., Fan, Y. X., et al. (2013). Effects of hydrogen-rich saline treatment on polymicrobial sepsis. *The Journal of Surgical Research, 181*(2), 279–286.

Lillo, R. S., & Parker, E. C. (2000). Mixed-gas model for predicting decompression sickness in rats. *Journal of Applied Physiology, 89*(6), 2107–2116.

Lillo, R. S., Parker, E. C., & Porter, W. R. (1997). Decompression comparison of helium and hydrogen in rats. *Journal of Applied Physiology*, *82*(3), 892–901.

Lin, M. T., & Beal, M. F. (2006). Mitochondrial dysfunction and oxidative stress in neurodegenerative diseases. *Nature*, *443*(7113), 787–795.

Lipinski, B. (2011). Hydroxyl radical and its scavengers in health and disease. *Oxidative Medicine and Cellular Longevity*, *2011*, 809696.

Liu, H., Colavitti, R., Rovira, I. I., & Finkel, T. (2005). Redox-dependent transcriptional regulation. *Circulation Research*, *97*, 967–974.

Liu, C., Kurokawa, R., Fujino, M., Hirano, S., Sato, B., & Li, X. K. (2014). Estimation of the hydrogen concentration in rat tissue using an airtight tube following the administration of hydrogen via various routes. *Scientific Reports*, *4*, 5485.

Liu, J., Wang, X., Shigenaga, M. K., Yeo, H. C., Mori, A., & Ames, B. N. (1996). Immobilization stress causes oxidative damage to lipid, protein, and DNA in the brain of rats. *The FASEB Journal*, *10*(13), 1532–1538.

Mandal, C. C., Ganapathy, S., Gorin, Y., Mahadev, K., Block, K., Abboud, H. E., et al. (2010). Reactive oxygen species derived from Nox4 mediate BMP2 gene transcription and osteoblast differentiation. *The Biochemical Journal*, *433*(2), 393–402.

Mano, Y., Kotani, T., Ito, M., Nagai, T., Ichinohashi, Y., Yamada, K., et al. (2014). Maternal molecular hydrogen administration ameliorates rat fetal hippocampal damage caused by in utero ischemia-reperfusion. *Free Radical Biology & Medicine*, *69*, 324–330.

Mates, J. M., Segura, J. A., Alonso, F. J., & Marquez, J. (2012). Oxidative stress in apoptosis and cancer: An update. *Archives of Toxicology*, *86*(11), 1649–1665.

Matsuda, M., & Shimomura, I. (2013). Increased oxidative stress in obesity: Implications for metabolic syndrome, diabetes, hypertension, dyslipidemia, atherosclerosis, and cancer. *Obesity Research & Clinical Practice*, 7(5), e330–e341.

Matsumoto, A., Yamafuji, M., Tachibana, T., Nakabeppu, Y., Noda, M., & Nakaya, H. (2013). Oral 'hydrogen water' induces neuroprotective ghrelin secretion in mice. *Scientific Reports*, *3*, 3273.

Miller, E. R., 3rd, Pastor-Barriuso, R., Dalal, D., Riemersma, R. A., Appel, L. J., & Guallar, E. (2005). Meta-analysis: High-dosage vitamin E supplementation may increase all-cause mortality. *Annals of Internal Medicine*, *142*(1), 37–46.

Motterlini, R., & Otterbein, L. E. (2010). The therapeutic potential of carbon monoxide. *Nature Reviews Drug Discovery*, *9*(9), 728–743.

Murphy, M. P. (1997). Selective targeting of bioactive compounds to mitochondria. *Trends in Biotechnology*, *15*(8), 326–330.

Murphy, M. P., & Smith, R. A. (2000). Drug delivery to mitochondria: The key to mitochondrial medicine. *Advanced Drug Delivery Reviews*, *41*(2), 235–250.

Nagata, K., Nakashima-Kamimura, N., Mikami, T., Ohsawa, I., & Ohta, S. (2009). Consumption of molecular hydrogen prevents the stress-induced impairments in hippocampus-dependent learning tasks during chronic physical restraint in mice. *Neuropsychopharmacology*, *34*(2), 501–508.

Nagatani, K., Nawashiro, H., Takeuchi, S., Tomura, S., Otani, N., Osada, H., et al. (2013). Safety of intravenous administration of hydrogen-enriched fluid in patients with acute cerebral ischemia: Initial clinical studies. *Medical Gas Research*, *3*(1), 13.

Nakao, A., Kaczorowski, D. J., Wang, Y., Cardinal, J. S., Buchholz, B. M., Sugimoto, R., et al. (2010). Amelioration of rat cardiac cold ischemia/reperfusion injury with inhaled hydrogen or carbon monoxide, or both. *The Journal of Heart and Lung Transplantation*, *29*, 544–553.

Nakashima-Kamimura, N., Mori, T., Ohsawa, I., Asoh, S., & Ohta, S. (2009). Molecular hydrogen alleviates nephrotoxicity induced by an anti-cancer drug cisplatin without compromising anti-tumor activity in mice. *Cancer Chemotherapy and Pharmacology*, *64*(4), 753–761.

Niki, E. (2009). Lipid peroxidation: Physiological levels and dual biological effects. *Free Radical Biology & Medicine*, *47*(5), 469–484.
Niki, E. (2014). Biomarkers of lipid peroxidation in clinical material. *Biochimica et Biophysica Acta*, *1840*(2), 809–817.
Ning, Y., Shang, Y., Huang, H., Zhang, J., Dong, Y., Xu, W., et al. (2013). Attenuation of cigarette smoke-induced airway mucus production by hydrogen-rich saline in rats. *PLoS One*, *8*(12), e83429.
Noda, K., Shigemura, N., Tanaka, Y., Kawamura, T., Hyun Lim, S., Kokubo, K., et al. (2013). A novel method of preserving cardiac grafts using a hydrogen-rich water bath. *The Journal of Heart and Lung Transplantation*, *32*(2), 241–250.
Oharazawa, H., Igarashi, T., Yokota, T., Fujii, H., Suzuki, H., Machide, M., et al. (2010). Protection of the retina by rapid diffusion of hydrogen: Administration of hydrogen-loaded eye drops in retinal ischemia-reperfusion injury. *Investigative Ophthalmology & Visual Science*, *51*(1), 487–492.
Ohsawa, I., Ishikawa, M., Takahashi, K., Watanabe, M., Nishimaki, K., Yamagata, K., et al. (2007). Hydrogen acts as a therapeutic antioxidant by selectively reducing cytotoxic oxygen radicals. *Nature Medicine*, *13*(6), 688–694.
Ohsawa, I., Nishimaki, K., Yamagata, K., Ishikawa, M., & Ohta, S. (2008). Consumption of hydrogen water prevents atherosclerosis in apolipoprotein E knockout mice. *Biochemical and Biophysical Research Communications*, *377*(4), 1195–1198.
Ohta, S. (2011). Recent progress toward hydrogen medicine: Potential of molecular hydrogen for preventive and therapeutic applications. *Current Pharmaceutical Design*, *17*(22), 2241–2252.
Ohta, S. (2012). Molecular hydrogen is a novel antioxidant to efficiently reduce oxidative stress with potential for the improvement of mitochondrial diseases. *Biochimica et Biophysica Acta*, *1820*(5), 586–594.
Ohta, S. (2014). Molecular hydrogen as a preventive and therapeutic medical gas: Initiation, development and potential of hydrogen medicine. *Pharmacology & Therapeutics*, *144*, 1–11. http://dx.doi.org/10.1016/j.pharmthera.2014.04.006.
Ono, H., Nishijima, Y., Adachi, N., Sakamoto, M., Kudo, Y., Kaneko, K., et al. (2012). A basic study on molecular hydrogen (H2) inhalation in acute cerebral ischemia patients for safety check with physiological parameters and measurement of blood H2 level. *Medical Gas Research*, *2*(1), 21.
Ono, H., Nishijima, Y., Adachi, N., Tachibana, S., Chitoku, S., Mukaihara, S., et al. (2011). Improved brain MRI indices in the acute brain stem infarct sites treated with hydroxyl radical scavengers, Edaravone and hydrogen, as compared to Edaravone alone. A non-controlled study. *Medical Gas Research*, *1*(1), 12.
Radi, R. (2013). Peroxynitrite, a stealthy biological oxidant. *The Journal of Biological Chemistry*, *288*(37), 26464–26472.
Reuter, S., Gupta, S. C., Chaturvedi, M. M., & Aggarwal, B. B. (2010). Oxidative stress, inflammation, and cancer: How are they linked? *Free Radical Biology & Medicine*, *49*(11), 1603–1616.
Ristow, M., & Zarse, K. (2010). How increased oxidative stress promotes longevity and metabolic health: The concept of mitochondrial hormesis (mitohormesis). *Experimental Gerontology*, *45*(6), 410–418.
Salganik, R. I. (2001). The benefits and hazards of antioxidants: Controlling apoptosis and other protective mechanisms in cancer patients and the human population. *Journal of the American College of Nutrition*, *20*(5 Suppl.), 464S–472S, discussion 473S–475S.
Schapira, A. H. (2008). Mitochondria in the aetiology and pathogenesis of Parkinson's disease. *Lancet Neurology*, 7(1), 97–109.
Schoenfeld, M. P., Ansari, R. R., Nakao, A., & Wink, D. (2012). A hypothesis on biological protection from space radiation through the use of new therapeutic gases as medical counter measures. *Medical Gas Research*, *2*, 8.

Schoenfeld, M. P., Ansari, R. R., Zakrajsek, J. F., Billiar, T. R., Toyoda, Y., Wink, D. A., et al. (2011). Hydrogen therapy may reduce the risks related to radiation-induced oxidative stress in space flight. *Medical Hypotheses*, *76*(1), 117–118.
Setsukinai, K., Urano, Y., Kakinuma, K., Majima, H. J., & Nagano, T. (2003). Development of novel fluorescence probes that can reliably detect reactive oxygen species and distinguish specific species. *The Journal of Biological Chemistry*, *278*(5), 3170–3175.
Shimouchi, A., Nose, K., Shirai, M., & Kondo, T. (2012). Estimation of molecular hydrogen consumption in the human whole body after the ingestion of hydrogen-rich water. *Advances in Experimental Medicine and Biology*, *737*, 245–250.
Shinbo, T., Kokubo, K., Sato, Y., Hagiri, S., Hataishi, R., Hirose, M., et al. (2013). Breathing nitric oxide plus hydrogen gas reduces ischemia-reperfusion injury and nitrotyrosine production in murine heart. *American Journal of Physiology. Heart and Circulatory Physiology*, *305*(4), H542–H550.
Smith, R. A., & Murphy, M. P. (2011). Mitochondria-targeted antioxidants as therapies. *Discovery Medicine*, *11*(57), 106–114.
Song, G., Li, M., Sang, H., Zhang, L., Li, X., Yao, S., et al. (2013). Hydrogen-rich water decreases serum LDL-cholesterol levels and improves HDL function in patients with potential metabolic syndrome. *Journal of Lipid Research*, *54*(7), 1884–1893.
Steinhubl, S. R. (2008). Why have antioxidants failed in clinical trials? *The American Journal of Cardiology*, *101*(10A), 14D–19D.
Sun, Q., Cai, J., Liu, S., Liu, Y., Xu, W., Tao, H., et al. (2011). Hydrogen-rich saline provides protection against hyperoxic lung injury. *The Journal of Surgical Research*, *165*(1), e43–e49.
Szabó, C. (2007). Hydrogen sulphide and its therapeutic potential. *Nature Reviews Drug Discovery*, *6*, 917–935.
Tanriverdi, H., Evrengul, H., Kuru, O., Tanriverdi, S., Seleci, D., Enli, Y., et al. (2006). Cigarette smoking induced oxidative stress may impair endothelial function and coronary blood flow in angiographically normal coronary arteries. *Circulation Journal*, *70*(5), 593–599.
Tormos, K. V., Anso, E., Hamanaka, R. B., Eisenbart, J., Joseph, J., Kalyanaraman, B., et al. (2011). Mitochondrial complex III ROS regulate adipocyte differentiation. *Cell Metabolism*, *14*(4), 537–544.
Tsukagoshi, H., Busch, W., & Benfey, P. N. (2010). Transcriptional regulation of ROS controls transition from proliferation to differentiation in the root. *Cell*, *143*(4), 606–616.
Turrens, J. F. (2003). Mitochondrial formation of reactive oxygen species. *The Journal of Physiology*, *552*(Pt. 2), 335–344.
van Berkel-Arts, A., Dekker, M., van Dijk, C., Grande, H. J., Hagen, W. R., Hilhorst, R., et al. (1986). Application of hydrogenase in biotechnological conversions. *Biochimie*, *68*(1), 201–209.
Vaziri, N. D., & Rodriguez-Iturbe, B. (2006). Mechanisms of disease: Oxidative stress and inflammation in the pathogenesis of hypertension. *Nature Clinical Practice Nephrology*, *2*(10), 582–593.
Walker, C. (2008). Antioxidant supplements do not improve mortality and may cause harm. *American Family Physician*, *78*(9), 1079–1080.
Wang, F., Yu, G., Liu, S. Y., Li, J. B., Wang, J. F., Bo, L. L., et al. (2011). Hydrogen-rich saline protects against renal ischemia/reperfusion injury in rats. *The Journal of Surgical Research*, *167*(2), e339–e344.
West, A. P., Brodsky, I. E., Rahner, C., Woo, D. K., Erdjument-Bromage, H., Tempst, P., et al. (2011). TLR signalling augments macrophage bactericidal activity through mitochondrial ROS. *Nature*, *472*(7344), 476–480.
Westerblad, H., & Allen, D. G. (2011). Emerging roles of ROS/RNS in muscle function and fatigue. *Antioxidants & Redox Signaling*, *15*(9), 2487–2499.

Wood, K. C., & Gladwin, M. T. (2007). The hydrogen highway to reperfusion therapy. *Nature Medicine, 13*(6), 673–674.

Xu, Z., Zhou, J., Cai, J., Zhu, Z., Sun, X., & Jiang, C. (2012). Anti-inflammation effects of hydrogen saline in LPS activated macrophages and carrageenan induced paw oedema. *Journal of Inflammation, 9*, 2.

Yoritaka, A., Hattori, N., Uchida, K., Tanaka, M., Stadtman, E. R., & Mizuno, Y. (1996). Immunohistochemical detection of 4-hydroxynonenal protein adducts in Parkinson disease. *Proceedings of the National Academy of Sciences of the United States of America, 93*(7), 2696–2701.

Yoritaka, A., Takanashi, M., Hirayama, M., Nakahara, T., Ohta, S., & Hattori, N. (2013). Pilot study of H(2) therapy in Parkinson's disease: A randomized double-blind placebo-controlled trial. *Movement Disorders, 28*(6), 836–839.

Yoshida, A., Asanuma, H., Sasaki, H., Sanada, S., Yamazaki, S., Asano, Y., et al. (2012). H(2) mediates cardioprotection via involvements of K(ATP) channels and permeability transition pores of mitochondria in dogs. *Cardiovascular Drugs and Therapy, 26*(3), 217–226.

Yu, P., Wang, Z., Sun, X., Chen, X., Zeng, S., Chen, L., et al. (2011). Hydrogen-rich medium protects human skin fibroblasts from high glucose or mannitol induced oxidative damage. *Biochemical and Biophysical Research Communications, 409*(2), 350–355.

Zhai, X., Chen, X., Shi, J., Shi, D., Ye, Z., Liu, W., et al. (2013). Lactulose ameliorates cerebral ischemia-reperfusion injury in ratsby inducing hydrogen by activating Nrf2 expression. *Free Radical Biology & Medicine, 65*, 731–741.

Zhang, Y., Sun, Q., He, B., Xiao, J., Wang, Z., & Sun, X. (2011). Anti-inflammatory effect of hydrogen-rich saline in a rat model of regional myocardial ischemia and reperfusion. *International Journal of Cardiology, 148*(1), 91–95.

Zhou, H., Fu, Z., Wei, Y., Liu, J., Cui, X., Yang, W., et al. (2013). Hydrogen inhalation decreases lung graft injury in brain-dead donor rats. *The Journal of Heart and Lung Transplantation, 32*(2), 251–258.

Zhou, R., Yazdi, A. S., Menu, P., & Tschopp, J. (2011). A role for mitochondria in NLRP3 inflammasome activation. *Nature, 469*(7329), 221–225.

Zhu, W. J., Nakayama, M., Mori, T., Nakayama, K., Katoh, J., Murata, Y., et al. (2011). Intake of water with high levels of dissolved hydrogen (H2) suppresses ischemia-induced cardio-renal injury in Dahl salt-sensitive rats. *Nephrology, Dialysis, Transplantation, 26*(7), 2112–2118.

AUTHOR INDEX

Note: Page numbers followed by "*f*" indicate figures, and "*t*" indicate tables.

A

Abboud, H. E., 295
Abe, K., 5, 13, 20, 40, 114, 170, 208–210, 212–214, 217
Abraini, J. H., 292
Abrous, D. N., 305
Acton, S. T., 104–105
Adachi, N., 295, 301, 304–305
Adams, M. W., 290
Adams, N., 274–275, 281
Adhikari, S., 201–202, 240–241
Agah, A., 214–216
Agarwal, R., 293, 308–309
Agbor, T. A., 170–172, 179*t*
Aggarwal, B. B., 293
Aggeler, R., 58–59, 61–62
Aggio, A., 293
Agrawal, N., 80–81
Ahern, P. P., 158–159
Ahmad, A., 186*f*
Ahmed, A., 235
Ahn, Y. S., 208–210
Ahrenholz, D., 199–200
Aihara, E., 176–178, 179*t*
Aittokallio, T., 148, 150
Akimoto, T., 148
Alasti, N., 214–216
Albensi, B. C., 214–216
Aldini, G., 4–5, 14–15
Aldrich, M. C., 148–149
Ali, B., 235
Allen, D. G., 293
Al-Magableh, M., 235
Alonso, F. J., 294
Altaany, Z., 81*t*
Alvarez, B., 46–47
Alvarez, C., 234–235, 274
Alves-Filho, J. C., 116
Amarnath, V., 6, 10
Ames, B. N., 293, 302, 305
Andersen, J. K., 293, 305
Andersen, J. N., 236
Anderson, M. L., 295
Andjelkovich, D. A., 6
Andrews, J. L., 32
Andriamihaja, M., 20, 170, 172, 234–235, 241–242
Andruski, B., 114–115
Ang, A. D., 150–151, 197
Ang, S. F., 200, 201–202
Angayarkanni, N., 146–147
Angulo, J., 293
Annis, D. S., 156–157
Ansari, R. R., 293
Anso, E., 294
Antelmann, H., 40, 59–60, 61, 64, 67, 71–72, 147, 161
Antoch, M. P., 32
Antonelli, E., 114–115, 119–120, 157–158, 173, 176, 208–210
Anuar, F., 199
Ao, G., 130–131
Appel, L. J., 295
Aracena-Parks, P., 86
Araque, A., 214–216, 217
Araujo, R. M., 176
Arb, C. V., 264
Armstrong, J. S., 6, 179*t*, 218, 219
Arnoni, C. P., 159
Arp, A. J., 148
Asano, Y., 304–305
Asanuma, H., 304–305
Ascenzi, P., 31–32
Asfar, P., 20–21
Asoh, S., 302, 309
Atanasiu, C., 172
Aukland, K., 291
Austin, R. C., 235
Azevedo, L. U., 176

B

Badhwar, A., 113–114
Badier, M., 292
Badings, H. T., 14

Bahaoddini, A., 153
Bai, A. P., 172–173
Bailey, S. M., 29, 32
Bain, J. A., 28–29
Baines, C., 116–117
Baird, A. W., 158–159
Balk, J., 272–275, 280–281
Ball, C. I., 235
Ball, C. J., 142
Ballinger, C. A., 22–23
Banerjee, R., 5, 13, 20, 40, 58, 148, 149–150
Banerjee, S. K., 210–212
Baranano, D. E., 222
Barathi, S., 146–147
Barbosa Junior, F., 159
Barcellos, L. F., 148–149
Barker, 214–216
Barnes, S., 46–47
Baross, J., 234–235
Basbaum, A. I., 197
Basell, K., 40, 59–60, 61, 64, 67, 71–72, 147, 161
Bashan, N., 294
Baskar, R., 151
Bassig, B. A., 148–149
Bateman, R. M., 291, 292
Baty, J. W., 237–238
Baue, A. E., 198
Bauer, A., 264
Bayir, H., 20–21
Beal, M. F., 293
Beard, K. F. M., 272–274
Beatty, J. K., 179*t*
Beauchamp, R. J., 6
Becher, D., 40, 59–60, 61, 64, 67, 71–72, 147, 161
Behar, T., 214–216
Belardinelli, M.-C., 5
Bell, E. L., 294
Beltowski, J., 212–213
Benamouzig, R., 172
Benavides, G. A., 20–21, 29–31
Benetti, E., 134
Benfey, P. N., 294
Benhar, M., 80, 156, 236–237, 238, 247–248
Benishin, C. G., 208–210, 212–213
Benjamin, I. J., 294
Benninghoff, B., 154
Bercik, P., 178
Beresina, O. V., 148–149
Berkowitz, O., 276–277, 280–281
Berliner, R. W., 291
Bernard, A., 156–157
Bernstein, C. N., 170–172
Bertlein, S., 46, 64–66, 150–151, 161
Bertolin, A., 147
Besner, G. E., 199–200
Bhatia, M., 5, 116, 148, 150–151, 196–197, 198, 199–200, 201–202, 212–213, 214–216, 240–241
Bhunia, A. K., 40, 81*t*
Bhushan, S., 20
Bian, J.-S., 11, 151, 234, 235
Bianchi, C., 134, 210–212
Biermann, J., 218–219
Bihari, A., 113–114
Bihari, R., 114–115
Billadeau, D. D., 156
Billiar, T. R., 293, 304–305
Bion, J., 198
Bir, S. C., 20, 146–147, 154
Birke, H., 272, 274–275, 276–277, 278, 280–281, 282–283, 282*f*
Biteau, B., 236
Bjelakovic, G., 293–294, 295
Bjorgo, E., 152–153
Blachier, F., 31–32, 170
Blackler, R. W., 119–120, 170–172, 176–178, 179*t*, 180–181, 182–184, 184*f*
Blackstone, E., 20–21, 31–32
Bloch, K. D., 20, 32
Block, K., 295
Blouin, J. M., 172
Bluestone, J. A., 159
Bo, L. L., 304–305
Boim, M. A., 159
Boitano, S., 156–157
Bolla, M., 180–181
Boltz, A., 272–274
Böltz, S., 46, 64–66, 150–151, 161
Bonnefoy, J. Y., 154
Boreiko, C., 6
Borges, K., 220–221
Bornstein, P., 156–157

Bos, E. M., 134–135, 139–140
Bosco, P., 293–294
Bosma, W., 235
Bosworth, C. A., 29–31
Bougaki, M., 150–151
Bouillaud, F., 31–32, 170, 234–235, 241–242
Bowen, J. D., 295
Bower, B. F., 291
Bowser, D. N., 214–216
Boyarskih, U. A., 148–149
Brach, T., 62
Bracher, A., 27–28
Bradford, M. M., 277–278
Braet, C., 208–213
Brain, S. D., 175–176
Brambilla, D., 293–294
Brancaleone, V., 114–116, 119–120, 157–158, 173, 175–176, 179*t*
Branco, L. G., 176
Brandt, U., 28–29
Braun, H. P., 155–156
Braun, R. D., 161–162
Breitner, J. C., 295
Brekke, J. F., 120–121
Brenner, S., 239
Breslin, J. W., 111
Brewer, A. C., 294
Bribriesco, A. C., 103
Briceno, J. C., 105
Brodie, A. E., 148, 153–154
Brodsky, I. E., 294
Brody, T. M., 28–29
Broell, J., 159
Broniowska, K. A., 236–237, 238, 247–248
Broome, J. D., 147
Brown, A. L., 161–162
Brown, E. J., 156–157
Brown, K., 218–219
Browning, L. W., 274–275, 281, 282–283, 282*f*
Brune, V., 150
Bruzzese, L., 152–153
Bu, D., 212–213, 221
Buchholz, B. M., 292, 304–305
Buckley, C. D., 175–176
Budai, B., 5, 6–7, 40, 64–66, 73
Bula, N., 183–184
Bulsiewicz, W. J., 196
Buras, J. A., 198
Burkhard, P., 280–281
Burnet, H., 292
Bus, J., 6
Busch, W., 294
Buschemeyer, W. C., 108–109
Butterfield, D. A., 5
Buxton, G. V., 307, 309
By, Y., 152–153
Byzova, T. V., 293

C

Cabrales, P., 105
Cacchioli, A., 153
Cai, J. H., 302, 306, 308–309
Cai, S., 20–21
Cai, W. W., 308–309
Cai, Y., 160
Calderwood, A., 234
Calfon, M., 151
Caliendo, G., 119–120, 158–159, 172–173, 176–181, 183–184
Calo, L., 234–235, 274
Calvert, J. W., 20, 29–31, 40, 133–134, 185, 218–219
Calzada, M. J., 156–157
Calzia, E., 20–21
Camargo, E. A., 176, 180–181
Campanini, B., 276–277
Campolo, M., 185–187, 186*f*
Cannon, M., 58–59, 61–62
Cantarella, G., 293–294
Cao, F., 306–307
Cao, K., 40, 170–172
Cao, Q., 40, 81*t*, 159–160
Cao, Y., 197
Capaldi, R. A., 58–59, 61–62
Carballal, S., 46–47
Cardaioli, E., 305
Cardinal, J. S., 292, 304–305, 307–308
Cargnoni, A., 293
Carini, M., 4–5, 14–15
Carlisle, R. E., 151
Carmona, M. C., 295
Carriere, A., 295
Carrieri, D., 290–291
Carroll, K. S., 40, 48*f*, 51*f*, 81–83

Carruthers, K. J., 200
Carson, D. A., 148, 153–154
Carvalheiro, H., 159
Cascio, M. B., 80–81
Cashel, J. A., 156–157
Castagliuolo, I., 196–197
Castellino, F., 147
Ceconi, C., 293
Centelles, M. N., 64–66, 148
Cerrito, P. B., 108–109
Chabli, A., 5
Chadefaux-Vekemans, B., 5
Chalon, B., 20–21
Chan, L., 200
Chan, M. V., 119–120, 158–159, 176, 180, 183–184
Chan, S. J., 208–210
Chandel, N. S., 294, 295
Chang, J. S., 148–149
Chang, L. C., 208–210
Chang, X., 212–213, 221
Chao, C., 148–149
Chattopadhyay, M., 152, 184–185
Chaturvedi, M. M., 293
Chaturvedi, P., 218–219
Chaudry, I. H., 198
Chauhan, A. K., 114–115, 173–174
Chaumontet, C., 172
Chavez-Chavez, R. H., 112
Chebib, M., 221
Chen, C. H., 307–308
Chen, C. P., 212–213, 219–220
Chen, G., 130
Chen, H. G., 306
Chen, J. S., 173, 272, 290
Chen, K. Y., 61, 64–66, 139
Chen, L., 307–308
Chen, M. H., 208–210, 218
Chen, M. J., 12, 219–220
Chen, R. Q., 218–219
Chen, W., 132–133
Chen, X., 131–132, 148, 218, 307–309
Cheng, J., 158–159
Cheng, P., 139
Cheng, R. Y., 152
Cheng, Y., 240–241
Cheng, Z. Y., 14–15, 170–172
Chenuel, B., 20–21
Chen,W., 185–187
Cheung, N. S., 6, 11, 12, 179*t*, 218, 219–220
Chiba, S., 153
Chilian, W. M., 105–106, 111
Chintalapani, S., 142
Chitoku, S., 304–305
Chokkalingam, A. P., 148–149
Chouchani, E. T., 236, 294
Christman, G. D., 86
Christopherson, K. S., 214–216
Chu, H., 222
Chu, S. H., 6, 11, 179*t*, 218, 219
Chuai, Y., 306–307
Chung, J. H., 81*t*, 161–162
Cini, C., 5
Cipollone, R., 31–32
Cirino, C., 175–176
Cirino, G., 14–15, 114–116, 119–120, 157–158, 173, 176–181, 183–184
Civale, L., 234
Claeyssens, M., 208–213
Clapham, D. E., 221
Clarke, P. H., 234–235
Clements, R. T., 134, 210–212
Coccia, R., 5
Cocheme, H. M., 294
Coffin, C. S., 172
Coffin, J. F., 22–23
Coimbra, T. M., 159
Colavitti, R., 294
Coletta, C., 20, 148–149, 185
Coley, D. M., 208–212
Collino, M., 134
Collins, D., 158–159
Collins, Y., 294
Colon, W., 81*t*
Condo, J., 152–153
Cong, B., 146–147
Conn, E. E., 274
Connor, K. E., 214–216
Cook, P. F., 234–235, 276–277, 280–281
Cooper, A. J., 218–219
Cooper, C. E., 20
Costa, J. J., 113
Costa, R. S., 159
Cotman, C. W., 220–221
Cotta-de-Almeida, V., 155

Cotter, T. G., 112–113
Crane, P. K., 295
Crapo, J. D., 27–28
Crawford, J. H., 29–31
Crespo, J. L., 272
Cribbs, R. K., 199–200
Crimi, E., 150–151
Crockett, S. D., 196
Cuadrado, A., 294, 308–309
Cueto, R., 22–23
Cui, W.-T., 234
Cui, X., 307
Cukier-Meisner, E., 176, 188
Culberson, C. R., 199
Cummings, J. H., 146–147
Cunha, F. Q., 159
Cunha, T. M., 176
Curello, S., 293
Cuzzocrea, S., 186*f*
Czaikoski, P. G., 116

D

Da Pozzo, P., 305
Da Silva, G. J., 119–120, 176, 183–184
da Silva, J. A., 159
Dabney-Smith, C., 274–275, 281
Dacosta, R. S., 108–109
D'Acquisto, F., 105, 175–176
Dagani, F., 219–220
Dai, H., 117–118, 137–138
Dai, W., 139
Dalal, D., 295
Dal-Secco, D., 176
Dao, J., 208–210
Darbor, M., 135–136, 176–178
Daut, J., 28–29
Davis, 104–105
Davison, M., 114–115
de Beus, M. D., 81*t*
de Boer, R. A., 134–135
de Kok, L. J., 235, 272, 274–275, 280–281
de la Motte, C., 102–103
De Marco, C., 5
de Palma, G., 182–183
de Paula, M. A., 176, 180–181
Dean, D. R., 272–274
Debiemme-Chouvy, C., 71
Debrabant, A., 102–103
Dekker, M., 290–291
DeKok, L. J., 262
Del Soldato, P., 152, 172–173, 178–180, 218–219
Delaney, S., 307
Delano, F. A., 112–113
Della Coletta Francescato, H., 159
Dellinger, R. P., 198
Dello Russo, C., 217
Dellon, E. S., 196
Delneste, Y., 154
Denadai-Souza, A., 176, 180–181
Denk, W., 103
Denou, E., 182–183
Denu, J. M., 58–59
Desideri, G., 293
Desikan, R., 237–238, 239
Devarie-Baez, N. O., 40–41, 42, 48*f*, 51*f*, 81–83
Devchand, P. R., 178–180
Dewhirst, M. W., 161–162
Di Meo, I., 149–150, 155–156, 281
Di Paola, R., 186*f*
DiCarlo, G., 214–216
Dicay, M. S., 150–151, 172, 178–180, 183–184
Dick, R. A., 4
Dick, T. P., 61–62, 63, 64, 73, 112–113
Dickhout, J. G., 151
Dickinson, D. A., 20–21, 29–31
Dinesh Kumar, S., 239–240
Ding, X. J., 235
Dingledine, R., 220–221
Dirksen, R. T., 86
Distrutti, E., 114–116, 119–120, 157–159, 172–173, 176, 187–188, 208–210
Dixon, J. E., 58–59
Docherty, N. G., 158–159
Doeller, J. E., 20–21, 29–31, 40, 133–134, 185, 218–219
Dogishi, K., 176–178
Doherty, J. J., 220–221
Dole, V. S., 114–115, 173–174
Dombkowski, R. A., 170
Donald, J. A., 170
Donatti, A. F., 176

Dong, J. H., 214–216
Dong, L., 218–219
Dong, X. B., 218
Dong, X.-J., 272
Dong, Y., 307
Doorn, J. A., 4
Doran, R., 120–121
Doran, S. F., 27–28
Doring, C., 150
Dorman, D. C., 234, 241–242
Douce, R., 276–277
Drabek, T., 20–21
Droge, W., 154, 293
Droux, M., 276–277, 280
Druesne-Pecollo, N., 293–294, 295
Du, J., 208–210, 212–213, 221
Du, P. F., 235
Du, S. X., 148–149
Dudek, A. M., 294
Duff, D. W., 212–213
Dufton, N., 179*t*
Dungey, A. A., 113–114
Dunn, G. D., 105–106
Dupre, L., 155
Durand-Gorde, J. M., 152–153
Durante, W., 117–118
Duy, C. N., 210–212

E

Eaton, S., 198
Eberl, T., 20–21
Eck, H. P., 154
Eckart, A., 105
Eckerle, S., 150
Eckert, C., 290–291
Egen, J. G., 102–103
Eggleton, P., 159
Ehmann, W. D., 5
Eichorn, M. E., 102
Eisaki, N., 276–277
Eisenbart, J., 294
Eisenmenger, S., 102
Ekundi-Valentim, E., 176, 180–181, 182
El Assar, M., 293
Elrod, J. W., 20, 29–31, 40, 133–134, 179*t*, 185, 218–219
Elsheikh, W., 119–120, 176, 183–185, 184*f*
Enders, G., 102
Endo, J., 294
Endorf, F. W., 199–200
Engler, R. L., 113
Enli, Y., 293
Enokido, Y., 208–210
Epstein, A., 148
Erdjument-Bromage, H., 80, 236–237, 238, 294
Erecinska, M., 20–21, 219–220
Erel, O., 293
Erion, M. D., 201
Ertl, A., 4
Esashi, Y., 275–276, 280–281
Esechie, A., 210–212
Esensten, J. H., 159
Espey, M. G., 151–152
Esposito, E., 186*f*
Esterbauer, H., 4–5, 6–7, 7*f*, 8*f*, 9*f*, 11*f*, 12*f*, 14, 14*f*, 15, 15*f*
Eto, K., 148
Evrengul, H., 293
Exner, M., 5, 7*f*, 8*f*, 9*f*, 11*f*, 12*f*, 14*f*, 15, 15*f*
Ezerina, D., 112–113

F

Fairfax, S. T., 120–121
Fan, H., 185–187
Fan, Y. X., 302, 306
Fanger, M. W., 147
Fanucchi, M. V., 20–21, 25–29, 31–32
Favier, A., 293–294, 295
Fay, W. P., 117–118
Fearnley, I. M., 236
Federico, A., 305
Feilen, N., 199
Fellenberg, K., 264
Fellin, T., 214–216
Feng, B., 276–277
Feng, J., 134, 210–212
Fernandez, E., 160
Fernandez, S., 27–28
Fernandez, Y., 295
Ferrandiz, C., 160
Ferrari, R., 293
Ferraz, J. G., 170, 172
Ferreira, G. B., 294
Ferri, C., 293
Ferri, L., 293

Ferris, C. D., 80, 222, 236–237, 238
Field, L., 152
Fijnheer, R., 170–172
Filipovic, M. R., 49
Filner, P., 262
Fingar, V. H., 108–109
Finkel, T., 293, 294
Fiorucci, S., 114–116, 119–120, 157–159, 172–173, 176–178, 179*t*, 180–181, 183–184, 187–188, 208–210
Fish, M., 199
Flannigan, K. L., 119–120, 170–172, 173–174, 176, 179*t*, 183–184, 184*f*
Fleming, C., 160
Florenzano, J., 176, 180–181
Flower, R. J., 115–116, 175–176
Fonseca, E., 160
Fontanari, P., 292
Forbes, T. L., 113–114
Formichi, P., 305
Forrester, M. T., 80, 156, 236–237, 238, 247–248
Foster, M. W., 236–237, 238, 247–248
Foussat, A., 156–157
Fox, B., 248
Francescato, H. D., 159
Francois, J. A., 277
Francom, D. M., 208–210, 212–213
Frazzon, A. P., 272–274
Frazzon, J., 272–274
Freed, C. R., 218
Freeman, B. A., 46–47
Freiberg,B.A., 155
Friedman, A. K., 217
Friedrich, B., 290
Fritsch, E. F., 277–278
Fritsch, J., 290
Fromonot, J., 152–153
Frossard, J.-L., 196–197
Fu, C., 40, 80, 81*t*, 151
Fu, D., 150–151, 197
Fu, Y., 305
Fu, Z., 307
Fuchs, C., 272–274
Fujii, H., 300–301, 304, 306–307
Fujimuro, M., 80–81
Fujino, M., 303
Fujita, K., 305
Fujita, Y., 305
Fukuda, R., 291, 292
Fukuda, S., 107–108
Fukushima, A., 272–275
Furne, J., 13, 146–147, 154, 208–210
Furtmuller, P. G., 179*t*

G

Gaboury, J. P., 96
Gadalla, M. M., 40, 80–83, 81*t*, 86, 155–156
Gagliardi, R. J., 220–221
Gaitonde, M. K., 279
Gallagher, P. J., 155
Gallicchio, M., 134
Gallus, G. N., 305
Gan, L., 294
Gan, Z. Y., 152
Ganapathy, S., 295
Ganong, A. H., 220–221
Gao, F., 306–307
Gao, L., 173
Garcia, B., 140–141
Garcia, I., 234–235, 272, 274
Garcia-Bereguiain, M. A., 208–210, 219–220
García-Mata, C., 234, 272
Gardette, B., 292
Gardette-Chauffour, M. C., 292
Garg, A. D., 294
Gaskin, F. S., 116–118
Gasparic, J., 146–147
Gazi, S. K., 40, 80–83, 81*t*, 155–156
Geller, D. A., 160
Gelman, A. E., 103
Gemici, B., 176–178
Geng, B., 208–210
Georgieff, M., 20–21
Ghersi-Egea, J. F., 218–219
Gholami, A. M., 264
Giacomelli, L., 147
Gibson, S. B., 294
Giles, G. I., 29–31
Gill, R. A., 235
Gill, V., 102–103
Gilligan, J. P., 201
Gilman, C. P., 86
Gilroy, D. W., 175–176

Giordano, M., 272–274
Giudici, R., 210–212
Giustarini, D., 146–147
Givvimani, S., 114–115, 208–212, 218–219
Gladwin, M. T., 309
Glawe, J. D., 146–147
Glode, L. M., 148
Gluud, C., 293–294, 295
Gluud, L. L., 293–294, 295
Gmunder, H., 154
Gnaiger, E., 20–21
Gobbi, G., 151, 153
Godo, N., 114
Goebel, U., 218–219
Goetz, A. E., 102
Goldman, S. A., 214–216
Goldwasser, E., 295
Gomez, T. S., 156
Gong, M., 235
Gonzalez, M. D., 158–159
Gooding, K. M., 159, 235
Goodwin, L. R., 208–210, 212–213
Goonasekera, S. A., 86
Gordon, J. I., 158–159
Gorin, Y., 295
Goto, Y., 179*t*
Gotor, C., 234–235, 272, 274
Goubern, M., 170
Gould, S., 149–150, 154, 156
Graham, D. G., 6, 10
Grande, H. J., 290–291
Granger, D. N., 112–113, 176–178
Granger, H. J., 105–106, 111
Grassi, D., 293
Gray, S. L., 295
Greenstock, C. L., 307, 309
Greer, P. K., 161–162
Greiner, R., 40, 59–60, 61, 64, 67, 71–72, 147, 161
Grieshaber, M. K., 31–32, 281, 282–283
Grisham, M. B., 112–113, 187–188
Groger, M., 141, 210–212
Groisman, A., 98–100, 101, 103, 106–108
Gruber, J., 239–240
Grushka, O., 183–184
Guallar, E., 295
Guan, F., 148–149
Guan, Q., 173
Guan, Y. Y., 240–241
Guerrero, A. T., 176
Guillot, C., 292
Guo, C., 135, 179*t*
Guo, H. H., 108–109
Guo, W., 14–15
Guo, Y., 185–187
Guo, Z., 218
Gupta, S., 170–172
Gupta, S. C., 293
Gutensohn, M., 274–275, 276–277, 278, 280–281
Gutierrez, E., 107–108
Gutierrez-Merino, C., 208–210, 219–220
Gutscher, M., 62
Gutteridge, J. M., 293, 297–298
Gwinn, W. M., 161–162

H

Haas, F. H., 264, 272, 274–275, 280–281
Hackam, D. G., 293–294, 295
Hagen, W. R., 290–291
Hagiri, S., 292, 307–308
Haidari, S., 105
Haig, A., 141
Haigh, R., 159, 248
Hajos, G., 208–213
Halassa, M. M., 214–216
Haldar, S. M., 236–237, 247–248
Hale, L. P., 161–162
Halliwell, B., 212–213, 219–220, 293, 297–298
Hamanaka, R. B., 294
Hamilton, S. L., 86
Hammad, S. M., 150
Hampton, M. B., 237–238
Han, A., 218
Han, D., 212–213
Han, H. Z., 306
Han, Y., 152, 160, 212–213, 221, 235
Hancock, J. T., 235, 236, 237–238, 239, 247–248, 254–256
Handy, D. E., 114–115, 173–174
Hanley, P. J., 28–29
Hanson, G. T., 58–59, 61–62
Haouzi, P., 20–21
Hara, M. R., 80–81
Hara, T., 158, 176–178

Hara, Y., 306, 307–308
Harding, H. P., 151
Harma, M. I., 293
Harris, A. G., 113–114
Harris, E. W., 220–221
Harris, K. A., 113–114
Harrison, J., 237–238, 239
Hart, D. A., 147
Hart, J., 235
Hasegawa, R., 275–276
Haselton, F. R., 102–103, 108–109
Haslett, C., 175–176
Hassanpour, M., 135–136, 176–178
Hataishi, R., 292, 307–308
Hattori, N., 305
Hatzfeld, Y., 272–274, 275–276, 280–281
Havryluk, O., 183–184
Hawkesford, M. J., 274–276, 280–281
Hawkins, H., 210–212
Hayashi, K., 179*t*
Hayashi, S., 176–178
Hayashida, K., 300, 300*f*, 304–305
Hayatdavoudi, G., 27–28
Haydon, P. G., 214–216, 217
Hayes, K. R., 32
He, B., 307–309
He, L. S., 156–157
He, Y. D., 235
Healy, M. J., 212–213
Heeg, C., 272–275, 276–277, 278, 280–281
Hegde, A., 150–151, 197, 199, 201–202
Hell, J. W., 214–216
Hell, R., 272–275, 276–277, 278, 280–281
Heller, E. A., 217
Hellmich, M. R., 148–149
Helman, W. P., 307, 309
Helton, T. D., 221
Hengartner, M. O., 239
Henson, D., 237–238, 239
Hercberg, S., 293–294, 295
Hermann, M., 5, 7*f*, 8*f*, 9*f*, 11*f*, 12*f*, 14*f*, 15, 15*f*
Hess, D. T., 80, 156, 236–237
Hidalgo, C., 86
Hildebrandt, T. M., 31–32, 149–150, 155–156, 281, 282–283
Hilhorst, R., 290–291
Hillebrands, J. L., 139–140
Hinks, J. A., 236
Hiraki, K., 13, 208–210, 212–213
Hirano, S., 153, 303
Hirayama, M., 305
Hirose, M., 292, 307–308
Hirota, S. A., 102–103
Hishiki, T., 294
Hoefgen, R., 262
Hoernstein, S. N., 274–275, 276–277, 278
Hofbauer, B., 196–197
Hoffman, N. E., 274
Hoffmann, I., 272
Hogenesch, J. B., 32
Hogg, N., 236–237, 238, 247–248
Hohenester, E., 280–281
Holbrook, N. J., 293
Holdorf, M. M., 274–275, 281
Holford, T. R., 148–149
Holzmann, B., 198
Hongfang, J., 146–147
Honzatko, A., 6
Hopkins, L., 280–281
Hori, T., 147, 154
Horvath, E. M., 134, 210–212
Hoschke, A., 208–213
Hosgood, S. A., 140
Hosp, F., 272–274, 276–277
Howard, J. G., 198
Hrycevych, N., 183–184
Hsu, A., 148
Hu, K. D., 235
Hu, L. Y., 130–131, 212–213, 219–220, 235
Hu, L.-F., 11, 234, 235
Hu, L.-Y., 235, 248
Hu, R. W., 172–173
Hu, S.-L., 235, 248
Hu, W. Z., 235
Hua, F., 81*t*
Hua, L. Y., 234
Hua, S. L., 234
Huang, B., 276–277
Huang, C. S., 308–309
Huang, H., 307
Hudspith, M. J., 219–221
Hughes, M. N., 64–66, 148
Hulbert, W. C., 6
Hunter, J. P., 140
Huxley, V. H., 110–111

Hwang, S. M., 120–121
Hyspler, R., 146–147
Hysplerova, L., 146–147
Hyun Lim, S., 304–305

I

Ichihara, M., 305
Ichinohashi, Y., 304
Ichinohe, A., 208–210
Ichinose, F., 20, 32
Ida, T., 147, 161
Igarashi, T., 300–301, 304, 306–307
Ignacy, T., 114–115
Ihara, H., 147, 161
Ikawa, T., 153
Imamura, K., 276–277
Imanaka, H., 276–277
Indrova, M., 146–147
Intaglietta, M., 105, 161–162
Irie, F., 214–216
Isbell, T. S., 20–21, 29–31
Ise, F., 179*t*
Isenberg, J. S., 151–152, 156–157
Ishdorj, G., 294
Ishibashi, T., 306, 307–308
Ishigami, M., 13, 208–210, 212–213
Ishii, H., 112–113
Ishii, I., 150–151, 197
Ishii, K., 5, 13, 40, 208–210, 212–213
Ishikawa, M., 291, 292, 295, 296, 296*f*, 298–300, 298*f*, 301, 304–305, 306–307
Ishizawa, K., 272–274, 275–276, 280–281
Isom, G. E., 31–32
Issekutz, A., 96
Itani, R. M., 178–180
Ito, M., 304, 305
Ito, T., 148
Ivanovic, Z., 161–162
Ivanovic-Burmazovic, I., 49
Ivins-O'Keefe, K., 151–152, 157
Iwasawa, K., 208–210
Iwata, S., 147, 154

J

Jacques, A. G., 272–274
Jaffrey, S. R., 80, 236–237, 238
Jaggar, J. H., 5
Jakopitsch, C., 179*t*
Jamain, S., 208–210
James, A. M., 236, 294
James, R. A., 234, 241–242
Jamshidi, N., 112–113
Jang, J., 212–213
Jansonius, J. N., 280–281
Jantzen, P. T., 214–216
Jaraki, O., 80
Jarem, D. A., 307
Jayagopal, A., 102–103, 108–109
Jazwa, A., 294, 308–309
Jeannin, P., 154
Jekel, H., 139
Jeng, A. Y., 201
Jeng, M. W., 147
Jenne, C. N., 103
Jeong, W., 59–60
Jerome, D. E., 151
Jez, J. M., 277, 280–281
Jhee, K. H., 148
Ji, M. H., 302, 306
Jia, J., 130–131
Jia, Y., 152
Jia-Ling, S., 6, 179*t*, 218, 219
Jiang, B., 40, 170–172
Jiang, C. X., 234, 235, 306
Jiang, H., 138, 151
Jiang, X., 170–172
Jin, H. M., 155–156
Job, D., 276–277
Johnson, E. L., 4, 14
Johnson, J. A., 294
Johnson, P. C., 161–162
Johnston, B., 96
Johnston, G. A., 221
Johnston, M. V., 86
Jones, G. E., 105
Jones, M. K., 178–180
Jordan, B., 12
Joseph, J., 294
Jost, R., 274–275, 276–277, 278, 280–281
Ju, Y., 40, 81*t*
Juarez, B., 217
Juergens, M., 280–281
Julia, A., 160
Julian, D., 148
Junemann, J., 274–275, 281, 282–283, 282*f*
Jung, S. M., 40, 81*t*

K

Kabil, O., 5, 13, 20, 40, 58, 148, 149–150
Kachko, I., 294
Kaczorowski, D. J., 292, 304–305
Kahn, P. C., 208–210
Kaila, K., 221
Kajimura, M., 114, 291, 292
Kajiwara, M., 212–213
Kajiya, M., 306
Kakinuma, K., 296, 299*f*
Kalani, A., 210–212, 218–219
Kalbfleisch, M., 114–115
Kaletta, T., 239
Kalogeris, T., 116–119, 136–138
Kalyanaraman, B., 294
Kamada, K., 116–118
Kamat, P. K., 210–212, 214–216, 218–219
Kamatani, N., 148, 153–154
Kamath, A. F., 114–115, 173–174
Kamhawi, S., 102–103
Kamimura, N., 302, 303*f*, 304–306, 309
Kamoun, P., 5, 114
Kanaumi, T., 208–210
Kaneko, K., 295, 301
Kang, K., 138–139
Kang, S. G., 161–162
Kang, Z., 302
Kanno, S., 153
Kapoor, S., 114–115
Karuppagounder, S., 40
Kashfi, K., 184–185, 292
Kashiba, M., 114
Katayama, T., 294
Kato, N., 304
Katoh, J., 307–308
Kaur, S., 149–150, 151–152, 154, 156–157
Kawabata, A., 176, 212–213
Kawai, D., 307
Kawai, T., 306
Kawamura, T., 304–305, 308–309
Keefer, L. K., 105–106
Keenan, C. M., 176–178
Keller, J. N., 12
Kelley, D., 234–235
Kemp-Harper, B., 235
Kenkel, I., 49
Kensler, T. W., 4
Kern, H. F., 196–197
Kern, W., 4
Kesse-Guyot, E., 293–294, 295
Keszler, A., 236–237, 238, 247–248
Kevil, C. G., 20, 146–147, 154, 156
Khakh, B. S., 214–216
Khan, M. S., 264
Khan, W. I., 170–172
Khatri, A., 150–151
Kida, K., 150–151
Kikuiri, T., 81*t*
Kim, B. N., 208–210, 212–213
Kim, C. H., 208–210
Kim, D. G., 208–210
Kim, H. R., 208–210
Kim, J. J., 170–172
Kim, J. M., 208–210
Kim, K. H., 212–213
Kim, M., 142
Kim, S., 40, 80–83, 81*t*, 155–156
Kim, S. F., 80–81
Kim, S.-O., 236–237
Kim, Y. W., 293
Kimblin, N., 102–103
Kimura, H., 5, 13, 20, 40, 114, 148, 170–172, 179*t*, 208–210, 212–216, 217, 218–219, 220–221, 222, 291
Kimura, K., 300, 300*f*, 304–305
Kimura, M., 176–178
Kimura, Y., 13, 20, 40, 170–172, 179*t*, 214–216, 218–219
King, A. L., 20, 29
King, M. R., 142
King, P. W., 290–291
Kipson, N., 152–153
Kirishi, S., 176
Kirk, M. C., 46–47
Kish, S. J., 5
Kiss, L., 210–212
Kisucka, J., 114–115, 173–174
Klegeris, A., 214–216
Kleipool, R. J., 14
Klimova, T. A., 294
Kloesch, B., 159
Kloosterhuis, N. J., 139–140
Knoferl, M. W., 20–21
Ko, T. H., 222
Kochanek, P. M., 20–21

Kodela, R., 152, 184–185
Koehl, M., 305
Koenitzer, J., 20–21, 29–31
Koenitzer, J. R., 20–21, 29–31
Koh, Y. H., 197, 199, 201–202
Koike, S., 40
Kokubo, K., 292, 304–305, 307–308
Kolluru, G. K., 156
Koltsova, E. K., 107–108
Kondo, K., 20
Kondo, T., 302
Kong, L., 105–106
Koningsberger, J. C., 170–172
Kopriva, S., 234, 272–274
Koritschoner, N. P., 160
Korthuis, R. J., 105–106, 110–111, 116–118
Kotani, T., 304
Kovacs, I., 238
Kovsan, J., 294
Koyama, M., 179*t*
Kraus, D. W., 20–21, 29–31, 40, 133–134, 185, 218–219
Kreimier, E. L., 29–31, 212–213, 219–220
Krenz, M., 116–117
Krishnan, N., 40, 80, 81*t*, 151
Krishnan, S., 160
Krombach, F., 102
Kruger, W. D., 148, 170–172
Kruse, C., 274–275, 276–277, 278, 280–281
Krüßel, L., 274–275, 281, 282–283, 282*f*
Krutzsch, H. C., 156–157
Krysko, D. V., 294
Ku, S. M., 217
Kubes, P., 96, 103
Kubo, S., 212–213
Kubota, M., 304
Kubota, S., 304
Kudo, Y., 295, 301
Kugler, P., 272–274
Kuhlmann, J., 276–277
Kuiper, P. J. C., 235, 272
Kumada, Y., 276–277
Kumagai, Y., 147, 161
Kumaran, S., 277
Kundu, S., 114–115, 208–212
Kupfer, A., 155
Kupfer, H., 155
Kurokawa, R., 303, 306, 307–308
Kuru, O., 293
Kusano, M., 272–275
Kusano, T., 275–276
Kutney, G. W., 67–69
Kuwano, Y., 107–108
Kuznetsov, A. V., 20–21
Kuznetsova, S. A., 156–157
Kvietys, P. R., 113–114
Kwak, M.-K., 4
Kwon, J., 59–60
Kwon, K. S., 246
Kwong Huat, B. T., 148

L

Labarre, J., 236
Laezza, F., 220–221
Lafouresse, F., 155
Lagreze, W. A., 218–219
Lahens, N. F., 32
Lahesmaa, R., 148, 150
Lai, C., 81*t*
Lallouchi, K., 5
Lam, F., 102–103
Lamattina, L., 234, 272
Lampel, M., 196–197
Lamy, L., 156–157
Lan, A. P., 218–219
Lan, Q., 148–149
Lan, Z., 140–141
Lancaster, J. R. Jr., 15, 40–41, 42
Lapatto, R., 148
Latino-Martel, P., 293–294, 295
Lau, H. Y., 150–151, 197, 199
Lavagno, L., 175–176
Le Moal, M., 305
Le Trionnaire, S., 185, 240–241
Leaderer, B., 148–149
Lee, B. S., 208–210
Lee, C., 59–60
Lee, C. P., 272–274
Lee, H.-G., 4, 10
Lee, H.-S., 196–197
Lee, J., 208–210, 212–213
Lee, J. K., 272–274
Lee, L. C., 172–173
Lee, S. R., 59–60
Lee, S. W., 240–241

Leemans, J. C., 139–140
Lefer, D. J., 5
Leffler, C. W., 5
Lehr, H. A., 98–100
Leiderer, R., 102, 113–114
Leite-Panissi, C. A., 176
Lemaire, C., 292
Lempereur, L., 293–294
Lenz, O., 290
Lerchenberger, M., 105
Leung, M., 108–109
Leuvenink, H. G., 139–140
Levin, E., 178–180
Levitt, M. D., 13, 146–147, 154, 208–210
Levonen, A. L., 148
Levy, M. M., 198
Lewandowski, K., 234
Lewis, M., 237–238, 239
Ley, K., 98–100, 101, 103, 104–105, 106–108
Li, B., 306–307
Li, D. F., 235
Li, F., 220–221
Li, G., 81*t*
Li, G. M., 302, 306
Li, H., 235
Li, J. B., 20–21, 235, 304–305
Li, L., 15, 20, 40, 150–151, 172–173, 179*t*, 180–181, 199, 212–213, 214–216, 235, 239–241, 272, 294
Li, M.-Y., 208–210, 234, 305–306
Li, P., 235
Li, Q., 15, 40–41, 42, 148–149
Li, R., 302
Li, S., 81*t*
Li, W., 103
Li, X. K., 235, 303, 305–306
Li, X. L., 132–133
Li, Y.-H., 235, 248
Li, Z., 138–139, 156–157
Li, Z. G., 235, 254–256
Li, Z. Q., 156–157
Liang, F., 135, 179*t*
Liang, X., 185–187
Liao, X., 218–219
Lieber, S. R., 274–275, 281
Lilley, K. S., 236
Lillo, R. S., 292
Lim, L., 149–150, 154, 156
Lin, C. W., 208–210
Lin, D., 4, 10
Lin, G.-H., 272
Lin, M. T., 293
Lin, Y.-T., 234
Lindermayr, C., 238
Linde-Zwirble, W. T., 198
Lipinski, B., 293
Lipke, P. N., 208–210
Lipscombe, D., 221
Lipton, S. A., 236–237, 247–248
Lisjak, M., 234
Liszt, M., 159
Liu, C., 303
Liu, D. Q., 306
Liu, H. W., 130–131, 274, 294
Liu, J., 185–187, 293, 305, 307
Liu, K., 302
Liu, P., 235, 248
Liu, Q., 4, 5, 10
Liu, S. Y., 302, 304–305
Liu, W. W., 140–141, 302, 307–309
Liu, X., 81*t*, 146–147
Liu, Y. H., 81*t*, 117–119, 130, 134, 136–138, 173, 210–212, 234, 235, 302
Liu, Z., 235, 248
Lo Faro, M. L., 159
Lobb, I., 114–115, 140–141
Lock, E. A., 218–219
Lominadze, D., 210–212
Long, E. K., 6
Lopez, A., 212–213
Lorenz, M., 105
Loscalzo, J., 114–115, 173–174
Lovato, S. J., 201
Lovell, M. A., 5
Lowicka, E., 212–213
Lu, H., 235
Lu, J., 199, 200
Lu, M., 11, 234, 235
Lu, P. D., 151
Lu, W., 234
Lu, Y., 5, 40, 170
Lu, Z. H., 108–109
Lund, J., 196
Lund, R., 148, 150

Luo, J. P., 235
Luo, X. N., 235
Luo, Y., 152, 160
Luquette, M. H., 199–200

M

Ma, D. F., 235
Ma, L., 178–180
Ma, V. Y., 200
Ma, Y., 138–139
Maarse, H., 14
Maas, F. M., 235, 272
MacAry, P. A., 201–202
Machide, M., 300–301, 304, 306–307
Macinkovic, I., 40, 46, 48*f*, 51*f*, 64–66, 81–83, 150–151, 161
Maekawa, Y., 304–305
Magee, E. A., 146–147
Maggiano, N., 217
Magierowski, M., 179*t*
Mahadev, K., 295
Mahle, K. C., 234, 241–242
Majima, H. J., 296, 299*f*
Maleki, M., 135–136, 176–178
Mali, V. R., 4–5
Malinverno, C., 153
Mallilankaraman, K., 170–172
Maltepe, E., 295
Manaenko, A., 307–308
Mancuso, C., 293–294
Mandal, C. C., 295
Maness, P. C., 290–291
Mani, S., 40, 81*t*, 235
Maniatis, T., 277–278
Manikandan, J., 199
Manko, A., 176–178
Mannino, D. M., 198
Mano, Y., 304
Mao, Z., 20–21, 25–27, 28–29, 31–32
Marchese-Ragona, R., 147
Mard, S. A., 135–136, 176–178, 183–184
Margreiter, R., 20–21
Marin, E. C., 159
Marioni, G., 147
Markesbery, W. R., 5
Marquez, J., 294
Marshall, D. C., 20
Marshall, H. E., 236–237
Marshall, J. C., 198
Martin, G. R., 150–151, 172, 178–180, 183–184
Martin, G. S., 198
Martin, W., 234–235
Martinez, E., 292
Martin-Romero, F. J., 208–210, 219–220
Marty, L., 62
Marutani, E., 150–151, 185–187
Maruyama, A., 272–274, 275–276, 280–281
Massberg, S., 102
Masuda, A., 305
Masutani, K., 309
Mates, J. M., 294
Mathai, J. C., 272–274
Mathieu, C. E., 295
Matsuda, K., 178–180
Matsuda, M., 305–306
Matsumoto, A., 236–237, 309
Matsunami, M., 176
Matsuyama, A., 276–277
Mattson, M. P., 5, 12, 214–216
Mauban, J. R., 120–121
Maulik, S. K., 210–212
Mawson, D., 235
Mazzalupo, S., 156–157
Mazzon, E., 134
McAvoy, E. F., 102–103
McCafferty, D. M., 114–115
McClintock, D. S., 294
McCormick, W., 295
McCurry, K. R., 304–305, 307–308
McDearmon, E. L., 32
McDonald, B., 102–103
McDougall, J. J., 114–115
McGeer, E. G., 214–216
McGeer, P. L., 214–216
McGowan, C. E., 196
McIntyre, J. O., 6, 10
McKnight, G. W., 172, 176–178
McKnight, W., 150–151, 172, 178–180, 182–184
McManus, B. E., 234, 241–242
Medani, M., 158–159
Meininger, C. J., 118–119, 136–137
Melendez, A. J., 199
Mempel, T. R., 103

Mencarelli, A., 114–115, 119–120, 157–159, 172–173, 176, 187–188, 208–210
Mendez, G., 20–21
Menezes, G. B., 102–103
Meng, G., 235
Meng, T. C., 236
Menger, K. E., 294
Menger, M. D., 98–100
Menini, E., 217
Menu, P., 294
Mesquita, F. P., 176, 180–181
Messmer, K., 113–114
Mestas, J., 98–100, 101, 103, 106–108
Metayer, C., 148–149
Metkar, S., 152
Metreveli, N., 210–212, 218–219
Meyer, A. J., 61–62, 63, 64, 73
Michaels, R. L., 214–216, 220–221
Michel, T., 80
Micheloni, C., 151
Mier, R. A., 105–106
Mikami, T., 301, 302, 305, 309
Mikami, Y., 40
Miles, A. M., 187–188
Miljkovic, J. L., 64–66, 150–151, 161
Miljkovic, J. Lj., 46, 49
Millen, B., 114–115
Miller, A., 235
Miller, B. H., 32
Miller, D. L., 239–240
Miller, E. R., 295
Miller, F. J., 27–28
Miller, J. M., 105, 274
Miller, T. W., 149–150, 151–152, 154, 156–157
Mills, R. W., 29–31
Mimoun, S., 172, 179*t*
Mineri, R., 149–150, 281
Mino, K., 276–277
Mirandola, P., 151, 153
Mishra, P. K., 208–212, 218–219
Missner, A., 272–274
Miyasaka, M., 112, 113
Miyashita, H., 304
Mizuno, H., 178–180
Mizuno, Y., 305
Mo, L. Q., 218
Modis, K., 20, 185
Mohammed-Ali, Z., 151
Mok, A., 140–141
Moltu, K., 152–153
Monks, C. R., 155
Monks, T. J., 218–219
Montfort, W. R., 156–157
Montine, T. J., 6, 10
Moochhala, S. M., 5, 116, 150–151, 197, 199, 200, 201–202
Moody, T. W., 152
Moore, K. P., 64–66, 148
Moore, P. K., 5, 12, 15, 20, 40, 116, 148, 150–151, 172–173, 197, 199, 201–202, 212–213, 219–220, 248, 272
Mora, J. R., 103
Morey, J. S., 150
Morgan, B., 112–113
Morhard, R. C., 20–21
Mori, A., 293, 305
Mori, T., 302, 307–308, 309
Moriga, Y., 276–277
Morizane, S., 294
Morris, J. C., 61, 64–66
Morrison, J. P., 20–21, 25–27, 28–32, 40, 133–134, 185, 218–219
Morrison, L. D., 5
Morrison, M., 20–21, 31–32
Mortenson, L. E., 290
Moss, M., 198
Motl, N., 20, 40, 58
Motta, J. P., 178, 179*t*
Motterlini, R., 291
Moulin, F. J., 234, 241–242
Moulton, V. R., 160
Moy, A. B., 111
Mu, W., 40, 80–83, 81*t*, 155–156
Mueller, A. L., 214–216
Mueller, C., 272–274, 276–277
Mueller, E. G., 86
Muellner, M. K., 5, 7*f*, 8*f*, 9*f*, 11*f*, 12*f*, 14*f*, 15, 15*f*
Mukaihara, S., 304–305
Muller, S. J., 274–275, 276–277, 278
Mullins, M. E., 80
Mullowney, C. E., 214–216
Munglani, R., 220–221
Munoz, L. E., 64–66, 150–151, 161
Muñoz, L. E., 46

Murai, K. K., 214–216
Murata, Y., 307–308
Murghes, B., 20
Murohara, T., 20
Muroi, E., 158
Murphy, M. P., 236, 294, 298–300
Murray, M. C., 152
Murray, T. V., 294
Muscara, M. N., 170
Musiani, P., 147
Mustafa, A. K., 40, 80–83, 81*t*, 86, 155–156
Mustafi, S. B., 294
Myers, M. P., 236
Myers, S. J., 220–221

N

Nagahara, N., 148
Nagai, T., 304
Nagai, Y., 208–210, 214–216, 217
Nagano, M., 148
Nagano, T., 296, 299*f*
Nagata, K., 301, 302, 305, 309
Nagatani, K., 302
Nagpure, B., 235
Nagy, A., 5, 6–7, 40, 64–66, 73, 179*t*
Nagy, P., 5, 6–7, 40, 59–60, 61, 64–66, 67, 71–72, 73, 147, 161
Nakabeppu, Y., 309
Nakahara, T., 305
Nakai, J., 120–121
Nakamura, H., 147, 154
Nakamura, M., 234–235
Nakamura, T., 236–237, 247–248, 275–276
Nakanishi, K., 276–277
Nakano, A., 103
Nakao, A., 292, 293, 308–309
Nakashima-Kamimura, N., 301, 302, 305, 309
Nakatsuka, A., 307
Nakaya, H., 309
Nakayama, K., 307–308
Nakayama, M., 307–308
Nambiar, M. P., 160
Namekata, K., 208–210
Nascimento, D. C., 116
Nath, C., 214–216, 218–219
Nath, N., 152
Natividad, J., 179*t*
Nava, R. G., 103
Navarrete, M., 214–216
Nawashiro, H., 302
Nedergaard, M., 214–216
Neeter, R., 14
Neisi, N., 135–136, 176–178
Nelson, D., 219–220
Neo, K. L., 240–241
Neoptolemos, J., 197
Nevalainen, O., 148, 150
Ney, L., 102
Ng, H., 235
Ng, L. F., 239–240
Ng, S. W., 150–151, 199, 201–202
Ng, Y. K., 208–210
Nguyen, L. N., 214–216
Nguyen, N. H., 199
Nguyen, Q. D., 208–213
Nicholls, P., 20
Nicholson, C. K., 20
Nicholson, M. L., 140
Niki, E., 307
Nikiforova, V., 262
Nikolova, D., 293–294, 295
Ning, Y., 307
Nishijima, Y., 295, 301, 304–305
Nishimaki, K., 291, 292, 295, 296, 296*f*, 298–300, 298*f*, 301, 302, 303*f*, 304–307, 309
Nisonoff, A., 147
Nissim, A., 159
Niu, X. F., 96
Noble, J. E., 262
Noda, K., 304–305
Noda, M., 309
Noelle, R. J., 147
Noever de Brauw, M. C., 14
Noguchi, H., 178–180
Noji, M., 272–276, 280–281
Nolz, J. C., 156
Norris, E. J., 199
Nose, K., 302
Notet, V., 20–21
Nourshargh, S., 94–98, 105
Novoa, I., 151
Nubel, T., 170
Nuss, G. W., 201
Nyirenda, M., 237–238, 239

O

O'Connell, P. R., 158–159
Ogasawara, Y., 5, 13, 40, 208–210, 212–213
Ogawa, F., 158
Ogier, V., 20–21
Oglesbee, D., 58–59, 61–62
Oh, G. S., 208–210
Oharazawa, H., 300–301, 304, 306–307
Ohno, M., 305
Ohsawa, I., 291, 292, 295, 296, 296*f*, 298–300, 298*f*, 300*f*, 301, 302, 303*f*, 304–307, 309
Ohta, S., 291, 301, 302, 303*f*, 305–306, 307, 308*f*, 309
Oikawa, A., 272–275
Oka, J., 40, 208–210, 214–216, 217
Okamoto, S.-I., 236–237, 247–248
Okazawa, H., 208–210
Okui, T., 176
Olah, G., 210–212
Oldenburg, B., 170–172
Olson, K. R., 29–31, 64–66, 146–147, 170, 184–185, 212–213, 219–220, 292
O'Neil, J. J., 27–28
O'Neill, L. A., 175–176
Ongini, E., 180–181
Ono, H., 295, 301, 304–305
Orlandi, S., 114–115, 119–120, 157–159, 172–173, 176, 187–188, 208–210
Orrico, M. I., 116
Ortiz, D., 105
Osada, H., 302
Osborne, J. A., 80
Ostrowki, R. P., 307–308
Osumi, K., 40
Otani, N., 302
Oter, S., 210–212
Otsuka, H., 20
Ottaviano, G., 147
Otterbein, L. E., 291
Ou, D., 130
Ouhara, K., 306
Ouyang, Q., 172–173
Ovadia, H., 294
Owen, C. S., 20–21
Owen, H. A., 274–275, 281
Ozeki, Y., 80–81

P

Pae, H. O., 208–210
Palaniyandi, S. S., 4–5
Palau, N., 160
Pálinkás, Z., 5, 6–7, 40, 59–60, 61, 64–66, 67, 71–72, 73, 147, 161, 179*t*
Pambianco, M., 151, 153
Pan, J., 40–41, 48*f*, 51*f*, 81–83
Pan, S., 138–139
Panda, S., 32
Pang, Y., 208–210
Papapetropoulos, A., 14–15
Papenbrock, J., 262
Pappin, D. J., 40, 80, 81*t*, 151
Pardue, S., 146–147
Parfenova, H., 5
Park, C.-M., 40, 48*f*, 51*f*, 81–83, 308–309
Parker, E. C., 292
Parks, D. A., 112–113
Parpura, V., 217
Pasini, E., 293
Pasquale, E. B., 214–216
Pastor-Barriuso, R. III., 295
Patel, H. D., 20–21, 29–31
Patel, M., 140
Patel, R. P., 20–21, 25–27, 28–32
Pattillo, C. B., 146–147
Paul, B. D., 40, 80, 81–83, 81*t*, 86, 150–151, 155–156
Pauleau, A. L., 62
Paulsen, C. E., 40
Pavalko, F. M., 155
Payne, A. H., 161–162
Pederzoli-Ribeil, M., 175–176
Peer, F., 113–114
Peery, A. F., 196
Peh, M. T., 239–240
Peiper, S. C., 108–109
Peiser, G. D., 274
Pendrak, M. L., 156–157
Peng, Z. F., 12, 219–220
Penicaud, L., 295
Perea, G., 214–216
Perluigi, M., 5
Perretti, M., 115–116, 175–176
Perrino, E., 146–147

Perry, A., 178–180, 185, 240–241
Perry, G., 4, 5, 10
Perry, S. F., 170
Peskar, B. M., 178–180
Peter, E. A., 146–147, 154
Peters, N. C., 102–103
Petersen, D. R., 4
Petri, B., 103
Peyton, K. J., 117–118
Pfeil, I., 150
Philbert, M., 218–219
Pickard, J., 104–105
Picklo, M. J., 6, 10
Pietrobon, D., 221
Pirker, K. F., 179*t*
Pittman, K., 102–103
Pizarro, A., 32
Polhemus, D. J., 20
Poole, D., 113
Popp, J., 6
Porter, W. R., 292
Pospelova, T. I., 148–149
Pospieszalska, M. K., 98–100, 101, 103, 106–108
Postlethwait, E. M., 20–21, 25–29, 31–32
Potier, L., 154
Potter, J., 148, 153–154
Prabhakar, N. R., 234–235
Prasad, P., 108–109
Pratico, D., 170–172
Prior, M. G., 212–213
Proebstl, D., 105
Pryor, W. A., 22–23
Puig, L., 160
Puneet, P., 199
Puthia, M. K., 200
Puukila, S., 40, 81*t*

Q

Qabazard, B., 239–240
Qi, J., 40, 170–172
Qi, Q. Y., 132–133
Qi, X., 185–187
Qi, Y., 208–210
Qian, L., 306–307
Qin, J., 212–213, 221
Qin, Z., 212–213
Qipshidze, N., 114–115, 210–212
Qu, C., 81*t*
Qu, K., 212–213, 219–220
Quinn, P. G., 161–162

R

Radermacher, P., 20–21
Radi, E., 305
Radi, R., 46–47, 293, 307–308
Ragazzoni, E., 217
Rahner, C., 294
Rai, S., 214–216, 218
Raina, A. K., 5
Raivio, K. O., 148
Rajasekaran, N. S., 294
Ramakrishnan, S., 146–147
Ramanathan, S., 156–157
Ramirez, M. V., 272–274
Ramnath, R. D., 199, 212–213, 214–216
Ransom, B., 214–216
Ransy, C., 20, 234–235, 241–242
Rao, F., 40
Rao, G. S., 280–281
Rauth, A. M., 108–109
Ray, J., 28–29
Read, K., 140
Redington, A. N., 20–21
Reed, D. J., 148, 153–154
Reed, T., 5
Reiffel, A. J., 142
Reiffenstein, R. J., 6, 208–210, 212–213
Reiner, J., 4
Ren, Y. K., 218–219
Renga, B., 176, 187–188, 208–210
Reuter, S., 293
Rey, F. E., 158–159
Rezessy-Szabo, J. M., 208–213
Rhee, S. G., 59–60, 246
Richardson, C. J., 146–147
Ridnour, L. A., 151–152
Riemenschneider, A., 262
Riemersma, R. A., 295
Rigoulet, M., 295
Rikitake, M., 306, 307–308
Rinaldi, L., 151
Risley, E. A., 201
Ristow, M., 294
Rivers-Auty, J., 150–151, 197

Rizzo, G., 114–115, 119–120, 157–158, 173
Rizzo, M. A., 120–121
Roberts, D. D., 151–152, 156–157
Rodarte, C., 111
Roderick, S. L., 276–277
Rodriguez-Iturbe, B., 293
Rodriguez-Manas, L., 293
Roediger, F., 12
Rogers, N. M., 156–157
Rohwerder, T., 283
Rojiani, A. M., 214–216
Romeo, M. J., 156–157
Romero, L. C., 234–235, 272, 274
Romero, P., 154
Rose, P., 40, 151, 272
Rose, R., 140
Ross, A. B., 307, 309
Rossi, R., 146–147
Rossoni, G., 146–147, 172–173
Rostain, J. C., 292
Roth, M. B., 20–21, 31–32, 239–240
Roth, S., 154
Roth, S. H., 6
Rother, M., 274–275, 276–277, 278
Rothman, S. M., 214–216, 220–221
Rovira, I. I., 294
Rowley, D., 198
Rudich, A., 294
Ruffet, M. L., 276–277
Ruifrok, W. P., 134–135
Rujescu, D., 208–210
Ruppert, T., 274–275, 276–277, 278, 280–281
Russ, P. K., 102–103, 108–109
Russell, M. J., 234–235
Russo, G., 176, 187–188
Ryan, B., 159
Ryan, J. C., 150

S

Saadat, M., 153
Saeed, A., 13, 146–147, 154, 208–210
Sagi, M., 153
Saito, K., 272–276, 280–281
Sakaguchi, M., 185–187
Sakamoto, C., 178–180
Sakamoto, M., 295, 301
Sakiyama, T., 276–277
Saksela, M., 148
Salganik, R. I., 295
Salmeen, A., 236
Salsi, E., 276–277
Salto-Tellez, M., 20, 150–151
Saluja, A. K., 196–197
Samami, A. A., 264, 276–277, 278
Sambrook, J., 277–278
Samhan-Arias, A. K., 208–210, 219–220
Sampaio, A. L., 175–176
Sampath, R., 155
Samstag, Y., 62
Sanada, S., 304–305
Sanchez-Madrid, F., 155
Sand, W., 283
Sang, H., 305–306
Sano, H., 234–235, 275–276
Sano, M., 294, 300, 300*f*, 304–305
Santagada, V., 119–120, 158–159, 172–173, 176–181, 183–184
Santos, K. T., 176, 180–181
Sanz, M. J., 96
Sanzgiri, R. P., 217
Saparov, S. M., 272–274
Sarelius, I. H., 108–109
Sarfeh, I. J., 178–180
Sasaki, H., 304–305
Sasakura, K., 185–187
Sathnur, P. B., 208–212, 218
Sato, B., 303, 306, 307–308
Sato, K., 306
Sato, N., 147, 154
Sato, Y., 292, 307–308
Sauvage, F. X., 71
Savani, R. C., 102–103
Sawa, A., 237–238
Sawa, T., 147, 161
Saxena, G., 218–219
Sayre, L. M., 4, 5, 10
Scallan, J., 110–111
Scelo, G., 148–149
Schallner, N., 218–219
Schantz, J.-T., 248
Schapira, A. H., 305
Schaur, R., 4, 6, 14
Schaur, R. J., 4, 15
Schelzig, H., 210–212

Scherrer-Crosbie, M., 20, 32
Scheuerle, A., 141
Schmid-Schonbein, G. W., 112, 113
Schmidt, A., 262, 272–274, 275–276, 280–281
Schnackerz, K. D., 280–281
Schoenfeld, M. P., 293
Scholz, N., 4–5, 6–7, 14
Schook, A. B., 32
Schreier, S. M., 5, 7*f*, 8*f*, 9*f*, 11*f*, 12*f*, 14*f*, 15, 15*f*
Schultz, T. W., 4, 14
Schumacker, P. T., 294, 295
Schwarzländer, M., 272–274
Schwer, C. I., 218–219
Sciaky, N., 155
Scimone, M. L., 103
Scott Isbell, T., 29–31
Scuderi, M. R., 293–294
Searles, R., 20, 32
Sechi, A. S., 155
Secundino, N., 102–103
Sediari, L., 176, 187–188, 208–210
Sedlak, T. W., 237–238
Segura, J. A., 294
Seike, T., 305
Sekiguchi, F., 176
Sekiya, J., 262
Seleci, D., 293
Sen, N., 40, 80–83, 81*t*, 86, 155–156
Sen, T., 40, 81–83, 81*t*, 86
Sen, U., 208–212
Sener, A., 140–141
Seo, T., 306, 307–308
Serhan, C. N., 175–176, 185–187
Setsukinai, K., 296, 299*f*
Shah Masood, W., 135, 179*t*
Shah, S. H., 146–147
Shahani, N., 237–238
Shan, C. J., 235
Shang, Y., 307
Shen, M., 139
Shen, W.-B., 234, 235
Shen, X. G., 146–147, 154, 156
Shen, Z. M., 208–210
Shi, D., 308–309
Shi, J., 308–309
Shi, M. M., 81*t*
Shibuya, N., 5, 13, 20, 40, 170–172, 208–210
Shigemura, N., 304–305, 309
Shigenaga, M. K., 293, 305
Shigetomi, E., 214–216
Shimamura, K., 148
Shimizu, K., 158
Shimmura, S., 304
Shimomura, I., 305–306
Shimouchi, A., 302
Shin, M. C., 199
Shinbo, T., 292, 307–308
Shinmura, K., 294, 300, 300*f*, 304–305
Shirai, M., 302
Shore, A. C., 159
Shrivastava, P., 201–202
Shuhendler, A. J., 108–109
Shui, B., 120–121
Shukla, R., 214–216, 218–219
Siau, J. L., 11
Sibille, E., 208–210
Sidhapuriwala, J. N., 150–151, 201–202
Sikka, G., 40, 81*t*
Silva, C. G., 159
Silva, M. J., 306
Silver, I. A., 219–220
Simon, D. I., 80
Simon, F., 141, 210–212
Simon, M. C., 295
Simonetti, R. G., 293–294, 295
Sio, S. W., 150–151, 200, 201–202
Sipes, J. M., 156–157
Sitkovsky, M., 198
Sivarajah, A., 134
Skinner, K., 235
Skovgaard, N., 29–31, 212–213, 219–220
Slaba, I., 102–103
Slavin, J., 197
Smith, C. G., 148
Smith, D. D., 5
Smith, K. J., 150
Smith, M. A., 4, 5, 10
Smith, R. A., 298–300
Smoliakova, I., 6
Snellinger, A. M., 86
Snijder, P. M., 134–135, 139–140
Snowman, A. M., 40

Snyder, S. H., 40, 80, 150–151, 155–156, 222, 236–237, 238
Sobocki, T., 152
Sobotta, M. C., 112–113
Sodha, N. R., 134, 210–212
Sofroniew, M. V., 214–216
Soldato, P. D., 218–219
Solgi, G., 135–136, 176–178
Song, G., 305–306
Song, W. J., 235
Song, X., 208–210
Soriano, R. N., 176
Souto, F. O., 116
Souto-Carneiro, M. M., 159
Souza, G. R., 176
Spahn, J. H., 103
Sparatore, A., 146–147, 152, 172–173, 218–219
Spector, J. A., 142
Sperandio, M., 104–105
Speroni, F., 276–277
Spiller, F., 116
Spiteller, G., 4
Spiteller, P., 4
Sponne, I., 20–21
Sponzilli, I., 153
Squadrito, G., 22–23
Squadrito, G. L., 27–28, 29–31
Sriram, K., 105
Srivastava, N., 234
Stadtman, E. R., 305
Staffieri, C., 147
Stahl, B., 141
Stamler, J. S., 80, 156, 236–237, 238, 247–248
Stanicka, J., 112–113
Stark, K., 105
Statile, J. L., 148
Stein, A., 20–21, 25–27, 28–29, 31–32
Stein, E. V., 149–150, 154, 156
Steinhubl, S. R., 293–294
Steinkellner, H., 5, 7*f*, 8*f*, 9*f*, 11*f*, 12*f*, 14*f*, 15, 15*f*
Steinlechner-Maran, R., 20–21
Steppan, J., 40, 81*t*
Stezoski, J., 20–21
Stokes, C. C., 214–216
Straume, M., 32
Struve, M. F., 234, 241–242
Stuiver, 235
Su, A. I., 32
Subramaniam, R., 12
Suematsu, M., 112, 113, 114, 291, 292
Sugimoto, R., 292, 304–305, 307–308
Sumagin, R., 108–109
Sun, H., 111, 217
Sun, J., 197
Sun, L. M., 148–149
Sun, Q., 302, 307–309
Sun, X., 138–139, 306, 307–309
Sun, X. J., 302, 306
Sundd, P., 98–100, 101, 103, 106–108
Sunico, C. R., 236–237, 247–248
Surette, M. G., 178
Sutter, T. R., 4
Suzuki, E., 208–210
Suzuki, H., 112–113, 148, 300–301, 304, 306–307
Suzuki, I., 281
Suzuki, M., 304–305
Suzuki, Y., 305
Svoboda, K., 103
Swarnkar, S., 214–216
Switzer, C. H., 152
Syer, S., 180–181, 182–183
Szabó, C., 14–15, 20, 40, 148–149, 178–180, 185, 210–212, 240–241, 291
Szczesny, B., 178–180, 185, 240–241
Szinicz, L., 234–235

T

Taber, S. W., 108–109
Tachibana, S., 304–305
Tachibana, T., 309
Tada, T., 275–276
Tai, C. H., 234–235
Takagi, T., 280–281
Takahashi, H., 272–274
Takahashi, K., 179*t*, 291, 292, 295, 296, 296*f*, 298–300, 298*f*, 301, 304–305, 306–307
Takaki, A., 307
Takanashi, M., 305
Takano, H., 214–216
Takashima, S., 208–210
Takasura, S., 179*t*
Takeshita, H., 153
Takeuchi, K., 176–178

Takeuchi, S., 302
Tallini, Y. N., 120–121
Tamaki, K., 300, 300*f*, 304–305
Tamaki, N., 307
Tamatani, T., 112
Tamizhselvi, R., 197, 199, 201–202
Tan, C. H., 20, 150–151
Tan, G., 138, 152, 160
Tanaka, M., 5, 208–210, 305
Tanaka, Y., 304–305, 309
Tang, H., 148–149, 152, 160
Tang, J., 235, 307–308
Tang, X. Q., 218–219
Tanizawa, K., 208–210
Tanriverdi, H., 293
Tanriverdi, S., 293
Tao, B. B., 155–156
Tao, H., 302
Tao, L., 20, 29–31, 40, 133–134, 185, 218–219
Tarr, J. M., 159
Tartakovsky, D. M., 105
Tas, A. C., 14
Tasken, K., 152–153
Taylor, E., 159
Taylor, J. D., 208–210, 212–213
Tazzari, V., 146–147, 152
Teixeira, S. A., 176, 180–181
Teklic, T., 234
Tempst, P., 80, 236–237, 238, 294
ten Tije, A., 108–109
Tenbrock, K., 160
Teng, X., 29–31
Thomas, D. D., 151–152
Thompson, C. B., 294
Thornton, J. D., 274–275, 281, 282–283, 282*f*
Throm, C., 272–274, 276–277
Tiacci, E., 150
Tian, M., 302, 306
Tian, X. L., 235
Tiberti, S., 293
Ticchioni, M., 156–157
Ticha, A., 146–147
Tiranti, V., 149–150, 281
Tirniceriu, A., 105
Tiveron, C., 149–150, 281
Togawa, T., 5, 208–210
Tokuda, K., 150–151
Toledano, M. B., 236
Tomei, C., 292
Tomura, S., 302
Tonks, N. K., 40, 80, 81*t*, 151, 236
Toohey, J. I., 67–69
Torii, Y., 275–276
Tormos, K. V., 294
Tota, S., 218–219
Tóth, I., 5, 6–7, 40, 64–66, 73
Touvier, M., 293–294, 295
Townsend, S. R., 198
Toyoda, Y., 293, 308–309
Tringali, G., 217
Trionnaire, S., 178–180, 185
Tristan, C., 237–238
Trzyna, W. C., 294
Tsai, A. G., 161–162
Tschopp, J., 294
Tseng, G. C., 208–210
Tseng, M., 108–109
Tsien, J. Z., 220–221
Tsien, R. Y., 58–59, 61–62
Tsokos, G. C., 160
Tsuchiya, Y., 147, 161
Tsugane, M., 40, 208–210, 214–216, 217
Tsukagoshi, H., 294
Tsung, A., 304–305, 307–308
Tu, C., 130
Tu, S., 236–237, 247–248
Turnbull, K., 67–69
Turrens, J. F., 293
Twiner, M. J., 150
Tyagi, N., 114–115, 210–212, 218–219
Tyagi, S. C., 210–212, 218–219

U

Uchida, K., 305
Uchida, T., 178–180
Udoh, U. S., 32
Ueda-Taniguchi, Y., 147, 154
Uhl, B., 105
Ullian, E. M., 214–216
Umemura, K., 13, 208–210, 212–213, 219
Unnikrishnan, S., 104–105
Untereiner, A., 235
Urano, Y., 296, 299*f*
Uttamchandani, M., 199

V

Vacek, J. C., 114–115, 210–212
Vadhana, P., 146–147
Vagianos, K., 170–172
Vairano, M., 217
Vajkoczy, P., 98–100
Valitutti, S., 147
van, Berkel-Arts, A., 290–291
van den Born, J. C., 134–135
van der Griend, R., 170–172
van Dijk, C., 290–291
van Dijk, M. C., 139
Van Dolah, F. M., 150
van Noesel, C. J., 150
vanBerge-Henegouwen, G. P., 170–172
Vandenabeele, P., 294
Vander, Heiden, M. G., 294
Vandiver, M. S., 40
Vasas, A., 5, 6–7, 40, 64–66, 73
Vasconcelos, Z., 155
Vaziri, N. D., 293
Velsor, L. W., 22–23
Verdial, F. C., 29–31, 212–213, 219–220
Verdu, E. F., 170–172, 179*t*
Verfaillie, T., 294
Verri, W. A., Jr., 176
Vestweber, D., 102
Vicente-Manzanares, M., 155
Vieira, S. M., 176
Villamaria, 142
Viner, N., 248
Visca, P., 31–32
Viscomi, C., 149–150, 155–156, 281
Visioli, O., 293
Vistoli, G., 4–5, 14–15
Vitale, M., 151
Vogt, J., 141
Voisin, M. B., 94–98, 105
Völkel, S., 282–283
Volle, C. B., 307
Vollmar, B., 98–100
Volpato, G. P., 20, 32
von Andrian, U. H., 103
von Bruhl, M.-L., 105
Vong, L., 150–151, 172, 175–176, 178–180, 182–184
Voronina, E. N., 148–149
Voropaeva, E. N., 148–149
Vreeswijk-Baudoin, I., 134–135
Vulesevic, B., 212–213

W

Wabnitz, G. H., 62
Wachter, U., 141
Wada, J., 307
Wada, K., 178–180
Waeg, G., 12
Wagner, K., 20–21
Wakabayashi, N., 309
Walker, C., 295
Wallace, J. L., 114–116, 119–120, 150–151, 157–159, 170, 172–173, 176–181, 179*t*, 182–184, 184*f*, 186*f*, 187–188, 210–212, 214–216
Walsh, C. T., 274
Walsh, J. J., 217
Wang, D., 138–139
Wang, E. A., 116–117, 149–150, 154, 156
Wang, F., 139, 304–305
Wang, G. L., 306
Wang, H., 178–180
Wang, J. F., 235, 304–305
Wang, L., 176
Wang, M. J., 117–119, 136–138, 155–156, 185–187, 218
Wang, P., 159–160
Wang, Q., 117–118, 137–138
Wang, R., 5, 20, 29–31, 40, 81*t*, 114, 116–117, 146–147, 154, 159–160, 170, 172, 234, 235
Wang, R. U. I., 5
Wang, S. H., 218, 235
Wang, T. T., 274
Wang, W. N., 306
Wang, W.-H., 272
Wang, X. Y., 173, 218–219, 293, 305
Wang, Y. W., 130–131, 132–133, 235, 292, 304–305, 307–308
Wang, Z. J., 199, 212–213, 214–216, 307–309
Wardclaw, A. C., 198
Warek, U., 272–274
Warenycia, M. W., 208–210, 212–213
Warke, V. G., 160
Warrilow, A., 274–276
Wartelle, C., 71
Waseem Akhtar, M., 236–237, 247–248

Watanabe, M., 272–275, 291, 292, 295, 296, 296*f*, 298–300, 298*f*, 301, 304–305, 306–307
Watanabe, Y., 147, 161
Waterhouse, C. C., 102–103
Wawrousek, K., 290–291
Way, J. L., 31–32
Webb, G. D., 234, 235
Wedmann, R., 46, 64–66, 150–151, 161
Wehland, J., 155
Wei, J., 131–132
Wei, Y., 307
Wei, Z.-J., 234, 235, 248
Weij, M., 139
Weiner, A. S., 148–149
Wells, J. V., 147
Wen, C., 218
Wenger, R. H., 295
Wenk, G. L., 214–216
Wesenberg, D., 274–275, 276–277, 278
West, A. P., 294
Westerblad, H., 293
Whatmore, J. L., 178–180, 185, 235, 240–241
Whiteford, J., 105
Whiteman, M., 6, 11, 12, 14–15, 20, 148, 150–151, 159, 179*t*, 218, 219–220, 235, 236, 240–241, 247–248, 254–256
Whitfield, N. L., 29–31, 212–213, 219–220
Wichterman, K. A., 198
Wier, W. G., 120–121
Wilson, D. F., 20–21
Wilson, I., 234
Wilson, K. E., 156–157
Wilson, L. G., 262
Wilson, M. T., 20
Wink, D. A., 151–152, 293
Winter, C. A., 201
Winter, D. C., 158–159
Winterbourn, C. C., 112–113, 237–238
Winyard, P. G., 159, 178–180, 185, 240–241
Wirtz, M., 272–275, 276–277, 278, 280–281, 282–283, 282*f*
Witko-Sarsat, V., 175–176
Wofsy, D., 159
Wohlgemuth, S. E., 148
Wolf, A., 276–277
Wong, B.-S., 6, 11, 179*t*, 218, 219
Wong, C. H., 103
Wong, F. L., 150–151, 197
Wong, P. T.-H., 11, 208–210, 212–213, 219–220, 234, 235
Woo, D. K., 294
Wood, J. L., 67–69, 70–71
Wood, K. C., 309
Wood, M. E., 248
Woodfin, A., 105
Woolley, J. F., 112–113
Workentine, M. L., 178, 179*t*
Wright, C. M., 86
Wu, F.-H., 272
Wu, J. F., 131–132, 235, 246, 248
Wu, L., 40, 81*t*, 170–172, 235
Wu, M. H., 111, 158–159
Wu, W. S., 148–149
Wu, X. Y., 20–21, 108–109
Wu, Z.-Y., 11, 81*t*

X

Xian, M., 40–41, 42, 185–187
Xiao, J., 148–149, 307–309
Xiao, L., 218–219
Xiao, Y., 130–131
Xie, H., 235
Xie, K. L., 306
Xie, L., 81*t*
Xie, X. H., 132–133
Xie, Z. Z., 81*t*
Xu, L., 139
Xu, R., 40, 81–83, 81*t*, 86
Xu, S., 235, 248
Xu, W., 111, 221, 302, 307
Xu, Z., 306

Y

Yamabe, T., 147, 154
Yamada, H., 305
Yamada, K., 304
Yamafuji, M., 309
Yamagata, K., 291, 292, 295, 296, 296*f*, 298–300, 298*f*, 301, 304–305, 306–307
Yamaguchi, H., 305
Yamaguchi, Y., 214–216, 275–276
Yamamoto, T., 291, 292
Yamanoue, T., 276–277
Yamazaki, S., 304–305

Yan, J., 160
Yan, X., 135, 179*t*
Yan, Z., 154
Yanaba, K., 158
Yanagi, K., 185
Yang, C. T., 152, 160, 218–219
Yang, G., 40, 81*t*, 151, 159–160, 170–172, 235
Yang, J., 208–210
Yang, K. S., 59–60
Yang, L., 235
Yang, R., 81*t*, 304–305
Yang, S. F., 274
Yang, W., 40, 170–172, 307
Yang, X., 218
Yang, Y. X., 114–115, 118–119, 136–137
Yang, Z., 212–213, 218–219, 221
Yao, M., 156–157
Yao, S., 305–306
Yarbrough, J. W., 4, 14
Yasin, M., 134
Yasojima, K., 214–216
Yasunaka, T., 307
Yazdi, A. S., 294
Ye, Y. K., 235
Ye, Z., 308–309
Yee, T. T., 112
Yennie, C. J., 307
Yeo, H. C., 293, 305
Yeum, K.-J., 4–5, 14–15
Yi, H., 280–281
Yin, J., 130
Yokota, T., 300–301, 304–305, 306–307
Yoritaka, A., 305
Yoshida, A., 304–305
Yoshida, M., 5, 208–210
Yoshizaki, A., 158
You, X., 131–132
Young, M., 32
Youssefian, S., 234–235
Yu, B., 20, 32
Yu, C., 173
Yu, G., 304–305
Yu, J., 290–291
Yu, P., 307–308
Yu, Y. S., 308–309
Yuan, L., 274–275, 281
Yuan, Y., 105–106, 111, 185–187
Yusof, M., 116–118
Yusuf, M., 148
Yutsudo, N., 305

Z

Zadak, Z., 146–147
Zakrajsek, J. F., 293
Zanardo, R. C., 114–116, 119–120, 157–158, 173, 175*f*, 179*t*
Zanoni, C. I., 176, 180–181
Zarbock, A., 98–100, 101, 103, 106–108
Zarse, K., 294
Zawieja, D. C., 105–106, 111
Zayachkivska, O., 179*t*, 183–184
Zeidel, M. L., 272–274
Zeng, F., 152, 160
Zeng, H., 151
Zeng, Q. T., 302, 306
Zeng, S., 307–308
Zeviani, M., 155–156
Zhai, X., 308–309
Zhan, J., 304–305, 307–308
Zhan, Y., 307–308
Zhang, C., 146–147
Zhang, D., 40–41, 42, 48*f*, 51*f*, 81–83
Zhang, G., 20–21, 159–160
Zhang, H., 116, 197, 199, 201–202, 234, 235
Zhang, J. H., 5, 32, 40, 120–121, 150–151, 170, 200, 302, 307
Zhang, L. J., 155–156, 305–306
Zhang, M. H., 308–309
Zhang, S. L., 235
Zhang, W. W., 146–147
Zhang, Y., 148–149, 151, 236–237, 238, 247–248, 307–309
Zhang, Z. J., 234
Zhao, B., 146–147
Zhao, C., 276–277
Zhao, H., 208–210
Zhao, J., 130, 208–210
Zhao, K., 40, 81*t*
Zhao, L., 306–307
Zhao, M., 138
Zhao, W., 5, 40, 170
Zhao, X. L., 235
Zhao, Y. Z., 235
Zheng, C.-J., 272

Zheng, D. D., 218
Zheng, J.-L., 235, 248
Zhi, L., 116, 199, 201–202
Zhou, H., 307
Zhou, J., 306
Zhou, R., 294
Zhou, W., 146–147, 218
Zhou, Y., 81*t*
Zhu, H., 131–132
Zhu, J. X., 114–115, 141
Zhu, W., 208–210
Zhu, W. J., 307–308
Zhu, X., 131–132
Zhu, Y. C., 155–156, 212–213, 214–216
Zhu, Y.-Z., 11, 14–15, 199, 212–213, 214–216
Zhu, Z., 306
Zhuang, Y. Y., 218–219
Zimmer, A. D., 274–275, 276–277, 278
Zinselmeyer, B. H., 103
Zollner, H., 4–5, 6–7, 14
Zuchtriegel, G., 105
Zuidema, M. Y., 117–119, 136–138
Zweifach, B. W., 113

SUBJECT INDEX

Note: Page numbers followed by "*f*" indicate figures and "*t*" indicate tables.

A

Acidic ninhydrin reagent, 260
Adenosine-5′-phosphosulfate (APS), 272–274
Adenosine triphosphate (ATP), 173, 174*f*
Adhesion molecule expression. *See* Chemokine expression
Air Liquide America Specialty Gases LLC, 21–22
Alexa-fluor 488 Gr1 antibody, 102–103
Antigen-presenting cells (APCs), 154
Anti-inflammatory actions
  acute and early phase preconditioning, 116
  after NaHS treatment, 116–117
  anti-adhesive effects, 116
  AnxA1 knockout mice, 115–116
  ATP-sensitive potassium (K_{ATP}), 115–116
  BKCa channels, 117–118
  CBS gene expression, 114–115
  CSE inhibition, 114–115
  eNOS knockout mice, 117–118
  exerts effects, 114
  heme oxygenase-1, 118–119
  intravital microscopy, 114–115
  leukocyte rolling and adhesion, 118–119
  mechanisms, 173–176
  3-mercaptopyruvate sulfurtransferase, 114
  molecular signaling mechanisms, 115–116
  NaHS preconditioning studies, 119
  neurovascular permeability, 114–115
  nonadhesive endothelial surface, 114–115
  pharmacologic HO inhibition, 118–119
  p38 MAPK inhibitors, 117–118
  septic mice, 116
  signaling mechanisms, 118–119, 118*f*
  treatment, 116
  vascular smooth muscle and endothelial cells, 114
  vasorelaxant effects of NO, 116–117
  WT animals, 115–116
Anti-inflammatory effects, H_2S
  angiogenesis, 173, 174*f*
  annexin-1, 175–176
  ATP, 173, 174*f*
  CSE, 173, 175*f*
  ICAM, 173
  leukocytes, 173
  macrophages, 175–176
  NaHS, 175–176
  vascular endothelium, 173, 175*f*
  visceral pain, 173, 174*f*
Anti-RP-1 antibody, 102–103
APS. *See* Adenosine-5′-phosphosulfate (APS)
ATP. *See* Adenosine triphosphate (ATP)

B

Bicinchoninic acid (BCA), 246
Biotin-HPDP, 84*t*
Biotin switch assay
  S-nitrosylation, 236–237
  thiols, 238
Biotinylation solution, 84*t*
Bis-Tris NUPAGE gel, 87
Blocking buffer, 84*t*
Bovine serum albumin (BSA), 246
  BSA–SOH, BSA–SSG and BSA–SSH, 46–47
  materials, 46
  mechanism, protein S-sulfhydration, 49
  MS and dot blot, 47–49
Brain synapses
  allicin, 210–212
  astrocytes, 214–216
  CBS, 208–210
  CNS, 212–214
  CSE, 208–210
  cytosol, 208–210
  dietary garlic, 210–212
  excitotoxic cell death, 214–216, 215*f*
  GABAergic synapse, 208–210
  GABA mediation, 221

Brain synapses (*Continued*)
glial cells, 217
gliotransmitters, 214–216
glutathione production, 214–216, 215*f*
LTP, 208–210
memory formation, 216–217
microglia, 214–216
3MST, 208–210
neurodegenerative disorders, 214–216
neuromodulator, 210–212
neuronal redox stress, 218–219
NMDA, 208–210, 220–221
postsynaptic signaling, 216–217
tripartite synapse, 214–217
BSA sulfenic acid (BSA-SOH), 46–47, 48*f*

C

Caenorhabditis elegans (C. elegans)
E. coli, 240
enzymes, 239–240
nematodes, 239
NGM plates, 240
worms, 239–240
Caerulein, 201–202
Calmodulin kinase, 222, 223*f*
Cardiovascular system
echocardiography, 133–134
ELISA studies, 133–134
H&E stain, 133–134
hypometabolism, 134–135
left anterior descending, 134
myocardial I/R injury, 133–134
myocardial ischemia, 133
NF-κB translocation, 134
nitrotyrosine stain, 134
reperfusion injury, 133
sulfide therapy, 134
troponin T levels, 134–135
TTC stain, 133–134
CAS. *See* Cyanoalanine synthase (CAS)
CBS. *See* Cystathionine β-synthase (CBS)
CD47 signaling
cytoskeleton, 155–156
ETHE1, 149–150
HPLC analysis, 146–147
interleukin-2, 147
leukocyte adhesion, 157–158
monobromobimane, 146–147
peripheral blood lymphocytes, 146–147
persulfides and polysulfides, 147
plasma sulfide, 146–147
SQRDL, 149–150
sulfane sulfur, 147
T-cells (*see* T-cell functions)
Cecal ligation and puncture (CLP), 116, 198–199
Cell injury
ethidium bromide and bisbenzimide, 113
fluorescent vital dyes bisbenzimide, 113
propidium iodide, 113
CellLight™ ER-GFP, 53, 54
Central nervous system (CNS)
ADT-treated animals, 130–131
amino acid neurotransmitters, 212–213, 213*f*
apoplexy and neurologic score, 130
bilateral vertebral arteries, 130
blood–brain barrier, 130–131
cerebral edema, 130–131
cerebral ischemia, 130
EPSPs, 213–214
hemoglobin, 212–213
iatrogenic interventions, 130
mechanical interventions, 130
middle cerebral artery occlusion, 130–131
MPO activity, 130–131
sulfane sulfur, 212–213
superoxide dismutase activity, 130
tetrazolium chloride, 130
thiosulfate, 212–213
Chemokine expression
fluorescent microspheres, 108–109
ICAM-1 expression, 108–109
microcirculation, 108–109
PECAM-1 fluorescence, 108–109
Cholecystokinin (CCK), 196–197
CK-MB. *See* Creatine-kinase-MB (CK-MB)
CLP. *See* Cecal ligation and puncture (CLP)
CNS. *See* Central nervous system (CNS)
Cold cyanolysis assay. *See* Sulfane sulfur
Creatine-kinase-MB (CK-MB), 28–29
CSE. *See* Cystathionine γ-lyase (CSE)
Cyanide (CN^-)-biotin
anhydrous DMF, 43
cyanoacetate-based reagent, 40–41
dry DMF, 43–44

1-ethyl-3-(3-dimethylaminopropyl) carbodiimide hydrogen chloride EDC, 43
flash column chromatography, 43–44
intracellular sulfenic acids, 54
Micro Bio-Spin™ P-6 gel columns, 48, 50
PBS, 54
preparation, 43*s*
Cyanoalanine synthase (CAS)
bisulfide formation, 275–276
cysteine, 255–256
KCN, 276
methylene blue assay, 275–276
OAS-TL, 280–281
reaction mixture, 259
spectrophotometry, 266
β-Cyano-L-alanine synthase, 266–267
Cyclooxygenase (COX), 178–180
Cystathionine β-synthase (CBS), 5, 170–172
Cystathionine γ-lyase (CSE), 5, 173, 175*f*
Cysteine
4-HNE, 12, 14, 15*f*
mimetic, 4–5
N-acetylcysteine, 8–9
Cysteine aminotransferase (CAT), 170–172, 171*f*
Cytochrome oxidase (COX), 254–255
mitochondria, 272
sulfide toxicity, 274–275
Cytoprotective actions, H_2S
COX-2 expression, 178–180
GI mucosal, 176–178
intestinal tissue, 178
leukocyte adherence, 176–178
NSAIDs, 176–178
ulcers healing, 178–180
Cytoskeleton
actin polymerization, 155
cysteine thiols, 156
ETHE1, 155–156
immunological synapse, 155
MTOC, 156
sulfhydration, 155–156
tubulin, 156

D

De novo formation, sulfane sulfur
cyanide prevention, 75*f*, 76
KCN, 73
procedure, 74–76
reagents and equipm, 74
Dichlorofluorescein (DCFH), 112–113
Dichlorofluorescein-diacetate (DCF-DA), 296, 297
6,8-Difluoro-4-methylumbelliferyl phosphate (DiFMUP), 59–60, 61
Dihydrorhodamine (DHR), 111–112
N,N-Dimethyl-*p*-phenylenediamine dihydrochloride (DMPD)
β-cyano-L-alanine synthase, 258
L-/D-DES, 258
3-(4,5-Dimethylthiazol-2-yl)-2,5-diphenyltetrazolium bromide test, 10
Disodium sulfide (Na_2S), 5, 6
Dithiothreitol (DTT), 81–83
Drugs, H_2S
aberrant crypt formation, 184–185, 184*f*
arthritis, 180–181, 181*f*
ATB-346, 180–181, 181*f*, 182
cardiovascular disease, 185–187, 186*f*
COX, 180–181
gastric prostaglandin synthesis, 183–184
GI toxicity, 184–185
hydroxybenzamide, 183–184, 183*f*
4-hydroxy-thiobenzamide, 183–184
IBD, 187–188
mesalamine, 187–188
naproxen, 180–181
neurodegenerative diseases, 185–187, 186*f*
NSAIDs, 180–181
proton pump inhibitors, 182–183
spinal cord injuries, 185–187
visceral analgesia, 188
DyLightTM405 conjugated streptavidin, 54

E

Electrophoretic mobility, 8
Elution buffer, 84*t*
Endotoxemia, 198, 198*f*
Enzymes activities
acidic ninhydrin reagent, 260
CAS, 257
β-cyano-L-alanine synthase, 266–267
cytochrome oxidase, 254–255

Enzymes activities (*Continued*)
equipments, 256
$FeCl_3$, 258
L-CS, 268–269
L-/D-DES, 261–262
signal transduction pathways, 254–255
SiR, 263–265
sulfur, 254–255
ESI-TOF-MS analysis, 47, 48
Ethylmalonic encephalopathy protein 1 (ETHE1)
Arabidopsis, 274–275
fatal metabolic disease, 281
GSSH, 274–275
SDO, 281
E3 ubiquitin ligase parkin, 80–81
Excitatory postsynaptic potentials (EPSPs), 213–214

F

Fenton/Weiss reaction, 293
Fluorescein isothiacyanate (FITC), 109–110

G

Gasses
advantages in toxicity, 292
cytochrome *c*, 292
hydreliox, 292
inhalation, 291
ischemia/reperfusion injury, 292
molecules, 291
myocardial nitro-tyrosine, 292
myocardium, 292
NO inhalation, 292
production, different enzymes, 292
Gastrointestinal (GI), 184–185
Gastrointestinal system
small intestine, 136–138
stomach, 135–136
Glutamate neurotransmission, 219–220
Glutathione (GSH), 4–5, 8–9, 14, 148
Glutathione persulfide (GSSH)
ETHE1, 274–275
SDO, 281–282
sulfur transferase, 274–275
Glyceraldehyde 3-phosphate dehydrogenase (GAPDH), 237–238. *See also* Bovine serum albumin (BSA)
Goldstein's reagent, 69, 70
GSH. *See* Glutathione (GSH)

H

Hemotoxylin and eosin (H&E) stain, 27–28
cardiovascular system, 133–134
hepatobiliary system, 138
musculoskeletal, 142
HEN buffer, 83*t*
HENS buffer, 84*t*
Hepatobiliary system
after I/R injury, 139
hepatic ischemia, 139
H&E stain, 138
in vitro cellular studies, 138–139
intravenous/intraperitoneal dose, 139
liver I/R injury, 138
NaHS, 138–139
PAG treatment, 138
PI3K-Akt1 pathway, 138–139
reperfusion injury, 138
Herein, 29–31, 32–33
H_2 gas concentration
H_2 sensor, 295–296
incorporation, 296*f*
measurement, 295–296
venous/arterial blood, 295
High-performance liquid chromatography (HPLC), 280
Histidine, 4
4-HNE. *See* 4-Hydroxy-2-nonenal (4-HNE)
4-HNE-DMA. *See* 4-Hydroxy-2-nonenal-dimethyl acetal (4-HNE-DMA)
H_2S. *See* Hydrogen sulfide (H_2S)
Human umbilical vein endothelial cells (HUVECs), 51, 52*f*, 53, 54
Hydreliox, 292
Hydrogen peroxide (H_2O_2)
anti-oxidants, 295
homeostasis, 294
and NO induces enzymes, 294
Nrf2 translocation, 294
physiological roles, 295
ROS production, 294
Hydrogen sulfide (H_2S)
Akt signaling, 31–32

anti-inflammatory actions
(*see* Anti-inflammatory actions)
brain synaptic remodeling (*see* Brain synapses)
CAT, 170–172, 171*f*
CBS, 170–172
CD47 (*see* CD47 signaling)
colitis, 172–173
cyanolysis reaction, 58–59
cytoprotective actions, 176–180
cytotoxicity of 4-HNE, 10–12
drugs, 180–188
electrons, 58
gaseous signaling molecules, 32–33
gene expression, 32
GI tract, 172
hyperhomocysteinemia, 170–172
hypometabolism, 20–21, 31–32
(*see also* Hypometabolism)
IBD, 170–172
in vitro systems, 20–21
inflammation, 196–202
L-Cysteine, 170–172
mammalian tissues, 20
in mice, 20–21
NaHS and Na_2S, 5–6
neuromodulation, 170
nocturnal laboratory rats and mice, 32
O_2 tension, 31–32
oxidation, 59
persulfidation, 58
polarographic sensor, 20–21
protein modification by 4-HNE
electrophoretic mobility shift, 8
immunoblotting, 8–9
protein thiols, 58
PTEN, 58–59
pyroxidal-5-phosphate, 170–172, 171*f*
reaction, 6–7
rescue molecule, 170
rhodanese activity, 31–32
roGFP2, 58–59
SQR, 172
sulfide pool, 59
thiol modifications (*see* Thiols)
thiosulfate in mitochondrion, 31–32
TNF, 172
use, 32–33
visceral pain, 176, 177*f*
4-Hydroxy-2-nonenal (4-HNE)
advanced lipid peroxidation end product, 14
biochemical effects, 4
brain, 13
2-butenal with hydrogen sulfide, 14
CBS gene, 5
C=O group, 4
cysteine, 14, 15*f*
disodium sulfide (Na_2S), 5–6
endogenousH_2S levels, 5
epoxidation, 4
GSH, 14
4-HNA preparation, 6
H_2S (*see* Hydrogen sulfide (H_2S))
hydrogen sulfide, 5
incubation, 13
lipophilic molecule, 15
neuromodulatory action, 5
pathogenesis, 4–5
preparation, 6
sequestering agents, 14–15
SH-SY5Y cells, 13
sodium hydrogen sulfide (NaHS), 5–6
TBST solution, 13
thiol compounds, 4–5
unsaturated aldehydes, 15
α,β-unsaturated aldehydes, 4
Western blot, 13, 14*f*
4-Hydroxy-2-nonenal-dimethyl acetal (4-HNE-DMA), 6
Hydroxyphenyl fluorescein (HPF), 296
Hypometabolism, 134–135
animals, 22
blood and tissues from H_2S-exposed rats, 29–31
exposure experiments, 22–25
gases and equipment, 21–22
lung histopathology assessment, 27–28
in mammals, 20–21
in mice, 20–21
physiological measurements, 25–27
in rats, 20–21
tissue and blood collection after H_2S exposure, 28–29

I

IAF. *See* Iodoacetamide fluorescein (IAF)
Iberiotoxin (IBTX), 136–137
Immunoblotting, 8–9
Inflammation
acute pancreatitis, 196–197
burn injuries, 199–200
caerulein, 201–202
carrageenan, 200–201
LPS, 198–199
lung injury, 201–202
substance P (*see* Hydrogen sulfide (H_2S))
Inflammatory bowel disease (IBD), 187–188
Intercellular adhesion molecule-1 (ICAM), 173
Intravital microscopy
application, 120–121
capillary density assessment, 113–114
cardinal signs, 94
Ca^{2+} sensor GCaMP2, 120–121
cell injury (*see* Cell injury)
chemokine and adhesion molecule expression (*see* Chemokine expression)
confocal/multiphoton imaging, 120–121
deleterious effects, 119–120
GFP and YFP mice, 120–121
hydrogen sulfide (*see* Hydrogen sulfide (H_2S))
in vitro and *ex vivo* approaches, 120
inflammatory stimulus, 94
leukocyte (*see* Leukocyte rolling, adhesion and emigration)
leukocyte/endothelial cell adhesive interactions, 119–120 (*see also* Leukocyte/endothelial cell adhesive interactions)
mesalamine, 119–120
microvascular permeability (*see* Microvascular permeability)
neutrophil-dependent mechanism, 94
neutrophil/endothelial cell adhesive interactions (*see* Neutrophil/ endothelial cell adhesive interactions)
postcapillary venular function, 119–120
reactive oxygen species (*see* Reactive oxygen species (ROS))
translational therapeutics, 119–120
Iodoacetamide fluorescein (IAF)
acrylamide electrophoresis, 237–238
C. elegans, 245, 245*f*
SDS, 244–245
I/R injury. *See* Ischemia-reperfusion injury (I/R injury)
Ischemia-reperfusion injury (I/R injury)
anti-inflammatory effects, 128, 142–143
arterial inflow, 128–129
BK channel signaling, 142–143
capillary and microvascular endothelium, 129
cardiovascular system (*see* Cardiovascular system)
cell death, 129
central nervous system (*see* Central nervous system)
endogenous antioxidant mechanisms, 129
gastrointestinal system (*see* Gastrointestinal system)
hepatobiliary system (*see* Hepatobiliary system)
hypotension, 128–129
inflammation, 128
musculoskeletal (*see* Musculoskeletal)
processes, 129
renal system (*see* Renal system)
respiratory system (*see* Respiratory system)
restoration, 128–129
thrombosis/ mechanical compression, 128–129
tissue ischemia, 128–129
Toll-like receptor activation, 129

L

L-Cysteine synthase (L-CL), 268–269
L-/D-cysteine desulfhydrase (L-/D-DES)
absorbance *vs.* concentration, 262
Bradford assay, 262
DMPD, 262
NaHS, 261
plant tissues, 261
spectrophotometry, 261
supernatant, 262
Leukocyte/endothelial cell adhesive interactions
abdominal midline incision, 100
Alexa-fluor 488 Gr1 antibody, 102–103
anti-RP-1 antibody, 102–103

color camera, 98–100
cremaster muscle, 100
custom-built animal board, 98–100
dorsal skin-fold chamber, 98–100
extravascular leukocytes, 101–102
ex vivo isolation/reinjection approach, 102
graphical representation, components, 98–100, 99*f*
imaging, 103
immersion objectives, 98–100, 101
knockout and transgenic mice, 100
LysM-eGFP mice, 102–103
microcirculation, 98–100
microvessels, 100
overstretching microvascular vessels, 100
physiologic factors, 98–100
platelet interactions, 102
superfusate pO_2, 101
surgical preparation, 100–101
thicker tissues, 102
transillumination, 101–102
Leukocyte rolling, adhesion and emigration
adherent and extravasa, 105
adherent cells, 104
autoperfused flow chamber systems, 106–108
characterization, 104
description, 103
extrinsic factors, 105–106
ex vivo microscopy, 105–106, 107*f*
in vivo environment, 105–106
interactions with endothelial cells, 106
intravital microscopic imaging, 105
microfluidic systems, 107–108
microparticle image velocimetry, 104–105
multicolor fluorescence spinning disk, 105
multiphoton confocal, 105
Newtonian fluid, 105
off-line analysis tools, 104–105
parenchymal cells, 105–106
postcapillary venules, 105
pro- and anti-adhesive molecules, 106–107
velocimetry approaches, 104–105
V_{WBC}/V_{RBC} provides, 104
wall shear stress and rate influence, 104–105
Li-COR Odyssey system, 86, 87
Lipopolysaccharide (LPS)
appendicitis and diverticulitis, 198
CLP, 198, 198*f*
endotoxemia, 198, 198*f*
glycolipid, 198
liver injury, 199
MPO, 199
sepsis, 198
substance P, 199
Long-term potentiation (LTP), 208–210
Lung histopathological scoring system, 27–28, 27*t*
γ-Lyase/cystathionine β-synthase, 292
Lysine, 4–5

M

Maleimide assay
Alexa Fluor 680-conjugated C2 maleimide, 86
considerations, 88
LiCOR Odyssey system, 86
Parkinson's disease, 88–89
physiological processes, 88–89
principles, 86
protocol, 87, 88
solutions and buffers, 86–87
sulfhydration and nitrosylation, 86
Male Sprague-Dawley rats, 22
Malondialdehyde (MDA), 307
Mass flow controllers (MFC), 23*f*
3-Mercaptopyruvate sulfurtransferase (3-MST), 80, 208–210
Methyl-methane thiosulfonate (MMTS), 81–83, 238
Methylsulfonyl benzothiazole (MSBT)
cell lysis, 51
methyl cyanoacetate, 41–42
MSBT-A, 42
PBS, 53
reaction with –SH and –SSH, 40–41
with thiol and persulfide substrates, 44–45
MFC. *See* Mass flow controllers (MFC)
Michael adducts, 4–5
Micro Bio-Spin™ P-6 gel columns, 46, 47, 48, 49, 50

Microtubule-organizing complex (MTOC), 156
Microvascular permeability
CCD camera, 109–110
confocal microscope, 110–111
FITC-albumin, 109–110
fluorescence intensity, 110–111
hydraulic conductivity, 111
hydrostatic pressure, 111
immune response, 109
inflammatory milieu, 111
interstitial fluid pressure, 109
intravital microscopic methods, 109–110
surface area changes, 109–110
vascular leakage assay, 110–111
volume flux, 111
water flux, 111
MitoTracker® Red CMXRos, 53, 54
Modified biotin switch assay
cell cultures, 83
detection, 81–83, 82*f*
dithiothreitol, 81–83
methylsulfonyl benzothiazole, 81–83
protocol, 84–86
reagents, 83
solutions and buffers, 83–84
sulfhydrated cysteines, 81–83
sulfhydration, 81–83
Western blotting, 81–83
Molecular hydrogen (H_2)
acute oxidative stress by ischemia/ reperfusion, 304–305
alga *Chlamydomonas reinhardtii*, 290–291
anti-inflammatory effects, 306
applications, 310
autoignition temperature, 290
chronic oxidative stress, 305
clinical examinations, 291
cyanobacteria, 290–291
cytoprotective effects, 291
diffusion, 304
drinking H_2-water, 309
energy metabolism, 305–306
gas concentration (*see* H_2 gas concentration)
highly reactive ROS, 296–298
H_2O_2 (*see* Hydrogen peroxide (H_2O_2))
hydrogenases, 290
hydroxyl radicals, 306–307
indirect reduction, oxidative stress, 308–309
inhalation, 301
mammalian cells, 291
maternal intake, 304
medical gasses (*see* Gasses)
oral ingestion by drinking hydrogen water, 301–302
peroxynitrite, 307–308
photosynthetic organisms, 290–291
putative catalyst, 310
rapid diffusion, 298–301
ROS (*see* Reactive oxygen species (ROS))
and saline injection, 302–303
therapeutic and preventive antioxidant, 291
MouseOx® Pulse Oximeter, 25
Musculoskeletal
bilateral gastrocnemius muscles, 142
deep venous thrombosis, 141–142
H&E stain, 142
hydrogen sulfide, 142
reconstructive surgery, 141–142
TNF-α levels, 142
Myeloperoxidase (MPO), 130–132, 133–134, 196–197

N

N-acetyl-L-cysteine (NAC), 154
NaHS. *See* Sodium hydrogen sulfide (NaHS)
Na_2S. *See* Disodium sulfide (Na_2S)
Nematode Growth Media (NGM)
C. elegans, 240
H_2S donor, 242
Neuronal redox stress, H_2S
cellular antioxidant, 218
cysteine formation, 218–219
glutathione, 218
NMDA receptors, 218
peroxynitrite, 219
ROS, 218–219
Neutralization buffer, 84*t*
Neutrophil/endothelial cell adhesive interactions
E- and P-selectins, 96–98

endothelial ligands, ESAM and CD99L2, 96–98
hydrodynamic dispersal forces, 96
ICAM-1 expression, 96–98
interstitial tissue, 94–96
ischemia and reperfusion (I/R), 96–98
leukocyte-endothelial interactions, 96–98
leukocytes, 96
myocardial cells, 96–98
neutrophil detaches, 96–98
process, 94–96
trafficking, 94–96, 95*f*
NF-E2-related factor 2 (Nrf2), 294
Nitrotyrosine stain, 134
N-methyl-D-aspartate (NMDA), 187, 208–210, 220–221
Nonsteroidal anti-inflammatory drugs (NSAIDs), 176–178

O

O-acetyl-L-serine sulfydrylase (OSSA), 264
O-acetylserine(thiol)lyase (OAS-TL)
CAS, 280–281
COX activity, 274–275
cysteine, 272–274
SAT, 276–280

P

Pancreatitis, acute
caerulein, 196–197
CCK, 196–197
CSE, 197
enzyme secretion, 196–197
lung injury, 196–197
MPO activity, 196–197
PAG, 197
substance P, 197
Parkinson's disease (PD), 305
Phosphatase and tensin homolog (PTEN) activity assay
dephosphorylation activity originates solely, 61
low pK_a cysteine, 61, 62*f*
principle, 59–60
procedure, 60–61
reagents and equipment, 60
recombination, 61
PHSS. *See* Polarographic hydrogen sulfide sensor (PHSS)
PI3K-Akt1 pathway, 138–139
Plant mitochondria, sulfide detoxification. *see* Sulfide detoxification
p38 MAPK inhibitors, 117–119
Polarographic hydrogen sulfide sensor (PHSS), 29–31
Polysulfides, H_2S
after preparation and dilution, 67
alkalinization, 64
chelators, 64–66
experiments, 66
H_2S(aq) solutions, 64–66
K_2S_x stock solutions, 66
preparation, 64
products, 64
PTEN activity/roGFP2 redox assay, 67
purest source of H_2S, 64–66
Sigma–Aldrich technical service, 64
sulfide salts and K_2S_x polysulfides, 67, 68*f*
Potassium (poly)sulfide (K_2S_x), 66, 68*f*, 70, 72*f*, 74*f*
Potassium thiocyanate (KSCN), 69, 70
Propargylglycine (PAG), 197
Protein modification, thiols
BCA assay, 244, 246
BSA, 246
C. elegans, 245, 245*f*
cysteine residue, 247
fluorescein, 246
IAF, 244–245, 246–247
nematodes, 245, 245*f*
peptides, 247
SDS-polyacrylamide, 244–245
toxicity assay, 244
tyrosine phosphatase, 246
PTEN-SBP-His
dispense, 61
pre-equilibrated Zeba gel filtration, 61
recombination, 60
Pyroxidal-5-phosphate (P5P), 170–172, 171*f*

R

Reactive oxygen species (ROS)
acute, 293
aerobic organism, 293

Reactive oxygen species (ROS) (*Continued*)
antioxidation therapy, 293–294
bicarbonate-buffered saline, 111–112
dichlorofluorescein, 112
Fenton/Weiss reaction, 293
hydroethidine, 112–113
I_{DHR}:I_{Bgrd} ratio, 111–112
intravital microscopy, 112–113
ionizing radiation, 293
microvasculature, 111–112
oxidant-sensitive probes *in vivo*, 112–113
pathologies, 293
produce, 293
superoxide anion radicals, 293
Renal system
bilateral nephrectomy, 140–141
ELISA assays, 140
hepatic I/R injury, 139–140
H_2S gas, 139–140
hypometabolism, 139–140
hypoperfusion and ischemia, 139
immunohistochemical stain, 139–140
intra-aortic balloon, 141
MPO stain, 140–141
NaHS solution, 141
neutrophilic infiltration, 140
proinflammatory and apoptotic markers, 141
renal I/R injury, 140
Syngenic Lewis rats, 140–141
TUNEL stain, 141
Respiratory system
alveolar-arterial gradient, 131–132
arterial inflow, 131
inflammatory injury, 132–133
I/R injury, 131
ischemia-reperfusion injury, 132–133
lung transplantation, 131–132
MPO assay, 131–132
NaHS treatment, 131–133
TLR-4 and NF-κB, 132–133
Rheumatoid arthritis (RA), 306
RNAlater® solution, 29
roGFP2-His
dispense, 63
pre-equilibrated Zeba gel filtration, 63
recombination, 62, 63
roGFP2 redox assay
disulfide bond formation, 63
DTT and diamide data, 64
oxidation and reduction kinetics, thiol/disulfide, 64, 65*f*
physiological conditions, 63
principle, 61–62
procedure, 63
reagents and equipment, 62–63

S

SAT. *See* Serine cetyltransferase (SAT)
Scott Specialty Gases, 21–22
SDO. *See* Sulfur dioxygenase (SDO)
SDS-PAGE sample buffer, 84*t*
Serine cetyltransferase (SAT)
Arabidopsis, 276–277
cysteine synthase, 276–277
His_6, 277
HPLC, 280
monobromobimane, 280
ninhydrin, 279
plant tissue, 277–278
protein extracts (PE), 278–279, 279*f*
purification, 278–279, 279*f*
S-Nitrosylation, 236–237
Sodium hydrogen sulfide (NaHS), 5, 6–7, 7*f*, 11, 11*f*, 12, 12*f*
Sodium (poly)sulfide (Na_2S_x), 66
S-sulfhydrated proteins
chemistry validation, small-molecule substrates
cyanoacetate with R-S-S-BT, 45–46
materials, 41–44
MSBT with thiol and persulfide substrates, 44–45
cysteine persulfides, 40
oxidative posttranslational modification, 40
physiological and pathophysiological processes, 40
"tag-switch" method (*see* Tag-switch method)
Streptavidin binding peptide (SBP), 60. *See also* PTEN-SBP-His
Sulfane sulfur
elimination
preincubation with cyanide efficiently, 73, 74*f*

principle, 71–72
procedure, 73
reagents and equipment, 72–73
prevention
de novo formation, polysulfides, 75*f*, 76
KCN, 73
procedure, 74–76
reagents and equipment, 74
principle, 67–69
quantitation
alkaline pH, 70–71
cyanolysis assay, 71, 72*f*
GYY4137 interferes, 71
procedure, 70
reagents and equipment, 69–70
thiol reductants, 71
UV/vis-absorption spectra, 71
Sulfhydration
biotin switch technique, 80
CBS and 3-MST, 80
growth factors, 80
hormones, 80
maleimide assay (*see* Maleimide assay)
modified biotin switch assay (*see* Modified biotin switch assay)
neurotransmitters, 80
and nitrosylation, 80–81
physiologic gasotransmitter, 80
protein S-sulfhydration, 80
Sulfide detoxification
APS, 272–274
CAS activity, 275–276
COX, 272
cyanide, 274
cysteine, 272–274, 273*f*
ETHE1, 274–275
GSSH, 274–275
homeostasis, 275
NSF1, 274
OAS, 276–280
SAT, 272–274
SDO activity, 281–283
Sulfide-quinone oxidoreductase, 31–32, 149–150
Sulfide quinone reductase (SQR), 172
Sulfite reductase (SiR)
cysteine, 264
ferredoxin, 264
homogenate, 264
OSSA, 264
Sulfur dioxygenase (SDO), 31–32
acetone, 283
chemolithotroph, 281
ETHE1, 281
GSSH, 281–282, 282*f*
HPLC, 281–282, 282*f*
oxygen consumption rates, 282–283
thiosulfate, 282–283
Systemic lupus erythematosus (SLE), 152, 160

T

Tag-switch method
bovine serum albumin and GAPDH
BSA–SOH, BSA–SSG and BSA–SSH, 46–47
materials, 46
mechanism, protein S-sulfhydration, 49
MS and dot blot, 47–49
design, 40–41
intracellular protein persulfides
immunoprecipitation, 51
Jurkat cell extracts, 50
materials, 49–50
intracellular S-sulfhydration by fluorescence microscopy
colocalization studies, 54
materials, 51–53
protocol, 53–54
T-cell functions
acute pancreatitis, 150–151
Akt/GSK3β pathway, 152
APCs, 154
asthma and allergy, 159–160
cobalt chloride, 152–153
colon and ovarian cancers, 148–149
GSH, 148
IBD, 158–159
lymphoblastoid cell lines, 153–154
lymphoid malignancies, 148–149
mammalian cells, 148
MEK inhibitor, 151–152
MHC, 148, 149*f*
mRNA expression, 148
NAC, 154

T-cell functions (*Continued*)
phospho-ERK (pERK), 151
PP2A, 152
proinflammatory, 150–151
psoriasis, 160
PTP1B, 151
renal injury, 159
rheumatoid arthritis, 159
SLE, 152, 160
sulfhydration, 150–151
TSP1, 148, 149*f*, 156–157
Tetrazolium chloride (TTC) stain, 133–134
Thiols
avidinhorseradish peroxidase, 238
C. elegans, 239–240
cell signalling, 235
cellular function, 234
desulfhydrases, 234–235
donor molecules, 240–241
electrophoresis, 239, 243
gasotransmitters, 234
mitogen-activated protein kinases, 240–241
NaHS, 242, 243*f*
NO, 238
protein modifications, 244–245
SH group, 236
toxicity estimation, 241–242
tyrosine phosphatases, 236
Thrombospondin-1 (TSP1), 148, 149*f*, 156–157
Trans-Blot® SD semi-dry blotter, 50
Transillumination intravital microscopy, 101
Tumor necrosis factor (TNF), 172
Tyrosine phosphatases, 236, 246

V

Viral antigen-free (VAF) rats, 22–23, 25–27

W

Wall shear stress and rate influence, 104–105

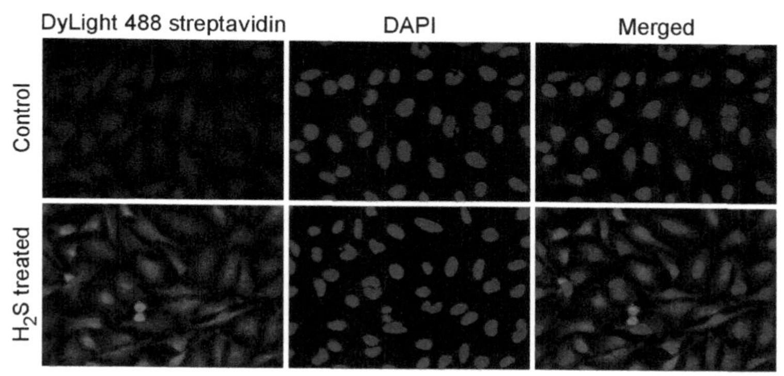

Chung-Min Park *et al.*, Figure 3 *In situ* labeling of intracellular S-sulfhydration by modified "tag-switch" assay. Cells (HUVEC) were either treated with 100 μ*M* Na_2S or not and then prepared for the microscopy, following the described protocol. Nuclei were stained with DAPI. Fluorescence microscopy was carried out using Carl Zeiss Axiovert 40 CLF inverted microscope, equipped with monochromatic RGB CoolLed light source (Andover, UK) and monochromatic AxioCam Icm1 camera. All experiments were performed at least in triplicate. Images were postprocessed in ImageJ software (NIH, USA).

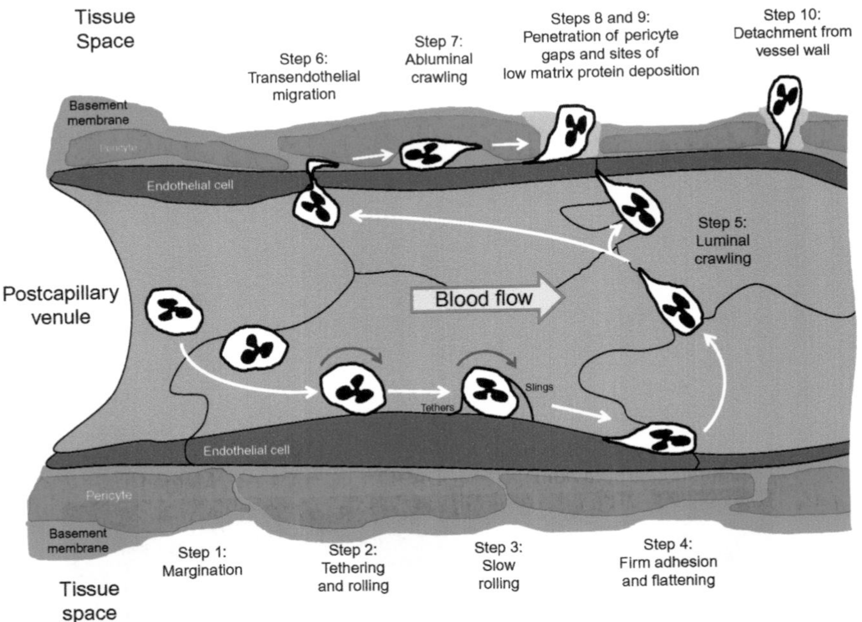

Mozow Y. Zuidema and Ronald J. Korthuis, Figure 1 Neutrophil trafficking to sites of inflammation involves 10 steps. Step 1: *Margination*. As neutrophils exit capillaries and enter the larger diameter postcapillary venular segment of the microcirculation, hydrodynamic forces move granulocytes to the vessel margins (margination). Step 2: *Tethering and rolling*. If appropriate adhesion molecules are expressed on activated endothelial cells and the marginated neutrophil, granulocytes are capture by adhesive interactions that mediate rolling of the leukocyte along the vessel wall. Step 3: *Slow rolling*. The rolling neutrophil monitors its local environment for the presence of activating factors that promote adhesion molecule expression, thereby enabling the leukocyte to further slow its rolling behavior by forming more weak adhesive interactions that are mediated by the selectins. As the cells roll along and interact with P-selectin on the endothelial surface, excess membrane on the surface of neutrophils is pulled out into long nanotubes (microvilli) that form tethers at the rear of the rolling leukocyte. These tethers eventually detach as the cell continues to roll, but do not retract. Instead, they persist and are slung in front of the rolling leukocyte, to interact again with P-selectin. The membranous nanotube is now referred to as a sling, which the neutrophil rolls over to again be retarded in its movement down the vessel wall as the sling transitions to a tethering function. Step 4: *Firm or stationary adhesion*. By establishing strong adhesive interactions mediated by integrin-dependent interactions with endothelial ICAM-1 that are upregulated by chemokines, the slowly rolling leukocyte progresses to firm adhesion. Step 5: *Luminal crawling*. Integrin-ICAM-1-dependent adhesion activates intracellular signaling pathways that induce cytoskeletal changes and polarization of the cell that lead to luminal motility. The crawling neutrophils move along interendothelial junctions in search of preferred routes for diapedesis and are often observed moving against the direction of flow in this exploration. Step 6: *Transendothelial cell migration*. Neutrophils cross the endothelial cell barrier by traversinginterendothelial junctions at preferential sites that overlie areas in the basement membrane that exhibit low matrix protein deposition. Occasionally, these inflammatory phagocytes cross the endothelial

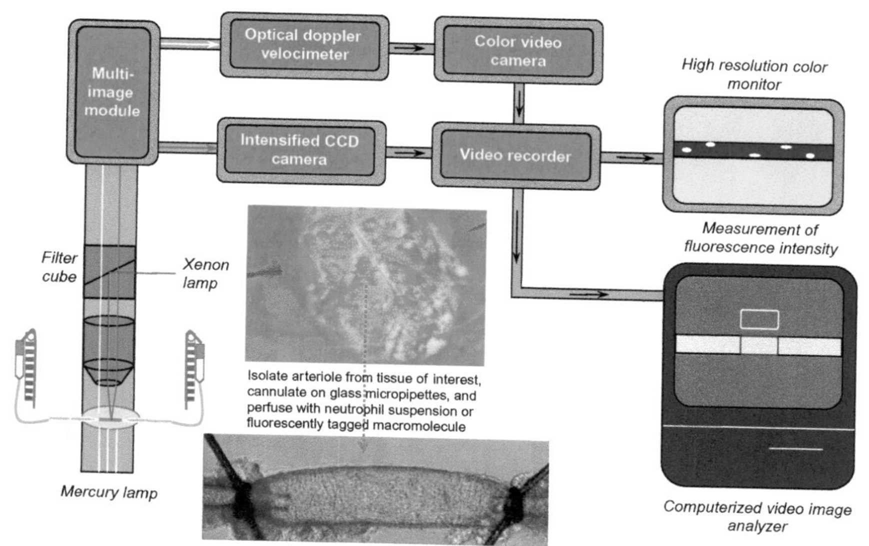

Mozow Y. Zuidema and Ronald J. Korthuis, Figure 3 Schematic representation of essential components for assessment of neutrophil interactions and venular permeability isolated, perfused single postcapillary venules using *ex vivo* microscopy. This approach uses the same components as shown in Fig. 2 for intravital microscopy. The major difference is that an isolated vessel perfusion chamber takes the place of the viewing pedestal animal board. For this preparation, a postcapillary venule is isolated from the tissue of interest, cannulated on glass micropipettes and perfused with neutrophils or fluorescently tagged macromolecules. The height of the perfusion reservoirs can be adjusted in equal and opposite directions by use of a pulley system to alter perfusion pressure and flow without altering midpoint pressure in the vessel. Neutrophil–endothelial cell adhesive interactions and venular permeability are determined as described in the text.

---

barrier by moving through cells in a transcellular route. Step 7: *Abluminal crawling*. Once through the endothelium, the diapedesing neutrophils crawl abluminally along pericyte processes, interacting with basement membrane structures at the same time. Steps 8 and 9: *Penetration of pericyte gaps and regions of low matrix protein expression in the basement membrane*. Abluminally crawling neutrophils breach the pericyte layer where gaps between these cells exist, which coincide with regions of low matrix protein deposition (depicted as a lighter shade of green) in the basement membrane. These steps occur more or less simultaneously and use different adhesion molecules to propel the leukocyte through these barriers. Step 10: *Detachment from the vessel Wall*. The last step in the process requires detachment of the migrating neutrophil from components of the vessel wall, which involves release of pseudopodial extensions from their sites of attachment to basement membrane components and adhesion receptors on pericytes and endothelial cells. Once this is achieved, the leukocyte follows directional cues provided by a chemotactic gradient and migrates through the tissue space towards inflammatory foci. *Modified from Voisin and Nourshargh (2013).*

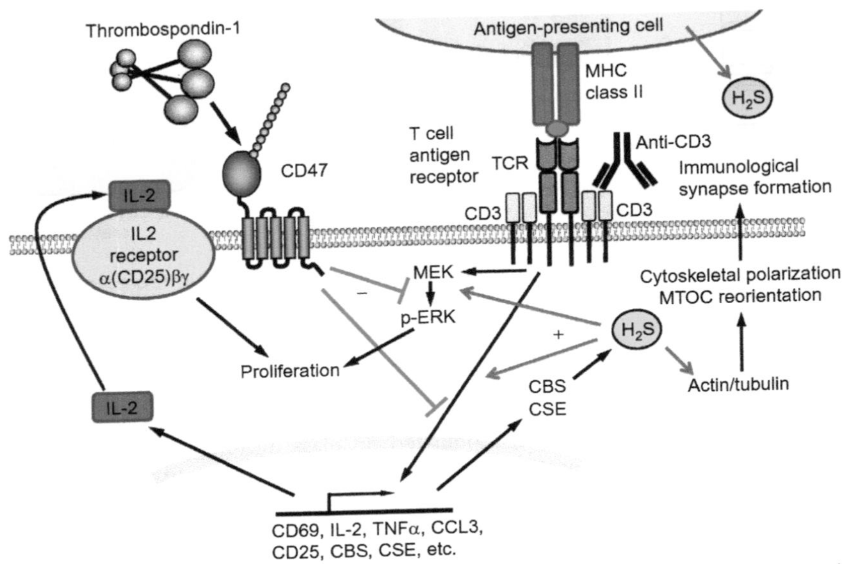

Sukhbir Kaur *et al.*, Figure 1 Enhancement of T cell activation by H_2S and inhibition by thrombospondin-1/CD47 signaling. T cell antigen receptor signaling is initiated by recognition of a specific peptide antigen in the context of MHC on an antigen-presenting cell. Optimal signaling to induce expression of T cell activation genes and proliferation requires expression of the H_2S biosynthetic enzymes CBS and CSE, but exogenous H_2S, probably in part from the APC, also enhances TCR signaling to activate the ERK MAP kinase pathway, modify actin and tubulin to induce reorientation of the microtubule-organizing center (MTOC) to facilitate formation of an immunological synapse with the antigen-presenting cell and induction of T cell activation genes. The latter include IL-2 that drives T cell proliferation and CD25 that encodes the α-subunit of the IL-2 receptor. Thrombospondin-1 signaling through its receptor CD47 inhibits these responses of T cells to exogenous H_2S and inhibits the induction of CBS and CSE during T cell activation.

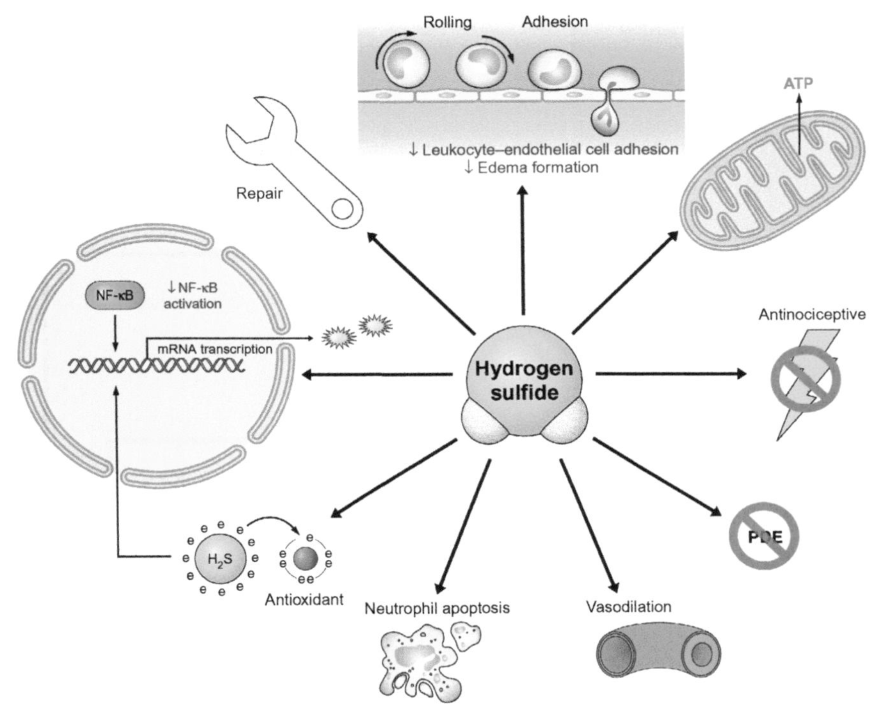

Burcu Gemici and John L. Wallace, Figure 2 Anti-inflammatory actions of hydrogen sulfide (H_2S). H_2S can affect many aspects of an inflammatory response, through many mechanisms. H_2S is a tonic inhibitor of leukocyte adherence to the vascular endothelium, limiting leukocyte extravasation and edema formation. Mitochondria can utilize H_2S as an electron donor in adenosine triphosphate (ATP) production, particularly during anoxia/hypoxia, and in doing so reduces generation of tissue-damaging oxygen-derived free radicals. Antinociceptive effects of H_2S have been demonstrated in models of visceral pain. By inhibiting phosphodiesterases (PDE), H_2S can elevate tissue cyclic guanylate monophosphate (GMP) levels, which can contribute to vasodilation. H_2S promotes resolution of inflammation through various mechanisms, including promotion of neutrophil apoptosis. The antioxidant actions of H_2S further reduce tissue injury. Several anti-inflammatory and antioxidant systems are activated by H_2S through its effects on transcription factors (including NfκB). Through multiple mechanisms, including induction of cyclooxygenase-2 expression and stimulation of angiogenesis, H_2S can promote repair of damaged tissue. *Reproduced from Chan and Wallace (2013), with permission of the publisher.*

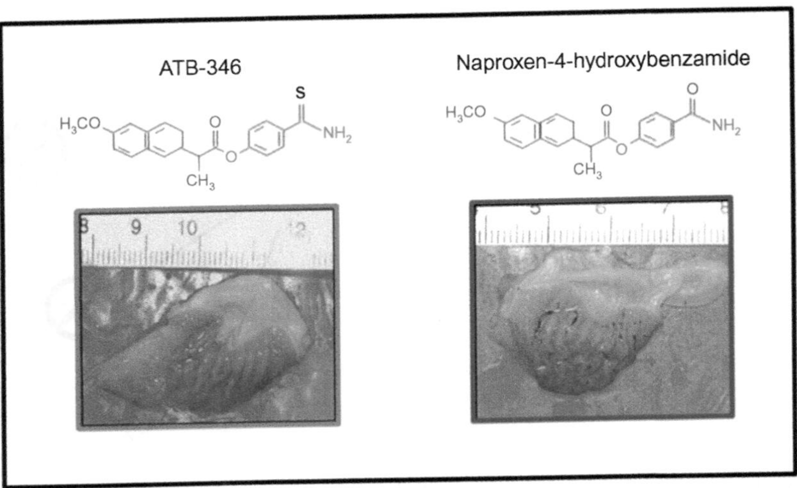

Burcu Gemici and John L. Wallace, Figure 6 The gastric-sparing property of ATB-346 is due to the ability of this drug to release H_2S. Administration of ATB-346 at a dose of 30 mg/kg did not elicit significant gastric damage, despite markedly suppressing gastric prostaglandin synthesis. However, administration of naproxen-4-hydroxybenzamide, which lacks the sulfur group and therefore cannot generate H_2S, elicits significant hemorrhagic damage in the stomach, with similar suppression of gastric prostaglandin synthesis as observed with the equimolar dose of ATB-346.

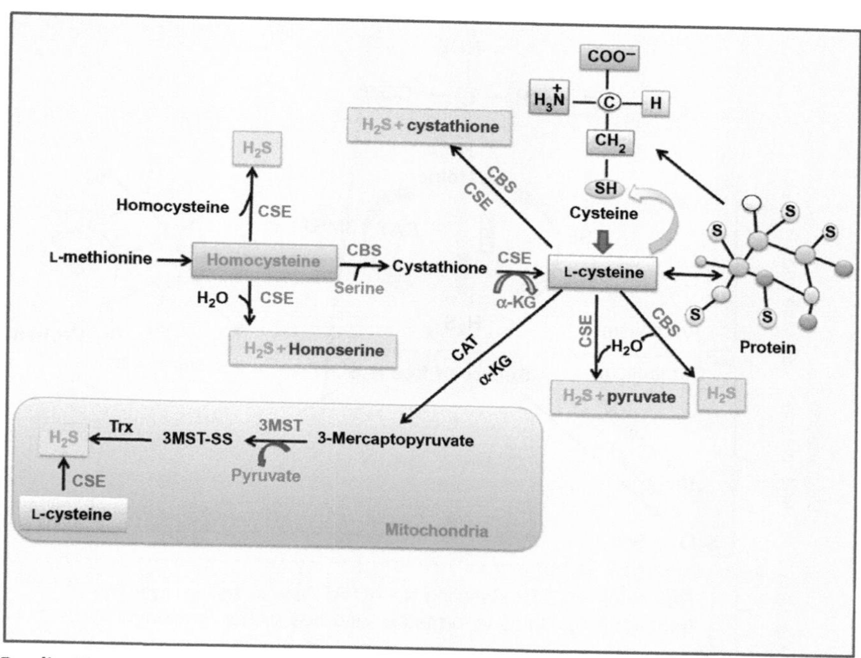

Pradip Kumar Kamat *et al.*, Figure 1 H_2S formation in cytosol as well as in mitochondria. The three enzymes CBS, CSE, and 3MST are responsible for H_2S generation in cells. CBS and CSE are usually present in cytosol while 3MST is present in mitochondria. The sources of cytosolic H_2S generation are L-methionine, homocysteine, cystathione, and L-cysteine. L-Cysteine also originates from a protein and cysteine and moreover also produces pyruvate and cystathione. In mitochondria, the source of H_2S synthesis are mercaptopyruvate and L-cysteine. CBS, cystathionine β-synthase; 3MST, 3-mercaptopyruvate sulfurtransferase; CSE, cystathionine γ-lyase; α-KG, α-ketoglutarate; CAT, cysteine aminotransferase; H_2S, hydrogen sulfide; TRX, thioredoxin.

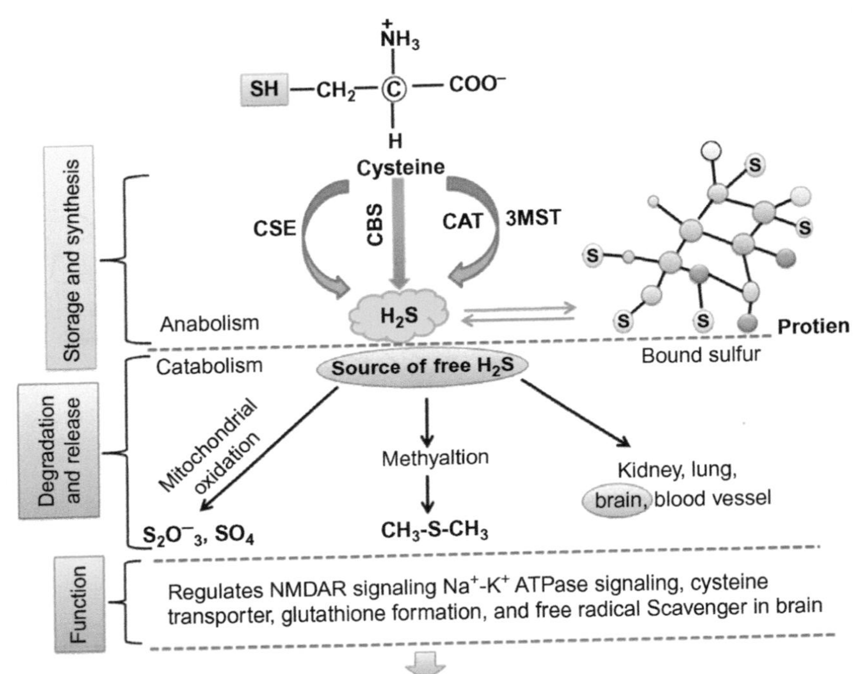

Pradip Kumar Kamat *et al.*, Figure 2 Diagrammatic representation shows the sources of H_2S and moreover anabolism and catabolism of H_2S. The usual sources of H_2S are blood lung, kidney, heart, and brain. Free H_2S gets converted into sulfate and thiosulfate through mitochondrial oxidation and sulfomethane via methylation which becomes a source of H_2S. Diagram also depicts the synthesis, storage, and function of H_2S in neuronal synapse. These free H_2S involved in NMDA receptor signaling, Na^+ K^+ ATPase signaling, and glutathione formation maintains and regulates neuronal function.

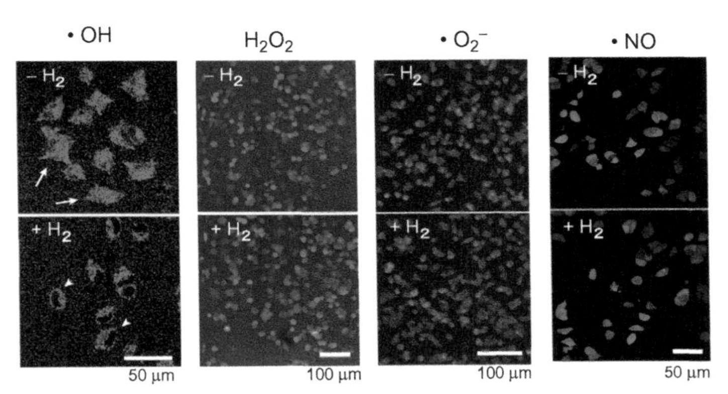

Shigeo Ohta, Figure 2 Selective reduction of reactive oxygen or nitrogen species by H_2 in cultured cells. PC12 cells were kept in medium with 0.6 m*M* H_2 (indicated by $+H_2$) or without H_2 (indicated by $-H_2$), and exposed to antimycin A or L-NAME (N^G-nitro-L-arginine methyl ester) to induce $^{\bullet}OH$, H_2O_2, and $^{\bullet}O_2^-$ (ROS) or $NO^{\bullet}$ (RNS) for 30 min. Each ROS or RNS was detected using flowing fluorescent dye; HPF, H_2DCF (2′,7′-dichlorodihydrofluorescein), MitoSOX, and DAF-2 DA (diaminofluorescein-2 diacetate) were used to detect $^{\bullet}OH$, H_2O_2, $^{\bullet}O_2^-$, and $NO^{\bullet}$, respectively. These representative fluorescence images obtained by laser-scanning confocal microscopy demonstrate the selective reduction of $^{\bullet}OH$ by H_2. *Adapted from Ohsawa et al. (2007) modified version of Fig. 1A, B and supplementary Fig. 1A, C, with permission from Nature Publishing Group.*

Edwards Brothers Malloy
Ann Arbor MI. USA
March 4, 2015